D0782078

Communities and Ecosystems

MONOGRAPHS IN POPULATION BIOLOGY

EDITED BY SIMON A. LEVIN AND HENRY S. HORN

Communities and Ecosystems

Linking the Aboveground and Belowground Components

DAVID A. WARDLE

PRINCETON UNIVERSITY PRESS

PRINCETON AND OXFORD

Copyright © 2002 by Princeton University Press
Published by Princeton University Press, 41 William Street,
Princeton, New Jersey 08540
In the United Kingdom: Princeton University Press, 3 Market Place,
Woodstock, Oxfordshire OX20 1SY
All Rights Reserved

Library of Congress Cataloging-in-Publication Data

Wardle, David A., 1963–
Communities and ecosystems : linking the aboveground and belowground
components / David A. Wardle.
p. cm. — (Monographs in population biology ; 34)
Includes bibliographical references (p.).
ISBN 0-691-07486-0 (alk. paper) — ISBN 0-691-07487-9 (pbk. : alk. paper)
1. Soil ecology. 2. Food chains (Ecology) I. Title. II. Series.

QH541.5.S6 W37 2002
577.5'7—dc21   2001055403

British Library Cataloging-in-Publication Data is available

This book has been composed in New Baskerville

Printed on acid-free paper. ∞

www.pup.princeton.edu

Printed in the United States of America

10 9 8 7 6 5 4 3 2

Contents

Acknowledgments

First I would like to thank the Life Sciences Editor of Princeton University Press, Sam Elworthy, for contacting me in the first place with the suggestion that I write this book. I also thank Sam and the other staff at Princeton University Press for their helpfulness during the book's preparation. I also thank Landcare Research at Lincoln, New Zealand, for institutional support while this book was being written. In particular, I would like to express my thanks to Kirsty Cullen and Karen Bonner at Landcare Research for preparing good copies of all the figures in this book.

I have been fortunate to have several colleagues and collaborators who have taught me a lot about plant, soil, and animal ecology. Special mention must be made of Gregor Yeates who has collaborated with me on a number of projects over the past twelve or so years and from whom I have learned much about soil ecology; and of Laurie Greenfield who as my undergraduate adviser in the mid 1980's led me into belowground ecology in the first place and who continues to challenge me to think more laterally about things than I would otherwise. Gary Barker, Marie-Charlotte Nilsson, and Olle Zackrisson have also each collaborated with me on a range of topics over much of the past decade and have each taught me much about various branches of ecology.

Duane Peltzer provided detailed comments on each draft chapter as it was prepared—not a small task. A number of colleagues provided very helpful comments on specific chapters most closely related to their own areas of expertise (not all of whom agree with absolutely everything I have written), i.e., Robert B. Allen, Jonathan M. Anderson, Richard D. Bardgett, Gary M. Barker, Peter J. Bellingham, Frank Berendse, David Coomes, Laurie G. Greenfield, J. Philip Grime, Andy Hector, Michael A. Huston, Juha Mikola, Heikki Setälä, Wim Van der Putten, Wendy Williamson, and Gregor W. Yeates.

Introduction

All terrestrial ecosystems consist of a producer subsystem and a decomposer subsystem. These components are obligately dependent upon one another, with the producers acting as the primary source of organic carbon for the system, and the decomposers being responsible for the breakdown of organic matter and the release and cycling of nutrients. Any approach to better understanding ecosystem functioning therefore requires explicit consideration of both these subsystems. Further, both the producer and decomposer subsystems involve consumer organisms, and as a result ecosystems include both a herbivore-focused food web (located largely, though not entirely, aboveground), and a detritus-based food web. The interactions that occur within each of these food webs, as well as between these food webs, play a major role in determining how ecosystems function. Although a combined aboveground—belowground approach is necessary for the adequate understanding of community- and ecosystem-level processes, most ecological work on aboveground organisms has traditionally been conducted without much explicit consideration of belowground organisms, and most soil biology has been carried out without much acknowledgment of what interactions and mechanisms occur aboveground.

In most terrestrial ecosystems, soils contain by far the greatest diversity of organisms present. On a global basis, the majority of organisms are invertebrates, and most of these spend at least a portion of their life cycle belowground (Ghilarov 1977; Giller 1996). One gram of soil probably contains several thousand bacterial species (Torsvik et al. 1994). It has been estimated that in the Andrews Experimental Forest of Oregon, U.S.A., there are probably around 8000 species of soil arthropods in contrast to only 143 species of aboveground vertebrates (Beard 1991). On a global basis, there are likely to be over a million species of fungi (Hawkesworth 1991) and a million species of nematodes, in contrast to 6000 species of reptiles and 9000 species of birds (May 1988).

Despite the sheer diversity of belowground organisms and the functional role of soil organisms in ecosystem-level processes (both locally and globally), most of the current body of ecological theory is based on synthesizing what aboveground and aquatic ecologists have found, and soil organisms have had a negligible impact on the development of this theory. The treatment of soil organisms in most general ecological textbooks is superficial and in some cases nonexistent. Quantitative literature syntheses and meta-analyses aimed at developing general principles with regard to food web theory (e.g., Cohen et al. 1990; Pimm et al. 1991) and the role of competition and/or predation in structuring ecological communities (e.g., Connell 1983; Schoener 1983; Sih et al. 1985; Gurevitch et al. 1992, 2000; Schmitz et al. 2000) all but ignore soil organisms. A recent review of keystone organisms (Power et al. 1996) makes no mention of arguably the world's most ecologically important keystone organism group, i.e., soil-dwelling nitrogen fixing bacteria. Another recent review, on biological stoichiometery "from genes to ecosystems" (Esler et al. 2000a), makes little mention of the large soil biological literature on the role of elemental ratios as drivers of decomposition and decomposer activity. Studies on belowground organisms make up only a small proportion (characteristically less than 3%) of papers published in the major ecological journals. As a result the current body of ecological literature and main general ecological concepts are representative of only a minority of the Earth's biota (Wardle and Giller 1996). Or, as stated by Ked-

2

dy (1989), "What would a body of ecological theory look like if Plantae, Fungi, Monerans, and Protistans played an appropriate role?" This remark could also be made just as strongly regarding soil invertebrates.

Further, even when the issue of the ecology of belowground organisms is acknowledged by ecologists at large, this is frequently without adequate reference to the vast soil biology literature. An extreme example of this involves a recent commentary in *Nature* titled "Ecology goes Underground" by Copley (2000), and which gives the incorrect impression that scientists are only now beginning to tackle the topic of soil biodiversity and functioning of the soil subsystem, through recently initiated projects in the U.S.A. and U.K. However, no mention is made of the significant body of work published by soil biologists over the past several decades precisely on this topic [see response to Copley (2000) by André et al. (2001), and authored by eight soil biologists from six countries].

Simultaneously, soil science, including soil biology, has tended to ignore what goes on aboveground; too often plants are seen by soil scientists merely as sources of carbon addition to the soil. As such, soil biology does not have a particularly strong theoretical framework, and few soil biologists have sought to apply concepts of ecological theory to their subdiscipline (see Wardle and Giller 1996; Ohtonen et al. 1997; Young and Ritz 1998). This is compounded by the practical problems of applying concepts developed for aboveground ecology to soil biota. First, soil organisms operate within the habitat created by the soil matrix (Coleman and Crossley 1995; Young and Ritz 1998), and there are major difficulties in studying soil organisms in situ because soil is opaque and has a very complex structure, both physically and chemically. Secondly, the extremely high diversity of soil organisms at microscopic spatial scales creates obvious problems in evaluating interactions of specific organisms with each other or with their environment. Thirdly, there are major taxonomic difficulties in identifying many components of the soil biota, especially when dealing with organisms of smaller dimensions, and probably over 90% of species of microflora and microfauna remain undescribed and unknown (Klopatek et al. 1992; Coleman and Crossley 1995).

In the last few years, there has been an increasing interest in exploring the interface of population-level and ecosystem-level ecology (see Vitousek and Walker 1989; Lawton 1994; Jones and Lawton 1995). This requires us to acknowledge the importance of both the aboveground and belowground compartments of terrestrial ecosystems as well as their interactions with each other. Despite this, there have been few attempts to bring together the widely dispersed literature on aboveground and belowground communities, or to interpret this in an ecosystem context. This is because aboveground and belowground ecology have traditionally developed largely independently of one another; the two subdisciplines usually publish in different journals and generally do not widely read each others' journals or bodies of literature. In this light it is perhaps unsurprising that often a finding that has been hailed as a breakthrough by one subdiscipline has long been common knowledge to the other.

This synthesis is aimed at both aboveground and belowground ecologists, with the aim of making both groups more aware of what goes on at the other side of the soil/surface interface. The primary goal of this book is to consider terrestrial communities and ecosystems from a combined aboveground-belowground perspective, to assess the feedbacks that exist between aboveground and belowground communities, and to interpret how these two components working in tandem govern ecosystem functioning and therefore the delivery of ecosystem services. A secondary goal is to consider aboveground-belowground linkages in the context of a changing global environment. This book is focused on concepts and issues, with examples described to illustrate these points wherever appropriate. In order to achieve this, wide use is made of both the general ecological literature and the soil biological literature.

Chapters 2–5 deal with interactions that are relevant to understanding aboveground-belowground linkages, and the consequences of these for ecosystem properties and processes. Chapter 2 considers the soil food web, including the biotic factors that regulate belowground organisms, their interactions at varying scales of resolution, and the consequences of these interactions for key belowground processes including those that regulate plant nutrient supply. Chapter 3 introduces plant communities, and evaluates how plant species effects and plant traits may be important in affecting both the soil biota and the processes that it regulates. Chapter 4 involves consideration of aboveground trophic interactions and food webs, and the consequences of these for the belowground subsystem, primarily as manifested through plants. Chapter 5 completes the circle by considering the means by which belowground organisms, their interactions, and soil food webs influence the growth and community composition of aboveground organisms. These four chapters in combination are intended to provide the basic tools for understanding how terrestrial communities and ecosystems function through explicit consideration of biotic interactions and aboveground—belowground linkages.

The next two chapters aim to apply the concepts developed in chapters 2–5 to two areas of current topical interest. Chapter 6 deals explicitly with the issue of biodiversity in an aboveground-belowground context, in terms of the factors that regulate this diversity, and with the consequences of diversity for the functioning of terrestrial ecosystems. Chapter 7 is focused on utilizing a combined aboveground-belowground approach to better understand the ecological consequences of the major drivers of anthropogenically induced global change. There is a considerable amount of research activity in the areas of both biodiversity and global change, as well as an immense production of

5

literature on these topics. Despite this, only a minority of studies have adopted a combined aboveground-belowground strategy in investigating many of the issues relating to these topics, despite the necessity of such an approach for understanding these issues in a true ecosystem-level context. Finally, chapter 8 aims to bring together a variety of underlying conceptual threads that weave their way throughout the preceding chapters.

CHAPTER TWO

The Soil Food Web

Biotic Interactions and Regulators

In any food web, the primary drivers are the autotrophs, which are responsible for determining the amounts of carbon that enter the system. However, in the case of the decomposer food web, the heterotrophic organisms are ultimately responsible for governing the availability of nutrients required for plant productivity. As a result, the plant and decomposer subsystems are in an obligate mutualism with one another, with each of the two components carrying out processes required for the long-term maintenance of the other.

Those organisms that comprise the decomposer food web span several orders of magnitude in terms of body size, and consist of the microflora (bacteria, fungi), microfauna (body width less than 0.1 mm, e.g., nematodes, protozoa), mesofauna [body width 0.1–2.0 mm, e.g., enchytraeids, microarthropods (mites, springtails)], and macrofauna (body width greater than 2 mm, e.g., earthworms, termites, millipedes). The primary consumers consist of the microflora, which are uniquely capable of directly breaking down complex carbohydrates in plant-sourced detritus and mineralizing the nutrients contained within. The secondary and higher-level consumers, i.e., the soil fauna, feed upon the microbes and each other, and therefore have immense importance in governing these microbial processes. Because of the vast ranges of body sizes of soil fauna, they determine soil processes at a range of spatial scales (fig. 2.1), and for

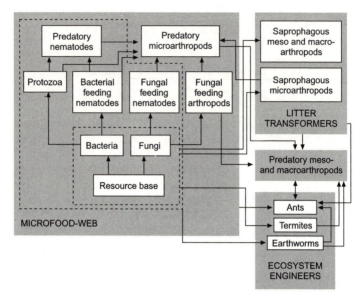

FIGURE 2.1. Structure of the soil food web. Only major groups of organisms and well-established linkages are shown. Arrows indicate direction of energy flow. The microfood-web, litter transformer, and ecosystem engineer categories are derived from Lavelle et al. (1995). Note that the groups of organisms represented in this diagram are only those that have been the most extensively studied; a more complete description of the feeding habits of other, less well-studied fauna, which may nevertheless be ecologically important (e.g., burrowing vertebrates, bacterial-feeding rotifers and tardigrades) is given in Petersen and Luxton (1982).

the purposes of this chapter will be considered as operating at each of three levels (*sensu* Lavelle et al. 1995), i.e., "microfood-webs" involving nematodes, protozoa, and those animals that directly feed upon them and each other; "litter transformers" including saprophagous mites and springtails and some macrofauna, which convert organic matter into organic structures (fecal pellets); and "ecosystem engineers" including earthworms and termites, which build physical structures that create habitats for smaller organisms.

In this chapter I will evaluate how soil food webs may be

driven by plant productivity, and evaluate the roles of specific types of biotic interaction (e.g., "bottom up" effects or resource limitation, "top down" effects or predator-regulated control, mutualism, facilitation) in governing the soil biota at the various scales of resolution identified by Lavelle et al. (1995). It will then be demonstrated how these interactions each govern the availability of the soil nutrient supply that is required for aboveground growth.

CONTROLS: TOP DOWN, BOTTOM UP, AND PRODUCTIVITY

To evaluate how net primary production (NPP) affects components of the soil food web we first need to consider the relative importance of top-down and bottom-up forces in regulating food web components. Much has been written about whether resource limitation (and hence competitive interactions) or predation is the primary control of different trophic levels in food webs, but soil food webs have made at best a very small contribution to the development of general theory about this topic. Before discussing how soil food webs conform to current theories I will briefly outline some of the main theories in this field.

The Hairston, Smith, and Slobodkin hypothesis (Hairston et al. 1960; hereafter referred to as HSS) predicts that for grazing food chains the importance of competition and predation alternates between trophic levels, so that producers are regulated by competition, primary consumers by predation, and secondary consumers by competition. A modification of this theory (Oksanen et al. 1981; Oksanen and Oksanen 2000) suggests that the relative importance of competition and predation is itself governed by primary productivity, so that primary consumers are more likely to be limited by predation in relatively productive than in unproductive systems through the increased importance of effects of secondary consumers. While the HSS hypothesis predicts a "green world" situation in which most plant productivity remains unconsumed due to top-down regulation of herbivores, the theory of Oksanen et al. (1981) predicts a green world along only subsets of gradients of NPP. An alternative theory, derived initially for marine systems (Menge and Sutherland 1976), predicts that top-down regulation of trophic level biomass becomes more important at lower trophic levels. Further challenges to the HSS theory stem from the view that green world phenomena actually result from the regulation of herbivore biomass by the low palatability of most of their potential food resources (White 1978). It has also been suggested that primary consumers may be regulated by both top-down and bottom-up forces; for example, Lawton and McNeill (1979) describe herbivores as being caught "between the devil and the deep blue sea," attacked by predators and parasites from the trophic level above, and also constrained by a nutrient-poor and well-defended food supply.

Evidence has been presented in support of each of these theories, and their relative abilities to predict what happens in the real world continue to be debated (e.g., Schoener 1989; Strong 1992; Polis 1999). Central to this issue is whether the available literature provides consistent evidence for competition or predation in regulating trophic levels. Schoener (1983) and Connell (1983) each synthesized much of the published available literature that reported the results of competition experiments, for various trophic levels and for various habitats. Although both these syntheses found that competition occurred in the majority of studies, their interpretations differed substantially. The analysis by Schoener (1983) appeared to support the HSS hypothesis at least in terrestrial systems, in that producers and secondary consumers competed in a higher proportion of situations than did primary consumers. In contrast, Connell (1983) proposed that much of this evidence was not persuasive, mainly because of the low numbers of competition studies on unexclosed herbivore populations that were available at

that time. The issue of top-down versus bottom-up regulation was also addressed by Sih et al. (1985), who, in contrast, considered published studies that involved assessment
of effects of manipulation for consumers, and found predation to occur in most cases. They claimed to find support of
the Menge and Sutherland hypothesis, in that effects of predation were increasingly more important in regulating lower
trophic levels, for both aquatic and terrestrial systems. However, there are two major problems with these kinds of
analyses. First, it is almost inevitable that studies finding significant treatment effects (e.g., effects of manipulating competitors or predators) are more likely to be published than
those that do not, meaning that the published literature
may contain a disproportionate frequency of studies identifying competition or predation relative to what occurs in
nature. Secondly, the analyses of Schoener, Connell, and Sih
et al. all used a "vote-counting" approach, in which only the
presence versus absence of an effect was noted for each
comparison, rather than the magnitude of the effect and
hence its importance. The second problem can be rectified
through the use of meta-analysis, in which the magnitude of
the effect is also taken into account. One such meta-analysis
(Gurevitch et al. 1992) considered field competition experiments published in seven journals over a ten-year period.
That study found little evidence for green world theories, in
that competition intensity was found overall to be greater
for herbivores than for primary producers or carnivores.
However, they concluded that the small amount of available
terrestrial arthropod data did appear to support the HSS
hypothesis. The issue of competition intensity has been debated especially intensely in the plant ecology literature,
particularly in relation to productivity or stress gradients
(Thompson 1987; Tilman 1988). Grime (1979) maintains
that competition should be more intense in productive environments, while Tilman (1988) maintains that competition
intensity should be important across such gradients even

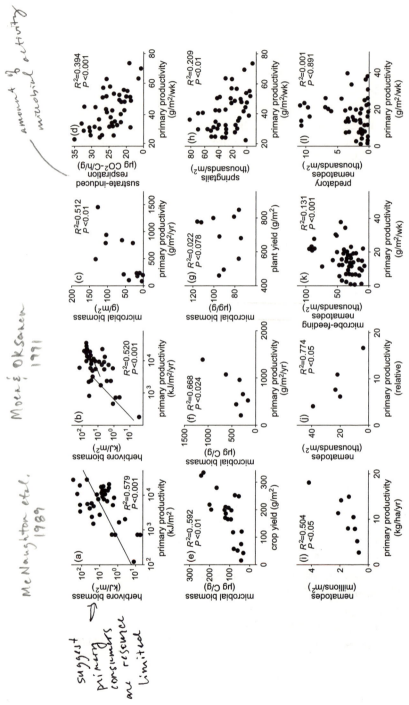

McNaughton et al., 1989

Moen & Oksanen 1991

amount of microbial activity

suggest primary consumers are resource limited

though the resources for which plants are competing should differ along the gradients. Results of experiments (e.g., Wilson and Tilman 1991; Campbell and Grime 1992; Keddy et al. 2000) and literature syntheses (e.g., Gurevitch et al. 1992; Goldberg and Novoplansky 1997) provide mixed support for both points of view and this issue remains unresolved. The specific issue of plant competition along stress gradients will be discussed further in chapter 3.

The issue of importance of top-down versus bottom-up effects can perhaps be most directly assessed by observing how trophic level biomasses vary along gradients of NPP. Here the evidence is again ambiguous, and a range of trophic responses to productivity has been observed (e.g., Oksanen 1988; Power 1992; Van de Koppel et al. 1996). However, it is clear that over large-scale productivity ranges at least, the biomass of aboveground primary consumers does increase with NPP, and this was shown in a study by McNaughton et al. (1989), which synthesized data from published studies representing a range of major terrestrial biomes (fig. 2.2a). However, Moen and Oksanen (1991), upon reanalyzing a

---

FIGURE 2.2. Biomasses or populations of consumer organisms in response to gradients of NPP, with emphasis on the decomposer subsystem. Consumer organisms considered are (a) aboveground herbivore biomass based on global data set (McNaughton et al. 1989); (b) aboveground herbivore biomass using modified version of the McNaughton et al. (1989) data set and with a two-part model fitted (Moen and Oksanen 1991); (c) soil microbial biomass in late successional North American ecosystems (D. Zak et al. 1994); (d) substrate-induced respiration (relative microbial biomass) in a New Zealand grassland (Wardle et al. 1995a); (e) soil microbial biomass in cropping fields in Alabama (Insam et al. 1991); (f) soil microbial biomass in coniferous forests in Oregon (Myrold et al. 1989); (g) soil microbial biomass in grass-sown plots in Rhode Island (Groffman et al. 1996); (h) springtail populations in a New Zealand grassland (Wardle et al. 1995a); (i) nematode populations in a New Zealand grassland (Yeates 1979); (j) nematode populations in Japanese forests (Kitazawa in Yeates 1979); (k) and (l) microbe-feeding and top predatory nematodes in a New Zealand grassland (Wardle et al. 1999b).

modified version of these data, showed that the entire data set could be best fitted by a two-part model (shown by the two lines in fig. 2.2b) in which when productivity exceeded 7000 kJ/m^2/yr herbivore biomass increased more slowly with increasing NPP than below that breakpoint. They interpreted this to mean that top-down effects of a third trophic level became important above this threshhold, consistent with the theory of Oksanen et al. (1981).

The decomposer food web differs from the aboveground (foliage-based) food web in that the primary consumers utilize dead rather than live plant material, and direct top-down effects of consumers on producers are therefore not possible, preventing the types of consumer-producer feedback that characterize foliage-based food webs. The question therefore arises as to how responses of decomposer food web components to gradients of NPP conform to theories developed for foliage-based food webs. To address this issue, I have collated ten data sets from eight studies in which biomasses or populations of consumers in decomposer food webs have been measured across a range of plant productivities (fig. 2.2c–l). In these studies NPP has been assessed only aboveground but this is interpreted as a surrogate of total ecosystem NPP. With regard to the biomass of the decomposer primary consumers (microflora or bacteria and fungi), data exist that suggest responses to NPP which are positive (Myrold et al. 1989; Insam et al. 1991; D. Zak et al. 1994; fig. 2.2c,e,f), negative (Wardle et al. 1995a; fig. 2.2d), or neutral (Groffman et al. 1996; fig. 2.2g). A similar pattern is apparent for secondary consumers; a positive response to NPP appears in the data of Yeates (1979) for total nematodes (most of which feed on bacteria and are therefore secondary consumers) and Wardle et al. (1999b) for microbe-feeding nematodes (fig. 2.2i,k), and a negative response in the data of Kitazawa in Yeates (1979) for nematodes and Wardle et al. (1995a) for springtails (fig. 2.2h,j). Tertiary consumers (predatory nematodes) were not found by Wardle et al. (1999b) to show a clear response to increasing NPP (fig. 2.21).

Why is such a range of responses to plant productivity observed for consumers in the decomposer food web? I identify four likely explanations. The first involves "hidden treatments" (*sensu* Huston 1997), or factors which covary with the treatments or gradients under consideration and which may independently drive the response variable. It is probably impossible to find a primary productivity gradient in which factors that may independently affect both autotrophs and consumers do not vary. For example, the relationships identified by McNaughton et al. (1989) and by Moen and Oksanen (1991) between NPP and herbivore biomass do not mean that a causative association need necessarily exist between the two; low herbivore biomass at the low end of the productivity gradient could be due to both herbivores and plants being controlled by a third factor, such as aridity or low temperature (the unproductive sites in their analyses included deserts and polar regions). In the case of the belowground studies depicted in fig. 2.2, it seems plausible that those abiotic factors varying along the NPP gradients could have independently affected both plants and consumers. The nature of these effects may be context dependent (and thus vary between situations and studies), yielding a range of possible relationships. Further, vegetation composition will itself vary across productivity gradients; domination by different plant functional types over different parts of the gradient will induce variation in consumer trophic level biomasses along the gradient (Wardle et al. 1999b). Secondly, the issue of whether top-down or bottom-up forces are likely to be important may in itself be context dependent; the data sets depicted in fig. 2.2c–l were from a range of differing environments and locations, and, given that a variety of patterns have been detected in herbivore-based food webs, it is perhaps unsurprising that the same would occur for belowground food webs. Thirdly,

15

while plants provide the primary consumers of decomposer food webs with carbon, they also directly compete with them for nutrients, and resource competition between plants and microbes has been frequently demonstrated (Okano et al. 1991; Kaye and Hart 1997). The direction of effects of increasing NPP on belowground consumers may be governed by which of two effects (stimulation of microbes by carbon addition, inhibition of microbes by resource depletion) dominates. Finally, relationships between NPP and soil consumer biomass may be confounded by the large effects that decomposer organisms may themselves have on plant growth through regulating nutrient supply (see chapter 5). In other words, while decomposer systems have often been regarded as donor controlled (see Pimm 1982), in the longer-term perspective the composition of the soil food web also affects plant production.

REGULATION BY RESOURCES AND PREDATION
IN SOIL FOOD WEBS

Because soil microfood-webs are characterized by organisms that engage in direct feeding interactions with each other, there is considerable amplitude for bottom-up and top-down forces to regulate component populations. Further, even though litter transformers and ecosystem engineers in the decomposer subsystem do not generally function as conventional predators, they can still be regulated by resource availability and predation. Despite their immense species diversity and key role in regulating ecosystem functioning, soil food webs have had a negligible impact on the ecological theories discussed so far in this chapter; none of the literature syntheses published by Connell (1983), Schoener (1983), Sih et al. (1985), or Gurevitch et al. (1992) to develop ecological theory about top-down versus bottom-up control include examples from decomposer communities. I will now discuss experimental evidence regarding the importance of bottom-up and top-down controls in regulating composition of soil food webs, and evaluate whether these webs conform to theories that were developed through use of data from other systems.

Bottom-Up Controls of Decomposer Microflora

The decomposer microflora, or the primary consumers of the decomposer food web, consists of two very different classes of organisms belonging to different kingdoms: fungi, which are sessile and intimately linked to the resource base; and bacteria, which are mobile and are active in the aqueous pore spaces within the soil. This split has profound trophic implications which will be discussed further on in this chapter, but I will first consider this trophic level (i.e., the "microbial biomass") as a whole. Figure 2.2 provides observational evidence that the microbial biomass increases with increasing NPP, consistent with what we would expect through bottom-up regulation, for three of the five cases depicted (fig. 2.2c,e,f versus fig. 2.2d,g). Four further lines of evidence that the microbial biomass *as a whole* is indeed bottom-up controlled are presented as follows.

1. Substrate Addition and Depletion

When resources in the form of readily available carbon (e.g., simple sugars) are added to the soil, there is almost always a rapid increase in microbial biomass (often within several hours), indicative of carbon limitation (e.g., Anderson and Domsch 1978; Nordgren 1992). While this is consistent with bottom-up control, it represents a short-term "pulse" in which microbial biomass increases within a shorter time frame than it takes for their predators to reach a sufficient level to exert top-down control. However, longerterm experiments (of several months duration) involving addition of glucose to field plots or soil in microcosms generally point to long-term stimulation of the microbial biomass by resource addition (Jonasson et al. 1996a,b; Mikola

17

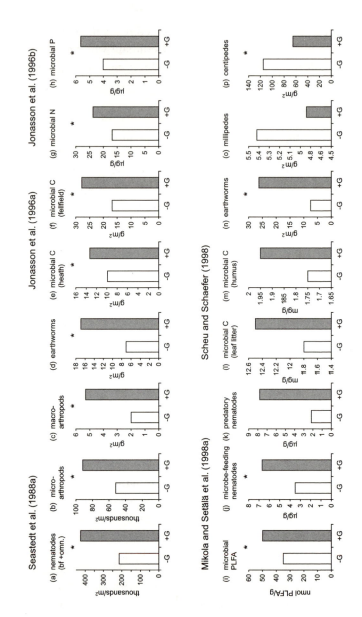

and Setälä 1998a; Mikola 1999; but see Scheu and Schaefer 1998) (fig. 2.3). Similar patterns are sometimes also observed following addition of other limiting resources to field plots, such as nitrogen and phosphorus (Jonasson et al. 1996a,b). Further, when available soil carbon is depleted, for example thorough long-term soil storage without resource addition, microbial biomass usually declines (Visser et al. 1984).

2. Rhizosphere Effects

In nearly all situations, the presence of actively growing plants enhances the soil microbial biomass, largely through the addition of relatively easily degraded carbon sources originating in the rhizosphere. Studies involving ^{14}C addition to herbaceous plants consistently find that a sizable proportion (sometimes over a third) of the added carbon ends up in the rhizosphere, and much of this is incorporated into the soil microbial biomass (e.g., Martens 1990; Johansson 1992). Microbial biomass in rhizosphere-associated soil is often several times greater than in corresponding nonassociated soil; it has been shown through the use of rhizoboxes (in which compartments of soil are separated by mesh at different distances from the roots) that the microbial biomass of soil located within a few millimeters of the root surface can be greatly stimulated (Klemendtsson et al. 1987; Youssef et al. 1989). Microbial biomass can also be

---

FIGURE 2.3. Effects of glucose addition on components of the decomposer food web, including primary consumers (panels e–i,m), secondary consumers (a,j), higher-level predators (k,p), and saprophages and earthworms (d,n,o). All studies involved field plots except for that of Mikola and Setälä (1998a) which used microcosms. For the Mikola and Setälä (1998a) study, only the final sampling date data for the experiment with three consumer trophic levels are presented. bf and omn represent bacterial feeders and omnivores. − G and + G indicate nonaddition and addition of glucose. * indicates that the glucose addition effect is significant at $P = 0.05$.

stimulated by enhancing plant photosynthetic rates (and hence allocation of carbon to roots) through increasing light levels, and shading of plants has been shown to reduce levels of microbial biomass (Harley and Waid 1956; Mikola et al. 2000). This pattern is not, however, universal, with only some atmospheric CO_2 enrichment experiments showing enhanced NPP through elevated CO_2 levels also finding corresponding increases in levels of microbial biomass (see chapter 7).

3. Resource Quality

As will be discussed in chapter 3, microbial biomass is elevated by improved resource quality (e.g., reduced carbon-to-nitrogen ratios, reduced levels of plant-derived secondary metabolites), and different plant species have differential effects on microbial biomass. Furthermore, microbial biomass in soils is clearly positively correlated with the resource status of soils (e.g., soil carbon and nitrogen concentrations) for most spatial scales and vegetation types (Wardle 1992). Such correlations are to be expected if the microbial biomass is bottom-up controlled.

4. The Importance of Competition

To show that bottom-up control is important in regulating microbial biomass requires evidence of competition among components of the biomass. There is clear evidence that competition among fungal species is common, and most studies that have explicitly tested for fungal competition have detected it. There are two mechanisms behind this competition, i.e., resource depletion ("exploitation competition") and interference (mainly through the production of antibiotics); although these mechanisms are not usually separated in competition experiments, both are probably important (Lockwood 1981; Wicklow 1981). There is a vast literature on investigating interactions between colonies of pairs of fungal species on agar plates (no doubt because of the ease of performing such experiments), and this is especially prevalent in the plant pathology literature in relation to the suppression of soil-dwelling plant-pathogenic fungi by other fungal species. While the majority of studies on agar find some form of inhibition between species (e.g., growth rate reduction; "deadlock" in which neither can grow into the zone occupied by the other), the approach is highly artificial and experiments that are performed on both agar and "real" substrates often yield entirely different competitive outcomes (Webber and Hedger 1986). Fungal competition experiments performed on "real" substrates, such as litter, soil, or wood blocks do, however, usually show competition to be important. One such approach (developed by Garrett 1963) involves adding inocula of two fungal species to litter to determine the effects of each species on litter colonization by the other, i.e., its "competitive saprophytic ability." Using such a design, Widden and Hsu (1987) found consistent competitive effects among most of the possible two-way combinations of five *Trichoderma* species, and Wardle and Parkinson (1992) identified strong antagonistic effects for all but one of the possible combinations among four coexisting soil fungal species. Fungal species may also reduce biomass production of each other in soil (Wardle et al. 1993) and experimental (Carreiro and Koske 1992; Wardle et al. 1993) and modeling (Schmit 1999) evidence also demonstrates that competition among fungal species can reduce their allocation of resources to reproduction.

While it is impossible to study competition of soil-inhabiting fungi without resorting to highly artificial, sterilized experimental systems, insights can be gained by considering fungi that colonize dead wood, because interactions between individual fungal colonies can be observed directly on their substrate. There is clear evidence from both observational and experimental work on dead wood that different fungal colonies interfere with one another, and that when hyphae of different colonies meet the interactions are often

21

highly antagonistic (Todd and Rayner 1978; Rayner and Todd 1979; reviewed by Boddy 2000). The experimental work of Holmer and Stenlid (1996), which considered competitive interactions among pairs of six wood-degrading fungi on wood blocks, found competition to be ubiquitous, and a clear competitive hierarchy among the six species emerged. Further, studies on wood decay fungi show antagonism between genetically different dikaryons of the same fungal species; Todd and Rayner (1978) found evidence of competition among genetically unrelated dikaryons of the wood decay fungus *Coriolus versicolor* with the intensity of interaction being less for more closely related dikaryons. While fungal species do not function as individuals or form distinct populations, it is nevertheless apparent that they show similar modes of competition as those groups of organisms that do.

Is competition in fungal communities important in all situations or is it context dependent? As stated earlier, plant ecologists have intensely debated whether or not competition is greater under more favorable (and less stressed) conditions. The same question can be applied to fungi. Only a handful of experiments have evaluated fungal competition intensity in situations of varying resource availability and they have given mixed results. Two studies, by Widden and Hsu (1987) and Carreiro and Koske (1992), compared fungal competition intensity on a low-quality and a high-quality substrate (pine versus maple litter, and oak litter versus dead moth wings, respectively); in both cases competition was intense on both substrates and there was no evidence that competition intensity was greater on the higher-quality resource despite the competitive balance between species being affected. Further, Wardle and Parkinson (1992) found, on both soil and agar, that competition intensity among pairwise combinations of four fungal species sometimes increased and sometimes decreased with increasing resource quality. However, Stahl and Christensen (1992) found in agar experiments considering interactions among seven fungal species that antagonism was actually greatest in high-resource media, and they concluded that fungi became territorial when colonizing resource-rich substrates but not in oligotrophic conditions. Despite this finding intense fungal competition clearly does occur in at least some harsh environments with low resource availability, such as discussed above for wood.

Despite the methodological difficulties in studying fungal competition, it would appear that fungal competition is important in the majority of situations, consistent with the other evidence I have presented that microbial biomass (which is usually dominated by fungal biomass) is strongly influenced by bottom-up forces. There is less available evidence with regard to the bacterial component of the microbial biomass, presumably because of methodological and taxonomic problems in investigating bacterial competition. Later in this chapter I will discuss the likely importance of competition within fungal versus within bacterial communities.

Bottom-Up Controls of Soil Fauna

Although the soil fauna consists of organisms that operate at vastly different spatial scales, there is clear evidence from a number of studies that bottom-up control can be important in regulating soil faunal groups, although there are also several studies that do not show this. Three of the five studies depicted in figure 2.2 do not show soil faunal populations to increase with NPP. Studies involving addition of basal resources to soil have, however, found associated increases in soil fauna, and given that soil fauna consists of secondary and higher-level consumers, these responses are therefore multitrophic. Positive responses of adding glucose to field plots were detected by Seastedt et al. (1988a) for a range of soil fauna with different trophic positions, although mixed responses to glucose addition were found by Scheu and Schaefer (1998) (fig. 2.3). In the microcosm experiments of Mikola and Setälä (1998a) (fig. 2.3) and Mikola (1999), in which model food webs were constructed, glucose addition stimulated microbial productivity and the biomass of the secondary consumer (microbe-feeding nematodes) but did not induce consistent increases in the tertiary consumer (predatory nematodes). Other studies have involved manipulation of field plots with organic matter, and although such studies can be confounded by physical effects of the resource addition (e.g., through enhanced soil moisture retention) they do provide relevant data about the physical effects of resource addition. For example, Dackman et al. (1987) found that addition of farmyard manure to soils also enhanced each of three consumer trophic levels: bacteria, nematodes (presumably mostly bacterial feeding), and nematode-trapping fungi. Similarly Brzeski and Szczech (1999) found large positive effects of sawdust addition to field plots on microbial-feeding nematodes over a seven-year period. In contrast, Yeates et al. (1999b) determined that sawdust addition to field plots did not stimulate the standing crop of microbe-feeding nematodes despite stimulating microbial production, but did stimulate top predatory nematodes over a seven-year period. In the same field experiment, Wardle et al. (1999c) identified positive effects of sawdust addition on springtails and some soil-associated macrofauna (notably spiders and opilionids) but only during some years of the experiment. Further, a study involving addition of a mix of organic substrates to field plots in a forest (Chen and Wise 1999) found all groups of soil meso- and macrofauna to be strongly stimulated by resource amendment, regardless of their trophic position (fig. 2.4).

It has been proposed that as the amount of basal resources increases food chain length also increases because sufficient carbon becomes available for fauna of higher trophic levels to exist (e.g., Kaunzinger and Morin 1998; Oksanen and Oksanen 2000; but see Post et al. 2000). With regard to the decomposer subsystem, Moore and de Ruiter

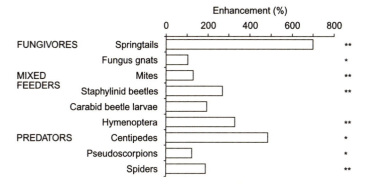

FIGURE 2.4. Effects of basal resource addition on major groups of soil fauna in a field plot experiment in an oak-maple forest in which plots were either amended with substrate (446 g/m^2 of additional detritus) or left intact. Only data for the final sampling date of the study (107 days after amendment) are included. Data are based on population densities. * and ** indicate that the enhancement effect by resource addition is significant at $P = 0.05$ and 0.01, respectively. Figure derived from results presented by Chen and Wise (1999).

(2000) found that food webs in cave sediments with carbon inputs of less than 1 g/m^2/yr consisted of only bacteria and protozoa, while sediments with greater amounts of inputs supported a wider range of microbe-feeding fauna and also predatory mites. However, the vast majority of soil types have a sufficient input of resources to support higher-level consumers, and there is little evidence of resource limitation influencing food chain length in most soil food webs, i.e., across the range of NPP that characterizes most of the world's ecosystems (cf. Whittaker 1975).

Evidence for bottom-up control of soil fauna, at least in the soil microfood-web, also arises from studies that have considered the rhizosphere and root zone of plants. Increases in bacterial biomass in the rhizosphere are frequently soon followed by large increases of bacterial-feeding protozoa (e.g., Clarholm 1985; Klemendtsson et al. 1987) and nematodes (e.g., Rønn et al. 1996). In the study of

25

Christensen et al. (1992) soil associated with freshly killed barley roots soon contained 80 times more protozoa and 30 times more bacterial-feeding nematodes than surrounding soil. These increases are, however, ephemeral and probably of too short a duration for higher-level predators to reach sufficient density to regulate this microfauna. Few studies have investigated fungivores in the rooting zone, although Parmelee et al. (1993) detected some significant rhizosphere effects of pine seedlings on total microarthropod populations. On a larger spatial scale, Wardle et al. (1999b) found that removing all phytomass from small grassland plots over a three-year period reduced some components of the soil fauna (e.g., springtails, earthworms, predatory nematodes) but not others (e.g., mites, microbe-feeding nematodes) despite reductions in soil microbial biomass.

Bottom-up regulation of soil fauna through resource (e.g., litter) quality is apparent in some studies and for some food web components, but not others (Wardle and Lavelle 1997; discussed in chapter 3). However, there is behavioral evidence that some soil animals will actively migrate to patches with improved resource quality and availability. For example, Griffiths and Caul (1993) presented experimental data that suggest that bacterial-feeding nematodes actively migrated to decomposing grass residues containing high bacterial populations, although protozoa did not. Meanwhile, Bengtsson et al. (1994) used experiments in which distinct patches were represented by vials connected by tubing to demonstrate that fungal-feeding springtails actively migrated over distances up to 40 cm to reach patches occupied by preferred fungal food sources (probably via fungal odor). Dispersal to new patches was greater when population densities were higher, food was less available, and individuals were in a feeding rather than in a molting phase. Further, Hall and Hedlund (1999) found that predatory mites (which fed on fungal-feeding springtails) were themselves attracted to patches occupied by fungi, indicating the presence of stimuli that attract mites to patches in which there is a greater chance of finding fungal-feeding prey.

Bottom-up regulation of soil biota should result in resource competition within trophic levels. Assuming that most faunas in soil food webs are generalist feeders, this seems likely, although significant resource partitioning may also be present due to the high diversity of microhabitats present in the soil (Anderson 1975; Ettema 1998; see also chapter 6). Only a few experimental studies have attempted to directly investigate competition among trophically equivalent soil faunal species. Theenhaus et al. (1999) found, upon combining two springtail species in microcosms, that a contramensural $(+,-)$ rather than a competitive $(-,-)$ relationship was detected, despite resource addition stimulating both species. However, in an experiment combining two earthworm species, Huhta and Viberg (1999) detected competitive reduction of both species. Studies on carabid beetles, many of which are top predators in soil food webs, reveal at least partial evidence of both competition and resource partitioning among species (Niemelä 1993). However, to date too few studies have been performed on soil faunal competition to be of much use in contributing to generalizations about top-down versus bottom-up regulation in soil food webs.

Control of Microflora by Predation

Direct predation of the soil microflora occurs mainly in the soil microfood-web. The nature of predation of the two components of the microflora, bacteria and fungi, differs tremendously. Bacteria, which inhabit the aqueous pores of the soil, are usually consumed whole by predation, while fungi, which are intimately associated with their substrate, are grazed by animals consuming hyphae growing on the colony surface.

Bacteria have few modes of resisting grazing and their main mode of defense is often simply by occupying soil pores that are too small to be accessed by their predators. Hassink et al. (1993) determined that grazing pressure by nematodes on bacteria was greater in soils which had a coarser texture and therefore a lower density of pores small enough for bacteria to avoid their predators. A greater density of bacteria therefore occurred in finer-textured soils. Evidence of top-down control of bacteria is also apparent from field experiments (Wardle and Yeates 1993; Wardle et al. 1995b) in which improved resource quality and amount had no effect on bacterial biomass but did cause an increase in higher trophic levels. Further field evidence for top-down control of bacterial biomass emerges from the studies of Santos et al. (1981) and Ingham et al. (1986), which found that large reductions in bacteria were caused by increases in their predators. In experimental microcosms, predation of bacteria by nematodes reduced their biomass in the study of Allen-Morley and Coleman (1989) while no consistent effects of nematodes on bacterial biomass were found by Mikola and Setälä (1998b), even though grazing increased bacterial turnover rate. Microcosm studies consistently find bacterial-feeding microfauna to greatly stimulate bacterial turnover rate and production (Stout 1980; Vreeken-Buijs et al. 1997; Mikola and Setälä 1998b) even when top-down regulation of the standing crop of bacterial biomass does not occur.

In contrast to bacteria, there is less consistent evidence for top-down suppression of fungal biomass by their grazers. Indeed, Hedlund and Augustsson (1995) found in microcosms that hyphal length of the fungus *Mortierella isabellina* was greatest when subjected to intermediate intensities of grazing by the enchytraeid *Cognettia sphagnetorum* (fig. 2.5). This study represents clear evidence of an overcompensatory growth response to grazing. Similarly, Hanlon (1981) found respiration of the fungus *Botrytis cinerea* to be maximal under intermediate levels of grazing by the springtail *Folsomia candida* (fig. 2.5). Exclusion of mesofauna, includ-

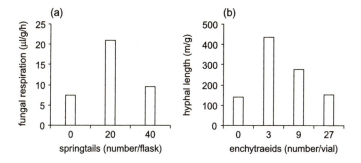

FIGURE 2.5. Stimulation of fungal activity or biomass by intermediate densities of grazers in microcosms (a) derived from Hanlon (1981), using data from only the final sampling date from figure 4 of that paper, and (b) derived from Hedlund and Augustsson (1995). In both cases values for the intermediate faunal density treatment are significantly different from those for both the lowest- and highest-density treatments at $P = 0.05$.

ing fungal grazers, reduced microbial biomass in field-placed litter bags in the study of Scholle et al. (1992), while in contrast Hedlund and Öhrn (2000) found that introduction of fungal-feeding springtails to microcosms reduced fungal biomass but stimulated their respiration.

It would appear from the available literature that bacteria are often regulated by top-down effects of their grazers while fungi frequently are not. This concurs with the hypothesis of Wardle and Yeates (1993) that bacterial biomass, but not fungal biomass, is predator regulated. I identify two reasons for this difference. First, bacterial feeders can ingest their prey whole, and their food has relatively few defenses against grazing, and low carbon-to-nitrogen and carbon-to-phosphorus ratios. However, fungi are morphologically relatively complex, contain a wide range of antifeeding (morphological and chemical) defenses, and have high carbon-to-nutrient ratios and therefore represent a poorer quality food source (Wardle and Yeates 1993). This explains why the biomass ratio of bacterial feeders to bacteria is much greater than that of fungal feeders to fungi (Wardle and Yeates 1993). Secondly, fungal species differ tremendously in resource quality and fungal feeders are therefore less generalist in their feeding habits than are bacterial feeders. This, combined with the likelihood that fungal biomass is usually resource limited (see earlier discussion), means that when a fungal species is reduced by grazing, unpalatable species are favored through competitive release. Experimental evidence points to unpalatable fungal species being promoted through selective grazing by springtails on palatable fungal species (e.g., Newell 1984; Klironomos et al. 1992). This type of biomass compensatory response to grazing in the fungal community may at least partially mitigate top-down control of fungal biomass that might otherwise occur.

Control of Soil Fauna by Predation

Regulation of the soil fauna through predation by other soil fauna appears to be common. Several studies have demonstrated top-down control of microbe-feeding microfauna. For example, Santos et al. (1981) found that selective removal of tydeid mites on litter caused large increases in bacterial-feeding nematodes. Similarly, microcosm studies have found large reductions of microbe-feeding nematode populations resulting from addition of top predatory nematodes (Allen-Morley and Coleman 1989; Mikola and Setälä 1998b) and predatory mites (Laakso and Setälä 1999a). Some mesofaunal groups have also been shown to be regulated by predation; two field studies (Kajak et al. 1993; Lawrence and Wise 2000) have both found convincing evidence that exclusion of spiders in the field (through small exclosure plots) induces large increases in springtail populations. Predator regulation of enchytraeids has been shown to occur in both microcosms (Huhta et al. 1998) and field litter bag studies (Nieminen and Setälä 1998).

Much remains unknown about predator regulation of larger soil fauna, although it is likely that the predators of these fauna consist largely of vertebrates (birds, lizards, and mammals) and live largely above ground. Indeed it is probably at this point that the foliage-based and detritus-based food webs meet (Polis 1994; Wardle 1995). Two studies show the likely importance of such effects. In the first, Wyman (1998) established field enclosure plots on the forest floor with and without salamanders, and found that salamanders significantly reduced densities of litter-dwelling millipedes, molluscs, beetle larvae, and dipteran larvae. In the second (Shvarts et al. 1997), data based on consumption rates by shrews in Eurasian forests suggested that they were likely to consistently ingest a substantial proportion (>20%) of the total soil invertebrate biomass stock, and it was predicted that these effects should therefore operate as an important determinant of invertebrate densities.

Trophic Cascades in Soil Food Webs

While there is general agreement that trophic cascades (in which predators induce effects that cascade down food chains and affect biomass of organisms at least two links away) can be important in some simple aquatic food webs, the issue of whether cascades are important or widespread in terrestrial communities has been a subject of much debate. Strong (1992) proposed that trophic cascades were "all wet," while Polis (1994) presented several reasons as to why cascades at the community level in terrestrial systems may be unimportant, including that food webs contain a high degree of omnivory, that the strengths of trophic linkages may often be too weak for cascades to occur, and that many food chains are subsidized by resources or energy channels which originate outside the food chain (see chapter 4). In contrast, Pace et al. (1999) suggested that trophic cascades are widespread regardless of ecosystem type, and cited several examples where terrestrial cascades clearly occur. Further, recent meta-analyses of studies in which terrestrial aboveground carnivorous arthropods have been manip-

	MANIPULATION	TROPHIC EFFECTS		PROCESS EFFECTS
Santos et al. (1981)	Reduce tydeid mites →	Increase bacterial feeding nematodes →	Reduce bacteria →	Reduce decomposition rate
Allen-Morley and Coleman (1989)	Add predatory nematodes →	Reduce microbe-feeding nematodes →	Reduce microflora →	Enhance C and N mineralization
Martikainen and Huhta (1990)	Add predatory mites →	Reduce microbe-feeding nematodes →	? →	No effect on decomposition
Kajak et al. (1993)	Reduce epigeic predators →	Increase saprophagous mesofauna →	Increase microflora →	Increase decomposition and N mineralization
Mikola and Setälä (1998a)	Add top predatory mites →	Reduce microbe feeding nematodes →	No effect on microbial biomass →	Reduce C and N mineralization
Wyman (1998)	Add salamanders →	Reduce invertebrates →	? →	Reduce decomposition rate
Laakso and Setälä (1999a)	Add predatory mites →	Reduce microbe-feeding nematodes →	No effect on microbial biomass →	Weak stimulation of C and N mineralization
Hedlund and Öhrn (2000)	Add predatory mites →	Presumably reduction of springtails →	Enhance hyphal length →	Enhance C mineralization
Lawrence and Wise (2000)	Exclude spiders →	Increase springtails →	? →	Increase decomposition rate

ulated (Schmitz et al. 2000; Halaj and Wise 2001) provide evidence suggesting that trophic cascades that reduce plant damage by plants are common, although the effects of cascades on plant biomass are generally weaker. However, whether these examples are representative of most terrestrial ecosystems remains to be demonstrated.

Are trophic cascades common in soils? Figure 2.6 outlines nine studies which help to address this question, and several of these provide evidence that manipulation of tertiary or higher-level consumers frequently induces cascades that affect the microflora. It is noteworthy that cascades have also been shown for both the bacterial-based energy channel (Santos et al. 1981; Allen-Morley and Coleman 1989), which operates as an aquatic system, and the fungal-based energy channel (Allen-Morley and Coleman 1989; Hedlund and Öhrn 2000), which operates as a terrestrial system. However, there are also examples in which cascades do not occur (Martikainen and Huhta 1990; Mikola and Setälä 1998b; Laakso and Setälä 1999a). In the study of Mikola and Setälä (1998b) the absence of a cascade occurred because although predatory nematodes reduced microbe-feeding nematodes, grazer-induced compensatory growth of the microflora was reduced when microbe feeders were reduced, resulting in no net change in microbial biomass. In other words, top predators indirectly affected microbial productivity, activity, and turnover, but not biomass.

A note of caution is, however, required with regard to these studies. First, they have mostly focused on the soil microfood-web (the soil biotic component in which direct feeding interactions are most apparent), and the findings

FIGURE 2.6. Summary of studies investigating trophic cascades in soil food webs and their consequences for decomposer processes. Santos et al. (1981), Kajak et al. (1993), Wyman (1998), and Lawrence and Wise (2000) involved predator manipulation using exclosures in the field and all others were performed in microcosms.

are therefore relevant to only a subset of the soil biota. Second, many of these studies have been performed in simplified microcosm systems in which consumers are presented with a limited range of prey, meaning that feeding linkages enforced in the experimental setup may be unimportant or weak in nature. The simplicity of microcosm food webs means that only species-level rather than community-level cascades can be demonstrated. In addition, omnivory may be common in real microfood-webs, particularly at higher trophic levels (Moore et al. 1988), meaning that the simple linear food chains present in artificially constructed food webs in microcosms may not necessarily dominate in real food webs. Indeed, Scheu and Falca (2000) and Ponsard and Arditi (2000) both used measurements of $\delta^{15}N$ in tissues from a range of soil faunal species to suggest that many arthropods in soil food webs feed across more than one trophic level, although the latter study suggested that the macroinvertebrate fauna broadly differentiates into two distinct trophic levels. However, the study of Santos et al. (1981) provides clear field evidence of a trophic cascade that influences bacterial densities on plant litter. Further, Wardle and Yeates (1993) and Wardle et al. (1995b) present field data suggesting that improvements in basal resource availability and quality stimulated microbial biomass because the enhancement of top predatory nematodes regulated microbe-feeding nematodes.

Multitrophic Considerations in Microfood-Webs

The material presented so far in this chapter provides clear evidence of bottom-up control of the microbial biomass and in particular the fungal compartment (usually representing the majority of the microbial biomass), several instances of top-down regulation of the bacterial compartment, and numerous examples of both top-down and bottom-up regulation of higher-level consumers. The question therefore arises as to whether this evidence supports theories developed regarding the importance of top-down versus bottom-up control of trophic level biomasses in other food webs. The presence of trophic cascades in soil food webs, if they are important, is consistent with both the HSS and Oksanen et al. (1981) theories. However, studies involving manipulations of basal resources provide only partial support for such theories. In the only direct multitrophic experimental test to date of these theories in soil food webs, Mikola and Setälä (1998a) added basal resources to microcosms containing microbe-nematode food webs occupying three consumer trophic levels. They found that both primary and secondary consumers responded positively to resource addition regardless of whether or not a top predator was included, and that biomass of the secondary consumer was governed by both top-down and bottom-up control. Both these findings are inconsistent with the HSS and Oksanen et al. (1981) theories, which predict top-down and bottom-up forces to regulate alternate trophic levels. That most studies show certain major soil faunal groups, in particular springtails, to be affected both by manipulation of predators and by manipulation of basal resources is also inconsistent with such theories.

Partial support for these theories does arise, however, from the studies of Wardle and Yeates (1993) and Wardle et al. (1995b), in which bacterial-based and fungal-based energy channels (*sensu* Moore and Hunt 1988) were compared. Correlational evidence from field plots with different resource manipulations (Wardle and Yeates 1993) showed that bacteria, bacterial-feeding nematodes, and fungal-feeding nematodes were regulated by predation, while fungi and top predatory nematodes were regulated by resources. This was also supported by data indicating that as the quality of sawdust placed in the field improved over time, only fungi and top predatory nematodes showed a consistent positive response (Wardle et al. 1995b). It is therefore apparent that in this system the fungal-based energy channel conforms to

35

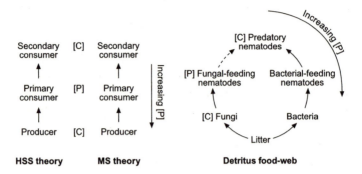

FIGURE 2.7. The soil microbial-nematode component of the soil microfood web in the study of Wardle and Yeates (1993) in relation to the Hairston, Smith, and Slobodkin (HSS) theory and the Menge and Sutherland (MS) theory. [C] and [P] indicate that competition (resource limitation) and predation, respectively, are likely to be the principal regulatory forces operating. From Wardle (1995). (Reproduced with permission from Academic Press)

the HSS theory while the bacterial-based energy channel conforms to the theory of Menge and Sutherland (1976) (fig. 2.7).

It is clear from the results of Wardle and Yeates (1993) that the bacterial-based and fungal-based energy channels behave very differently, and this is consistent with the evidence stated earlier in this chapter that fungi usually respond to resources while bacteria are frequently strongly affected by manipulation of their predators. It is also noteworthy that bacterial-based energy channels, which operate in an aquatic environment, conform to a theory that was developed from an aquatic perspective (Menge and Sutherland 1976), while fungal-based energy channels, which are based on sessile organisms, conform to a theory that was developed for terrestrial communities (i.e., the HSS theory). The findings that predation regulates microbes in the soil aqueous environment more than in the nonaqueous environment, and that a higher proportion of bacterial biomass than fungal biomass is consumed by microbe-feeding fauna, make an interesting parallel with what has been found for foliage-based food webs. Here, evidence from literature syntheses (Cyr and Pace 1993; Cebrian 1999) indicates that herbivores consume a much higher proportion of foliage in aquatic than in terrestrial systems, and that top-down regulation of phytomass is more likely to be important in aqueous habitats. Further, the high palatability of bacteria relative to fungi presumably results in part from bacteria having much lower carbon-to-nitrogen and carbon-to-phosphorus ratios than do fungi; this parallels the results of a recent literature synthesis of foliage-based food webs (Esler et al. 2000b) showing that basal trophic levels in aquatic food webs have much lower values for these ratios than do basal trophic levels in terrestrial food webs.

LITTER TRANSFORMERS, ECOSYSTEM ENGINEERS, AND MUTUALISMS

Soil fauna that operate at different spatial scales differ considerably in the nature of effects that they exert on the soil microflora. Animals in the microfood-web enter direct predator-prey relationships with their food sources. However, those mesofauna that operate as "litter transformers," and macrofauna that operate as litter transformers and ecosystem engineers, have effects on the soil microflora which are generally less direct, and which do not fit easily into classical theories relating to top-down and bottom-up regulation in food webs.

Litter Transformers

Litter transformers operate by consuming plant debris, and egesting this material as fecal pellets which are made up of much smaller particles than the ingested material. This egested material has a vastly modified pore size distribution, is frequently more compacted, and has a greatly enhanced moisture retention ability (Swift et al. 1979). This

37

transformed litter is highly favorable for microbial growth, particularly that of the bacterial component. The result is a vastly modified bacterial-to-fungal ratio; for example, Hassall et al. (1987) found bacterial density in fecal pellets produced by isopods to be enhanced by over one hundred times, and fungal density by only three times, compared to that present in fresh preingested litter. Differences in bacterial and fungal responses appear to occur because compaction of detritus into pellets can be detrimental to fungal growth (Lavelle 1997). The relationship between the fauna and microflora is mutualistic because increased microbial activity in the fecal pellets results in elevated rates of decomposition and nutrient mineralization, and reingestion of pellets by the fauna results in their improved nutrition. This system functions as an "external rumen," and is of particular importance when faunas consume litter with little prior microbial conditioning and in nutrient-poor situations (Wardle and Lavelle 1997). In such situations animals prevented from reingesting their fecal pellets can be very adversely affected (Hassall and Rushton 1982).

The Macrofaunal Gut as a Microbial Environment

Soil fauna classified by Lavelle et al. (1995) as "ecosystem engineers" can interact with soil microflora via the "external rumen" relationship described above (e.g., Hamilton and Sillman 1989), but in most cases the main focus of mutualism between the animals and microbes actually occurs within the faunal gut, which operates as an environment for microbial activity. In the case of earthworms, free-living microflora are promoted by favorable moisture and pH conditions in the gut, and a readily available substrate in the form of intestinal mucus consisting of low-molecular-weight carbon sources (Martin et al. 1987). Mucus production has been identified in the gut zone of several earthworm species, although species differ in the amounts of mucus produced (Trigo et al. 1999). In turn the microflora produce cellulases which break down organic matter entering the gut and release nutrients to the earthworm. However, passage through the earthworm gut usually stimulates bacteria rather than fungi (e.g., Daniel and Anderson 1992), and there is evidence of digestion of ingested fungal hyphae (e.g., Edwards and Fletcher 1988; but see Wolter and Scheu 1999) and microfauna (e.g., Bonkowski and Schaefer 1997) by earthworms. In other words, earthworms appear to act as mutualists with some subsets of the soil microfood-web and as predators of others.

The arthropod gut also serves as a habitat for microflora. There is some evidence (e.g., Borkott and Insam 1992) that even arthropods as small as springtails may have a resident symbiotic gut microflora, but this phenomenon appears to be widespread for larger soil arthropods including cockroaches, millipedes, and isopods (Bignall 1984). The role of mutualisms involving resident gut microflora is especially apparent for termites, which house a diverse array of bacteria and (in the case of lower termites) specialist cellulose-digesting protozoa (Breznak and Brune 1994). These resident organisms are essential for the digestion of complex carbohydrates, particularly those from ingested dead wood. Termite guts host a range of functional groups of microflora, including nitrogen fixing bacteria, which enable them to acquire new nitrogen not present in their ingested substrates (Bentley 1984).

Formation of Physical Structures

In addition to developing mutualistic interactions with microflora in their gut cavity, those soil organisms regarded as "ecosystem engineers" create physical structures which provide a modified habitat for soil microflora and other soil fauna, especially those that operate at much smaller spatial scales. Earthworms which form middens, casts, and burrows, termites which form mounds and galleries, and ants which form nests all qualify as "allogenic" ecosystem engineers as

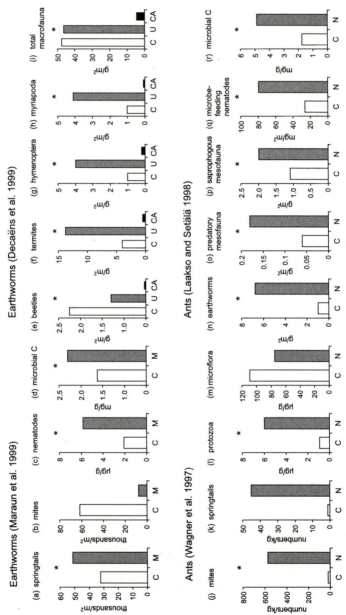

Earthworms (Maraun et al. 1999)

(a) springtails
(b) mites
(c) nematodes
(d) microbial C

Earthworms (Decaëns et al. 1999)

(e) beetles
(f) termites
(g) hymenoptera
(h) myriapoda
(i) total macrofauna

Ants (Wagner et al. 1997)

(j) mites
(k) springtails
(l) protozoa
(m) microflora

Ants (Laakso and Setälä 1998)

(n) earthworms
(o) predatory mesofauna
(p) saprophogous mesofauna
(q) microbe-feeding nematodes
(r) microbial C

defined by Jones et al. (1994), that is, those that alter the physical state of materials. Earthworms are frequently observed to stimulate microflora in the structures that they create (Satchell 1983; Brown 1995) through improving aeration and moisture conditions, secretion of mucus, and enhanced microbial accessibility to resources (e.g., by burial of plant litter). It would appear that disturbance-adapted microbes are particularly adapted to earthworm activity; for example, McLean and Parkinson (2000a) found evidence that faster-growing fungal species were favored by high densities of the forest-dwelling earthworm *Dendrobaena octaedra*. Soil fauna is also frequently enhanced in structures created by earthworms, for example in middens (e.g., Maraun et al. 1999) (fig. 2.8a–d), in soil underlying casts (e.g., Decaëns et al. 1999) (fig. 2.8e–i), and in the walls of burrows (Tiunov et al. 2001). This may result either from improved physical structure of the environment such as improved porosity, or through stimulation of organisms in lower trophic levels (e.g., microflora) (Brown 1995). However, earthworm effects on soil biota are not always positive. Some species of earthworms produce highly compacted casts which adversely affect microbial activity (e.g., Martin 1991). Adverse effects of earthworms on populations of microfauna also sometimes occur (Yeates 1981; Hyvönen et al. 1994), although it is unclear whether this is due to alteration of pore size distribution, modification of organic matter by passage through the earthworm gut, competition for the same food resources (i.e., bacteria), or direct predation on microfauna. Compacted structures created by some earthworm species may

FIGURE 2.8. Effects of structures created by ecosystem engineers on other components of the soil biota. Treatment labels are C, control soil; CA, earthworm cast material; M, earthworm middens; N, ant nest material; U, soil under earthworm casts. Data from Decaëns et al. (1999) are from the pasture site in their study only. The other studies are all from forests. * indicates that effects of ecosystem engineers are significant at $P = 0.05$.

reduce larger soil fauna (e.g., arthropods) through reducing the density of larger pores. For example, Decaëns et al. (1999) found a much lower density of macroinvertebrates in casts created by the anecic earthworm *Martiodrilus caramiguensis* than in the underlying soil (fig. 2.8e–i), and selection for soil fauna with smaller mean body sizes.

Structures constructed by soil arthropods also affect soil biota. Termite mounds are usually constructed to resist decomposer activity; they are composed of tightly compacted mineral particles and frequently also undigested resistant organic material sourced from fecal pellets. The result is that the organic matter content of termitaria can be up to 50% greater than in surrounding soil (Lavelle 1997). However, "fungus gardens" maintained by some termite species, in which fungi are cultivated by termites on a fungus comb, represent a mutualistic association in which the fungus degrades lignin and cellulose, enabling the termites to utilize plant material more completely than is otherwise possible (Hyodo et al. 2000). In addition, structures created by mound-forming ants can greatly stimulate both microbial and decomposer faunal densities. Wagner et al. (1997) and Laakso and Setälä (1998) both found that mounds created by ants contained much higher densities of most soil faunal groups than did surrounding soil (fig. 2.8j–r). Possible mechanisms for the stimulation include more constant and favorable microclimatic conditions in the mound, continual supply by ants of fresh organic material, and ant-induced reductions of predatory macrofauna (Laakso and Setälä 1998). Further, Laakso and Setälä (1997) found that populations of the earthworm *Dendrobaena octaedra* were greatly enhanced in the mounds of red wood ants (*Formica aquilonia*) and that they were not preyed upon by ants. They suggested that the ant-earthworm association was mutualistic, in that the benefits to ants of maintaining earthworm populations in the mounds (i.e., prevention of mounds from being overcome by molds and fungal pathogens, allowing greater longevity of the mounds) presumably outweigh the nutritional benefit to ants through consuming the earthworms.

Significance of Mutualisms in Soil Food Webs

In the aboveground and aquatic food web literature, most theory is based on the premise that competition and predation are the primary interactions driving communities. In the soil microfood-web this appears to be the case and microfood-webs conform reasonably well to some of these theories. However, it is apparent that as faunal body size increases, faunal effects on microbes become increasingly positive, and mutualisms become more common. Saprophytic mesofauna mainly undergo mutualistic interactions with microflora in the "external rumen" while for larger soil fauna these relationships occur largely in the gut cavity. Classical food web theories (including those discussed earlier in this chapter) spectacularly fail to account for these sorts of relationships, or for habitat modification by larger organisms for smaller ones. Although theoretical models predict that close mutualisms in food webs are destabilizing and therefore should not persist (Pimm 1982), mutualisms are probably ubiquitous with regard to the decomposer biota (as well as decomposer-plant and other microbial-plant associations; see chapter 5), and are integral for the functioning of terrestrial ecosystems (Wall and Moore 1999). Mutualism must therefore be placed alongside predation and competition as a major primary interaction in soil food webs.

THE FUNCTIONALITY OF SOIL FOOD WEBS

The soil microbial biomass is directly responsible for the decomposition of organic substrates and the mineralization of nutrients required for plant growth, and highly specialized subsets of the bacterial component of this biomass are also responsible for fixing atmospheric nitrogen and both chemoautotrophic and heterotrophic nitrification and

43

denitrification. Greater rates of these processes are usually observed when conditions are favorable for microbial activity and a greater biomass of microbes is therefore present (e.g., Flanagan and van Cleve 1983; Beare et al. 1991). Given that the magnitude of the microbial biomass is frequently influenced by NPP (fig. 2.2), it is to be expected that more productive situations which support greater microbial biomasses will also show greater rates of carbon and nitrogen mineralization. D. Zak et al. (1994) found for a range of late successional ecosystems in North America that more productive sites had both greater microbial biomass and enhanced rates of mineralization. Similarly, Zagal (1994) found in controlled glasshouse conditions that reducing NPP of barley plants through lowering light intensity reduced rates at which ^{15}N and ^{14}C were incorporated into the microbial biomass, and ultimately the rates of nitrogen mineralization by microbes. However, the composition of the microbial biomass is also important in governing soil processes. There is evidence from both selective biocide experiments in the field (e.g., Beare et al. 1992) and microcosm experiments with artificially created communities (e.g., Ingham et al. 1985) that fungi and bacteria can differ tremendously in their effects on carbon and nutrient release, and fungal species differ immensely in the rates that they mineralize added substrates (e.g., Chen and Ferris 2000).

Process Regulation by Predation

Predator-prey interactions in the microfood-web can have major effects on nutrient mineralization, given that microbe feeders are capable of consuming up to 30–60% of annual microbial production (Persson et al. 1980). Grazing of bacteria by nematodes and protozoa greatly stimulates microbial productivity and turnover, and microfaunal grazers often excrete large amounts of excess nutrients derived from their prey, especially in nutrient-rich situations (Freckman

1988; Darbyshire et al. 1994). Both these mechanisms are powerful determinants of decomposition and nutrient mineralization. Grazing of fungi by fungal feeders can either stimulate or reduce rates of decomposer processes depending upon nutrient conditions and grazing intensity (and therefore whether overcompensatory fungal growth occurs) (Seastedt and Crossley 1980; Hanlon 1981; see also fig. 2.5). While disruption of hyphae by grazers can increase nutrient release, enhancement of fungal growth due to grazing (e.g., Hedlund and Augustsson 1995) can also result in net nutrient immobilization. It would appear that grazing is more important in the bacterial energy channel than in the fungal channel for releasing plant-available nutrients. This is consistent with the discussion earlier in this chapter proposing that bacteria (with few defenses) are regulated largely by predation while fungi (which can be well defended) are regulated more by resource availability. If this theory is correct then increased resource availability (e.g., NPP) is more important in governing fungal-driven decomposer processes while predation is more important in affecting bacterial-driven processes. As a result of this, nutrient dynamics associated with the bacterial-based energy channel are much more rapid than those associated with the fungal-based energy channel (Coleman et al. 1983).

Higher-level predators also affect decomposition processes. Multitrophic controls of decomposition are apparent in those studies which have considered trophic cascades; all but one of the nine studies depicted in figure 2.6 found that manipulation of tertiary or higher-level consumers induced effects on carbon and/or nitrogen mineralization through altering microbial activity, although these effects could be either positive or negative. This effect has been demonstrated for a wide range of predators, including predatory nematodes, mites, spiders, and salamanders. There is also theoretical support for the view that higher-level predators influence carbon and nitrogen mineralization (Zheng et al.

1997, 1999). Zheng et al. (1999) used two mathematical models to investigate effects of food chain properties on decomposition rate: a Lotka-Volterra model, which found that decomposition rate could either increase or decrease with addition of higher trophic levels; and a donor-control model, which predicted that addition of higher trophic levels would only increase decomposition rates. However, the latter of these two predictions is inconsistent with several of the results depicted in figure 2.6.

Litter Transformers and Ecosystem Engineers

Arthropods which operate as "litter transformers" usually enhance mineralization through the external rumen mutualism with microbes described earlier. However, there are also instances in which net immobilization of nutrients occurs. For example, Van Wensem et al. (1993) found that addition of isopods to microcosms resulted in enhanced release of nitrogen from partially decayed litter, but not from fresh litter in which immobilization of nitrogen by microbes presumably occurred. Further, Teuben (1991) determined that addition of isopods to microcosms appeared to have a buffering effect, in that isopods often reduced availability of nutrients (particularly ammonium) when available levels were otherwise high, and enhanced availability when available levels were low.

Stimulation of microbial activity by "ecosystem engineers" such as earthworms frequently greatly stimulates decomposition and nutrient mineralization (Scheu 1994; Lavelle 1997), and this has been shown to result both from burial of plant litter and from formation of casts that place microorganisms and organic matter in intimate contact (Cortez and Bouché 1998). However, enhancement of decomposition and nutrient availability in structures created by earthworms is often very transient, and in the longer term these structures may actually protect organic matter against decomposition. For example, Martin (1991) and Lavelle and

Martin (1992) found for the tropical earthworm *Millsonia anomala* that while rapid mineralization of organic matter occurs during digestion and cast formation, in the scale of days to weeks the newly mobilized nutrients are immobilized in the microbial biomass. As the structures age and become more compacted decomposition is actually retarded, so that one year after egestion the casts have an 11% higher content of coarse organic debris than control soil (Martin 1991). Structures created by termites (e.g., termitaria) often also contain organic matter that is protected against microbial activity and therefore decomposition (Lavelle 1997).

Food Web Structure and Body Size Considerations

The structure of soil food webs is clearly important in governing decomposition and mineralization processes. This is apparent from studies that have fitted models to field data on dynamics of food web components and carbon and nitrogen mineralization, to determine the relative contributions of different faunal groups to mineralization (e.g., Hunt et al. 1987; Paustian et al. 1990; de Ruiter et al. 1993a). Although analyses of these food webs demonstrates that different functional groups of fauna are important in driving these processes in different situations (de Ruiter et al. 1994), they consistently point to the importance of faunal composition as a determinant of belowground processes. Further, despite these faunal compositional differences across ecosystems, Verhoef and Brussaard (1990) concluded upon reviewing the available literature that the faunal contribution to nitrogen mobilization in soils is rather constant, and is usually around 30% in both natural (e.g., forest, prairie) and managed ecosystems.

The importance of composition of soil faunal functional groups in regulating processes is also apparent through the results of manipulative experiments. In addition to those experiments described earlier which involve top predator manipulation, several experiments which involve manipulation of composition of those faunas that interact with microflora (either as predators or as mutualists) also point to the importance of soil faunal composition in driving processes. Figure 2.9 summarizes the results of five studies in which food webs were made increasingly complex by including increasingly larger organisms in the food webs. It is apparent that in four of these cases addition of faunas with larger body sizes (either mesofaunas or macrofaunas) were capable of causing effects on processes that were in excess of what was observed in the microfood-web consisting of only microbes and their microfaunal predators. In addition to demonstrating the effects of animal body size on processes, these results also provide evidence that animals which undergo mutualisms with microflora (i.e., litter transformers, ecosystem engineers) are capable of mineralizing a subset of the organic matter pool that those animals which prey directly upon the microflora cannot.

STABILITY AND TEMPORAL VARIABILITY

Until a decade ago it was widely maintained in the literature that food webs were likely to consist of short chains (typically three to four links), have a low incidence of onmivory and connectance, and lack within-habitat compartments. This was based on mathematical models which predicted that complexity destabilized food webs (May 1973) and that these properties were therefore required for food webs to remain sufficiently stable to persist (e.g., Pimm 1982; Cohen et al. 1990; Pimm et al. 1991). Synthesis and analysis of 113 published food webs were used to support these predictions (Cohen et al. 1990; Pimm et al. 1991). None of these published food webs adequately considered the soil subsystem (Wardle 1991) and it became increasingly apparent that soil food webs did not conform to these predictions. First, soil ecologists working with decomposer food webs recognized that omnivory was extremely common

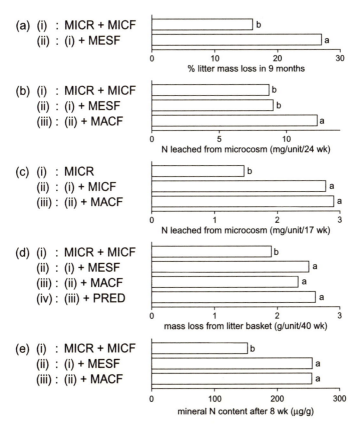

FIGURE 2.9. Effects of adding animals of increasing body size on decomposer-mediated processes in five studies. MICR indicates microflora, MICF microfauna, MESF mesofauna, MACF saprophagous macrofauna, and PRED predatory macrofauna. For each study, bars topped by different letters are significantly different at $P = 0.05$; bars are arranged in order of increasing trophic complexity. Results depicted are from (a) Vossbrink et al. (1979); (b) Coûteaux et al. (1991); (c) Alphei et al. (1996); (d) Setälä et al. (1996); (e) Vedder et al. (1996). Studies (a), (d), and (e) are field or mesocosm experiments in which mesh sizes were varied to exclude animals of differing body sizes while the other two are microcosm studies based on artificially created food webs. In both microcosm studies microflora consisted of nematodes and protozoa; in the study of Coûteaux et al. (1991) mesofauna consisted of collembola and the macrofauna included isopods; in the study of Alphei et al. (1996) macrofauna consisted of earthworms. Only data from the ambient CO_2 treatment from Coûteaux et al. (1991) are presented.

(Walter et al. 1991), that within-habitat compartments ("energy channels") were a fundamental component (Moore and Hunt 1988), and that food chains frequently consisted of up to seven links even when only the microfood-web was considered (Ingham et al. 1986; Hunt et al. 1987). Secondly, despite possessing these "destabilizing" features, soil food webs are clearly highly resilient to disturbance and usually recover entirely from major disturbances such as drying, freezing, and fumigation within a matter of days.

Although soil food webs did not feature much in the writings of many of those developing food web theories at the time, food web researchers working in other systems began to question the view that complex food webs with high omnivory and long food chains were indeed too unstable to persist (e.g., Polis 1991; Martinez 1992; Hall and Raffaelli 1993). In particular, those compilations of food webs used for testing theories about food web properties were increasingly becoming recognized as oversimplifications of the real world, and more comprehensive studies of food webs in estuaries (Hall and Raffaelli 1993) and deserts (Polis 1991) proposed that food webs were more complex than earlier work suggested. Indeed Polis (1991) suggested that the fact that cataloged food webs used for developing theory "depict so few species, absurdly low ratios of predators on prey and prey eaten by predators, so few links, so little omnivory . . . argues strongly that they poorly represent real biological communities" and that "contrary to strong assertions by many theorists, patterns from food webs of real communities generally do not support predictions arising from dynamic and graphical models of food web structure."

Complex food webs may be more stable if interaction strengths are generally weak (May 1973). It is now increasingly recognized that the distributions of interaction strengths in food webs are highly skewed and that food webs normally consist of mostly weak interactions and a few strong ones; this is supported by both experimental (e.g., Paine 1992) and theoretical (e.g., Berlow et al. 1999) evidence. This appears to be the case in soil food webs (De Ruiter et al. 1995; Wardle et al. 1995b). It has also been suggested through theoretical models that a high incidence of mainly weak interactions promotes stability of complex natural food webs (McCann et al. 1998). This appears likely for soil food webs; De Ruiter et al. (1995) found through applying models to empirical data from several soil food webs that food web interactions, while mostly weak, were organized in such a way as to promote the stability of micro-food-webs.

The apparent stability of soil food webs may arise not just from generally weak interaction strengths but also because donor-controlled food webs may be inherently more stable (De Angelis 1992) and because of the buffering nature of the physical structure of soil (Wardle 1995). However, the stability of any food web must ultimately be driven by the traits or attributes of the dominant species present, an issue that has been mostly overlooked in food web research. In soils, resilience to perturbations of microfood-web organisms is high because most organisms can grow very rapidly after disturbance, and resistance is high because most microfood-web components can remain dormant or inactive for lengthy periods. For example, during soil drying (probably the most common disturbance in most soils), many food web components can persist in a state of anhydrobiosis (Whitford 1989) and can reactivate rapidly upon rewetting.

The issue of how soil food web structure and composition affect stability has been subjected to very little experimentation. However, Allen-Morley and Coleman (1989) conducted a microcosm experiment in which soil microfood-webs of different structures were subjected to a disturbance involving freezing. They found that species-specific characteristics of the component organisms, rather than food web structure, was the main determinant of resilience following disturbance. Further, the results of two literature meta-analyses point to faunal body size as a key trait affecting stability of soil food webs. In the first, Bengtsson (1994) synthesized data from 67 previous studies in coniferous temperate forests for which temporal population data (of between 1 and 23 years depending on the study) for various soil faunal groups were presented. In this synthesis, temporal predictability for faunal groups declined (i.e., temporal variability increased) as the mean body size of taxa increased (Spearman's rank correlation coefficient = -0.73; $P < 0.01$). In the second (Wardle 1995), which synthesized 106 studies investigating effects of a disturbance (soil cultivation) on soil biota, organism size and resistance to disturbance were found to be negatively correlated ($r = -0.86$, $P = 0.006$) (fig. 2.10). These pieces of fragmentary evidence point to community composition and to traits of dominant organisms in soil food webs in governing stability, a theme that will be revisited in further chapters.

Finally, given that the decomposer subsystem is ultimately driven from above ground, the issue emerges as to whether the amount of resource input to the soil affects food web stability. This issue remains unexplored for soil biota except through the use of theoretical models; the analyses of De Angelis (1992) predict greater food web stability with greater NPP, and Moore et al. (1993) performed model simulations which suggest that soil food webs in more productive situations are more resilient although the effect of productivity on resilience across the productivity range that represents most of the Earth's ecosystems is not large. There is also evidence that stability of the soil microbial biomass is enhanced by increased resource availability; a meta-analysis of 58 previously published studies (Wardle 1998a) found that temporal variation of microbial biomass was significantly negatively correlated with soil carbon content across these studies (partial $r = -0.451$, $P < 0.001$). Further, Wardle et al. (1999b) found temporal variability of soil carbon mineralization in field plots to be less when soil carbon and nitrogen levels were higher. Given that the dynamics of components of the soil food web (and food web responses to external abiotic perturbations) have a pivotal role in gov-

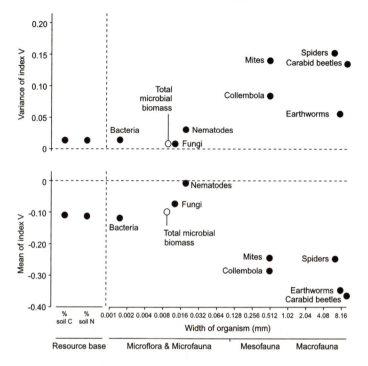

FIGURE 2.10. Resistance of components of decomposer food webs to disturbance (conventional tillage of soil) in relation to mean body size, based on
a meta-analysis of 106 studies. The index *V* becomes increasingly negative
with decreasing resistance to disturbance. From Wardle (1995). (Reproduced with permission from Academic Press)

erning soil processes (Whitford 1989), a better understanding of factors that govern soil food web stability (including
community composition, food web structure, ecosystem productivity, and nutrient availability) is essential for evaluating
more fully how soil food webs drive ecosystem function.

SYNTHESIS

A better understanding of the linkages between the
aboveground and belowground subsystems, and the consequences of these for ecosystem function, requires us to con-

sider the interactions that occur in soil food webs and how these regulate nutrient availability. It is apparent that a variety of interactions are important but at different spatial scales. In the soil microfood-web there is much evidence for both top-down and bottom-up control of trophic groupings being important, but with the bacterial-based and fungal-based energy channels being governed by very different mechanisms. In particular, fungi appear to be regulated mainly by resources and bacteria mainly by predation; the bacterial-based energy channel (which operates mainly in the aqueous component of the soil) appears to function in a manner characteristic of aquatic systems, while the fungal-based energy channel (which is based on sessile microflora) functions in a manner typical of many terrestrial systems. Microfood-web structure, and the interactions which occur within the web (including trophic cascades) are clearly important determinants of decomposer processes.

While the soil microfood-web generally conforms to ecological theories based on competition and predation, these theories fail when interactions involving larger saprophagous soil fauna (litter transformers and ecosystem engineers) are considered. Here, soil animals generally enter mutualisms with microbes through a variety of mechanisms. For litter transformers mutualisms take place mostly in the "external rumen" (fecal pellets) while for ecosystem engineers they occur in the animal gut cavity. These larger organisms also create a variety of physical structures that profoundly affect microbial activity. Mutualistic interactions are ubiquitous in soils and are key determinants of soil processes, and they must therefore be considered alongside competition and predation as being primary interactions occurring in soil.

A recurrent issue in this chapter is the ecological importance of food web composition, which is ultimately a function of the ecophysiological traits of the component organisms. With regard to the soil fauna, a key trait is clearly body

54

size, and this has been shown throughout the chapter to be a major determinant of the types of interaction that fauna undergo with microbes (predation to mutualism to symbiotic mutualism), decomposer processes, and temporal variability. This suggests that a primary determinant of soil functioning is the relative contributions of the microfood-web, litter transformer, and ecosystem engineer systems to the soil. Finally, while the nature of biotic interactions in soil food webs and their effects on decomposer processes continue to be well studied, much remains unknown about the temporal dynamics of soil food web components and food web stability. However, such knowledge is critical to better understanding how soil food webs affect key processes against the natural background of temporal variability of conditions that occurs in the belowground environment.

CHAPTER THREE

Plant Species Control of Soil Biota and Processes

In chapter 2 I considered how NPP and basal resource inputs may influence soil food webs and how this could in turn affect soil processes that ultimately determine plant growth. However, plant species differ tremendously with regard to the quality of organic matter that they return to the soil, and this in itself has important consequences for both the soil biota and the processes that it regulates. Although trophic dynamic theories (including those discussed in chapter 2) frequently consider the magnitude of NPP as the main means of bottom-up control of consumer trophic levels, it may be that the quality of resources, and differences between plant species with regard to this, is the ultimate driver for consumer organisms (White 1978; Polis 1999), and therefore consumer-driven processes.

Ecologists are becoming increasingly aware that plant species differ with regard to their effects on ecosystem properties (Hobbie 1992; Lawton 1994), including nutrient cycling. While the issue of how plant species and vegetation composition influences ecosystem properties is frequently identified as a major recent area of advancement in ecology, these sorts of effects have long been appreciated by soil biologists. In an important but frequently overlooked series of experiments, Handley (1954, 1961) presented clear evidence that the shrub *Calluna vulgaris* differed from other plant species because its leaf material contained tannin-protein complexes which retarded nitrogen mineralization and the nutrient acquisition and growth of *Betula pendula* seedlings. Further, the mull and mor theory of Muller (1884) discussed later in this chapter points to the role of plant species in affecting forest soils. And, in a longer time scale perspective, farmers have been aware for millennia of the benefits to soil fertility and ecosystem productivity of inclusion of the plant functional group consisting of nitrogen fixing legumes.

While there is an appreciation by ecologists that plant species differ in their effects on nutrient cycling, and differ in terms of rates of mineralization of carbon and nutrients in their tissues, there is less appreciation of the role played by the soil biota. Plant species effects on these ecosystem-level processes must ultimately arise from effects that plant species have on soil organisms. In this chapter, I will consider how plant species affect soil biota and the consequences of this for key soil processes that they regulate. These effects will then be interpreted in terms of how plant species differ in their ecological strategies and ecophysiological traits.

PLANT SPECIES EFFECTS ON SOIL BIOTA

Plant species differ in their effects on soil biota not just through differences in NPP and the amounts of basal resources returned to the soil, but also through influencing the quality of resources available to belowground organisms. Species differ in terms of the proportion of fixed carbon added in soluble or simple forms to the soil (mainly via the rhizosphere), the extent to which they deplete available nutrients in the soil, and the chemical composition of leaf and root litter that is produced. Components of soil food webs that are bottom-up controlled are likely to be responsive to these differences (cf. chapter 2).

Responses of Soil Microflora

On a global basis, soil microbial biomass can vary considerably across vegetation types, and this part reflects the qual-

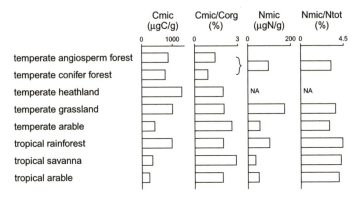

FIGURE 3.1. Soil microbial biomass carbon (Cmic) and nitrogen (Nmic) values for a range of temperate and tropical ecosystem types, derived from a literature synthesis of microbial biomass values from 112 published studies. Corg and Ntot represent the concentrations of organic carbon and total nitrogen in soil. NA means not available. From table 5 of Wardle (1992).

ity of resource input to the soil from the dominant plant species present. In temperate areas, soils in coniferous forests generally have lower values for microbial biomass and the biomass carbon-to-organic-carbon ratio than do angiosperm forests or grasslands (fig. 3.1). This is probably reflective of the poorer quality of litter returned by coniferous trees (see Wardle 1992). Further, tropical rainforests tend to support higher amounts of microbial biomass per unit organic matter than do other tropical vegetation types, which may be due in part to litter quality effects (fig. 3.1).

Plant species effects on microbial biomass can also be detected at much more localized spatial scales within habitats. Soils planted with different herbaceous or tree seedling species often support vastly different quantities of microbial biomass (e.g., Bradley and Fyles 1995; Groffman et al. 1996; Priha et al. 1999; fig. 3.2a,b,d), and different plant species grown in the same soil can either stimulate or inhibit microbial biomass relative to unplanted soil (e.g., Bardgett et al.

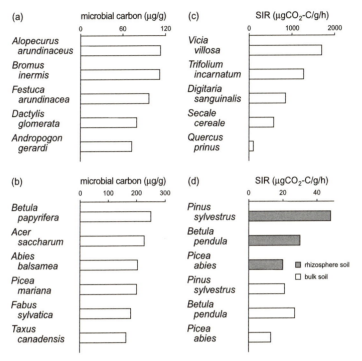

FIGURE 3.2. Examples of how plant species differ with regard to the levels of microbial biomass that they support. (a) Microbial biomass in soil from grassland plots (Groffman et al. 1996); (b) microbial biomass in pots planted with seedlings of different tree species (Bradley and Fyles 1995); (c) substrate-induced respiration (SIR; a relative measure of active microbial biomass) on leaf litter of different plant species (Beare et al. 1990); and (d) SIR in rhizosphere and bulk soil in pots planted with seedlings of three tree species (Priha et al. 1999).

1999a). Plant species differ in their effects on microbial biomass in a manner that cannot be explained solely in terms of intrinsic productivity differences across species, meaning that either the quality of resource input or the nature of nutrient depletion must therefore differ sufficiently across plant species to influence microbial biomass. For example, Wardle and Nicholson (1996) determined the regression relationship between belowground productivity and microbial biomass for each of ten herbaceous plant species grown in pots, and found that while these two variables were significantly associated for all ten species, the slopes of the regressions differed substantially across species. Further, the studies of Wheatley et al. (1990) and Bardgett et al. (1999a) both found that while microbial biomass in pots differed according to which herbaceous plant species were grown, these differences were not related to differences in root productivity across species.

The response of microbial biomass to plant species effects is especially apparent when leaf litter is considered. Although relatively few studies have sought to quantify microbial biomass on litter, there is evidence that important differences exist across plant species due to interspecific variation in leaf litter quality. For example, Beare et al. (1990) determined that there was nearly a tenfold range of active microbial biomass densities on leaf litter of several plant species monitored under the same conditions (fig. 3.2c). Similarly, Wardle (1993) found that microbial biomass densities on litter collected from a range of forests in New Zealand varied over a fourfold range.

There are also important differences between plant species with regard to the compositions of the microbial communities that they support. Bardgett et al. (1999a) found clear differences in microbial phospholipid fatty acid (PLFA) profiles (indicative of microbial community structure) for soils planted with different herbaceous species, and Grayston et al. (1998) showed through the use of BIOLOG plate counts that soils from under four different herbaceous plant species each supported different microbial communities. Plant species also differ in the composition of the microbial communities that occur on their root surfaces and in their rhizosphere zones. For example, Aberdeen (1956) determined that saprophytic fungal community composition on root surfaces varied considerably among different herbaceous species, and Marilley and Aragno (1999) used molecular techniques to show that bacterial community structure in the root zones of *Lolium perenne* and *Trifolium repens* differed markedly. Quality of leaf litter is a powerful determinant of fungal community composition because fungal species differ in the complexity of resources that they can degrade, and litter from different tree species often supports vastly differing fungal communities (e.g., Widden 1986).

Microbial species interactions can also be influenced by plant species effects, especially in the case of fungi which are bottom-up regulated and therefore influenced by competitive relationships. In controlled experimental conditions, Aberdeen (1956) showed that the competitive outcome between pairs of fungal species in the plant root zone differed according to plant species. Similarly, with regard to plant leaf litter, Widden and Hsu (1987) found that the competitive balance between pairs of *Trichoderma* species was strongly influenced by which litter type they were growing on.

Soil Microfood-Web Responses

Different vegetation types support vastly different populations of consumer organisms in the soil microfood-web, indicative of multitrophic effects. Ecosystems dominated by plant species that produce poorer-quality resources typically support lower populations of nematodes (Yeates 1979) and protozoa (Bamforth 1973), although coniferous forests with characteristically poor litter quality frequently support higher populations of many microarthropod groups than do angiosperm forests (Petersen and Luxton 1982; Wallwork 1983). Ingham et al. (1989) analyzed the microfood-web of three vegetation types under a comparable climatic regime in Colorado and Wyoming, United States, and found that the bacterial energy channel (bacteria and bacterial-feeding nematodes) dominated in alpine meadow and shortgrass prairie sites, while the fungal energy channel (fungi and mi-

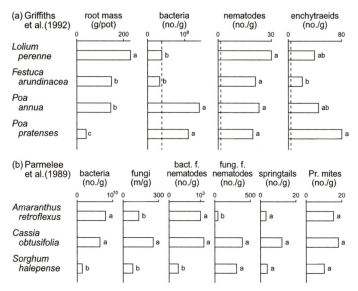

FIGURE 3.3. Plant species effects on consumer trophic levels in microfood-webs. (a) Root biomass and concentrations of consumer organisms in rhizosphere soil for an experiment in which plots were planted with different grass species. Only data for low-nitrate treatment are included. Dotted line represents populations in unplanted bulk soil. (b) Populations of consumer organisms on litter from three weed species differing in resource quality. Only data for the Griffin site are included. Bact. f. and fung. f. indicate bacterial feeding and fungal feeding, respectively. Pr indicates Prostigmatid. For both (a) and (b) bars within each subgraph that are topped with the same letter are not statistically significant at $P = 0.05$.

croarthropods) dominated in a lodgepole pine (*Pinus contorta*) forest site. These basic food web differences were presumably determined by differences in the quality of input of basal resources. At more localized scales, plant species effects can be important determinants of consumers in the microfood-web. For example, Griffiths et al. (1992; fig. 3.3a) found that populations of secondary consumers (enchytraeids, and nematodes, the majority of which feed on bacteria) in the plant rooting zone differed tremendously across four species of grasses. These differences could not be explained in terms of differences of root mass production across the species. Similarly, Wasilewska (1995) found that pots sown with different grass species mixtures and monocultures supported different nematode community assemblages.

Litter quality is sometimes a strong determinant of population sizes of microfood-web fauna, and populations of these fauna can differ across litter from different plant species (e.g., Parmelee et al. 1989; Blair et al. 1990; Hansen 1999; fig. 3.3b). However, different components of the microfood-web show vastly differing responses to resource quality depending upon whether they are top-down or bottom-up regulated. For example, Wardle et al. (1995b) observed in decomposing sawdust over a three-year period that although resource quality improved markedly over that time, only fungi and top predatory nematodes showed a corresponding increase. Microbe-feeding nematodes and bacteria did not respond consistently to the improved resource quality, presumably because they were regulated by top-down control. The response of soil organisms to plant species differences in quality of resource input reflects bottom-up regulation, and it is therefore to be expected that different components of the soil food web will respond differently to resource quality effects, with the fungal-based energy channel being more sensitive than the bacterial-based energy channel (see chapter 2).

Responses of Litter Transformers and Ecosystem Engineers

While the soil microfauna respond indirectly to litter quality via the microflora that they consume, litter-transforming arthropods ingest litter directly and are therefore more likely to be strongly influenced by differences in litter quality across species. Leaf litter under different tree species supports vastly differing communities of arthropods, due in large part to interspecific differences in litter quality (e.g., Badejo and Tian 1999; Pereira et al. 1998). Litter choice experiments indicate that saprophagous arthropods preferentially consume leaf litter that has high nutrient concentrations and low levels of phenolic compounds (e.g., Nicolai 1988).

Larger soil fauna, including ecosystem engineers, also show clear preferences for some resource types above others. For example, Saetre (1998) found upon mixing humus from *Betula pendula* (high quality) with that from *Picea abies* (low quality) that the earthworm *Aporrectodea caliginosa* required at least 25% *B. pendula* humus in the mixture in order to remain active throughout the 14-week duration of the experiment. In a field experiment, Tian et al. (1993) applied residues of five different crop and tree species to the soil surface, and found that earthworms, termites, and ants all showed clear preferences for certain residues above others; both earthworms and ants preferred those residues with the lowest carbon-to-nitrogen ratios. Similarly, Hendriksen et al. (1990) found that litter bags containing residues of different plant species contained vastly different populations of detritus-feeding earthworms at about 50 days after placement (fig. 3.4). Litter-feeding earthworms also prefer consuming litter from particular plant species; for example, Satchell (1967) found that individuals of *Lumbricus terrestris* preferentially fed upon litter with high nitrogen concentrations (fig. 3.4) and low levels of polyphenolics.

Organisms Directly Associated with Plant Roots

While saprophytic organisms and their consumers which are loosely associated with plant roots often respond to differences between plant species, plant species effects on soil organisms which are intimately associated with their roots can be particularly strong. Associative nitrogen fixing bacteria such as *Azospirillum* and *Azotobacter* are often stimulated only in the root zones of certain grass species (particularly some C4 grasses). Symbiotic nodule-forming strains of nitrogen fixing bacteria are favored only by particular plant spe-

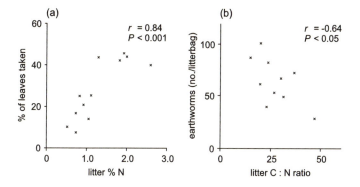

FIGURE 3.4. Earthworms in relation to resource quality. (a) Palatability of leaf litter from 12 tree species to the earthworm *Lumbricus terrestris* in feeding tests, in relation to litter nitrogen content. Each point represents a different tree species. Data from Satchell (1967). (b) Populations of detritus-feeding earthworms in litter bags of ten different litter types from seven tree species, in relation to litter carbon-to-nitrogen ratio. Each point represents a different litter type. Derived from Hendriksen et al. (1990).

cies or varieties which are capable of producing specific lectins that the bacteria can recognize. Many mycorrhizal fungal species are able to be hosted by only certain plant species; ectomycorrhizal, arbuscular, and ericoid mycorrhizal fungi all have a specific range of host plants and plant functional types that they can colonize (see Francis and Read 1994; Smith and Read 1997). Within these three mycorrhizal types, there is often considerable specificity of fungus-plant associations. For example, Högberg et al. (1999) demonstrated that in two Swedish boreal forests different ectomycorrhizal fungal species known to be associated with three different tree species generally had $\delta^{13}C$ values which the $\delta^{13}C$ values of the host plant. Further, different plant species that form associations with arbuscular mycorrhizae vary tremendously in the number of hyphae that they support (e.g., Kruckelmann 1975), presumably in part because of differences between species in the amounts of resources that they allocate to the fungal component.

65

Soil organisms that have negative effects on plant growth, such as invertebrate root herbivores and fungal pathogens, show varying levels of specificity and are therefore stimulated to varying degrees by the presence of certain plant species (e.g., House 1989; Bateman and Kwasna 1999). There are several instances in which differential effects of these organisms on different plant species have been shown to have important consequences for plant demography and community structure, and this will be discussed in chapter 5.

Mechanisms Involved

The responses of soil biota to differences between plant species emerges through a number of mechanisms. These include interspecific differences in the quantities of resources produced (NPP), quality of resources produced, competitiveness against microorganisms for nutrients, and formation and modification of habitats. (Further mechanisms involving plant species effects on disturbance regimes and spatial and temporal variability will be covered later in this chapter.) Based on the discussion in chapter 2, differences among species in NPP should have important consequences for those organisms regulated by bottom-up control. The same is true for differences in resource quality provided by different species. The level of stimulation of organisms in the rooting zone is clearly regulated by the extent to which soluble carbon is released into the rhizosphere. Further, plant litter types that have high carbon-to-nitrogen ratios and high levels of phenolics and structural carbohydrates (e.g., lignin, cellulose) tend to support low levels of microbial biomass (e.g., Beare et al. 1990) and to be much less palatable than other litter types to saprophagous invertebrates (e.g., Satchell 1967; Hendriksen 1990).

Plant species differ in the extent to which they compete for nutrients with microbes (Kaye and Hart 1997) and this has the potential to influence levels of decomposer biota.

Plant species vary in their nutrient-foraging strategies, with some foraging over much coarser spatial scales than others (Campbell et al. 1991), and with species varying in their rates of nutrient uptake per unit root mass and rate of short-term root production in response to localized patches of nutrient enrichment (e.g., Grime 1994; Fransen et al. 1999). These differences in nutrient uptake strategies between species may affect both the depletion of nutrients otherwise available for microbes, and the heterogeneity of nutrient availability at a spatial scale relevant to the soil microfood-web. However, much remains unknown about the consequences for microbes of localized depletion of nutrients by plant roots.

While the discussion this far has focused mostly on how plant species affect soil biota through influencing nutrient availability, plant species effects are also manifested through alteration and creation of habitats including ecosystem engineering (*sensu* Jones et al. 1994). There are several examples of these effects, which operate at a vast range of spatial scales. Plant species differ in their effects on soil structure, for example through production of cementing materials, anchorage by roots, and creation of pores through root penetration (see Angers and Caron 1998); these effects all have important consequences for soil organisms. Further, plant species differ in the extent to which they create and maintain a litter and humus layer; mor-forming forest plant species produce a deep layer of humus and undecomposed plant litter resulting in domination by the fungal-based energy channel and a high density of microarthropods, while the habitat created by mull-forming plant species is more conducive for the bacterial-based energy channel and earthworms. The structural attributes of plant debris, which varies considerably among species, also affect its suitability as a habitat for soil biota. For example, Hansen (1999) found that dead leaves of *Quercus rubra* contained microhabitats in woody petioles and within the leaf planes that promoted high populations of endophagous oribatid mites (i.e., those that burrow into plant material), while dead leaves of *Acer saccharinum* and *Betula aniensis* supported lower populations of endophages. Trees also produce woody debris, and woody substrates produced by different tree species often support vastly different assemblages of both basidiomycete fungi and saprophagous fauna. In forested ecosystems, many tree species support epiphytic plant and lichen communities which themselves produce detritus for decomposer biota. In tropical and warm temperate rainforests large epiphytic communities which produce "suspended soils" can form in tree crowns, and these soils can support both high levels of microbial biomass (Vance and Nadkarni 1990) and a diverse range of decomposer fauna, including many species that are adapted for life in forest canopies (Walter and Behan-Pelletier 1999, Winchester et al. 1999).

LINKS AMONG PLANT SPECIES, SOIL BIOTA, AND SOIL PROCESSES

Given that plant species vary in their effects on soil organisms, it is perhaps to be expected that the processes that the soil organisms regulate will also be responsive to plant species effects. Several studies have shown that the magnitude of various microbially driven processes in soils (e.g., carbon and nitrogen mineralization, nitrification, and denitrification) vary in response to plant species, through differences among species in the quality of resource input to the soil. For example, Wedin and Tilman (1990) found that nitrogen mineralization in soil from under monocultures of five perennial grass species varied by a factor of ten, and the magnitude of mineralization was negatively related to root lignin content and root carbon-to-nitrogen ratio. Similarly, Steltzer and Bowman (1998) found in an alpine meadow that nitrogen mineralization and nitrification rates were consistently greater under patches of the grass *Deschampsia* than under the forb *Acomastylis,* and this was consistent with differences between species with regard to litter nitrogen and phenolic concentrations. Plant species removal experiments in the field point to differences among species in their effects on both soil CO_2 release (Wardle et al. 1999b) and CH_4 release (e.g., Verville et al. 1998). The influence of plant species on soil processes is also apparent from the vast plant decomposition literature, and several studies have found litter decomposition rates to be related to various measures of "litter quality" including concentrations of nitrogen, phosphorus, lignin, and polyphenols (as well as various ratios of these), as well as levels of sclerophylly and soluble carbohydrates (see, e.g., Swift et al. 1979; Taylor et al. 1989; Berg and Ekbohm 1991; Palm and Sanchez 1991; Heal et al. 1997; Gallardo and Merino 1999).

Plant species effects on ecosystem processes must ultimately result from their effects on those soil organisms that drive the processes. Beare et al. (1991) found for litter from seven plant species that the rates of decomposition were closely related to the magnitude of the microbial biomass present on the litter prior to decomposition. Further, Flanagan and van Cleve (1983) determined for a range of forest types differing in quality of resource input that annual mineralization of nitrogen and phosphorus was closely positively correlated with the amounts of fungal hyphae present in the soil. However, other studies have found varying degrees of matching between the effects of plant species on microbial biomass and effects of species on soil processes (see, e.g., Wheatley et al. 1990; Vinton and Burke 1995; Priha et al. 1999). Plant species effects on processes are not necessarily correlated with their effects on microbial biomass because plant species can differ in their effects on the level of activity per unit microbial biomass (Wardle et al. 1998a), presumably through interspecific differences in the quality of resource input. The role of higher-level consumers (soil fauna) is also important. With regard to the microfood-web, Griffiths (1994) showed that predator-prey relationships between protozoa and bacteria were likely to result in significant mineralization of nitrogen only if the bacterial prey had a carbon-to-nitrogen ratio of under 5. Further, Darbyshire et al. (1994) determined that excretion of mineral nitrogen and phosphorus by a soil ciliate was significant only when it was fed with bacteria with a high concentration of these nutrients. These two examples suggest that plant-species-induced differences in substrate quality are likely to affect processes driven by microfood-web interactions. Similarly, mineralization of carbon and nitrogen through the effects of feeding activity of fungal-feeding and litter-transforming arthropods (and consequent effects on the microflora) is governed by the resource quality of the substrate (e.g., Hanlon 1981; Teuben 1991). With regard to earthworms, Saetre (1998) found that *Aporrectodea caliginosa* thrived and stimulated soil carbon mineralization in high-quality humus (from *Betula pendula*) but not in poor-quality humus (from *Picea abies*).

The study of Saetre (1998) suggests that certain plant species may select for soil biota that facilitates decomposition of their own residues. There is some evidence suggesting that this is indeed the case. For example, Hunt et al. (1988) collected litter from three plant species, each of which was dominant in a different habitat, and performed a reciprocal litter transplant experiment in which all litter types were decomposed in all three habitats. All three litter types decomposed faster than expected in the habitat from which they were sourced relative to the other two habitats (fig. 3.5a). In an agrosystem, Cookson et al. (1998) placed litter bags containing litter from three species (*Triticum aestivum, Hordeum vulgare,* and *Lupinus albus*) into plots that had previously been either amended or not amended with *T. aestivum* residues. Only *T. aestivum* litter decomposed significantly more rapidly on the previously amended plots, suggesting possible selection for a soil community adapted to decomposing *T.*

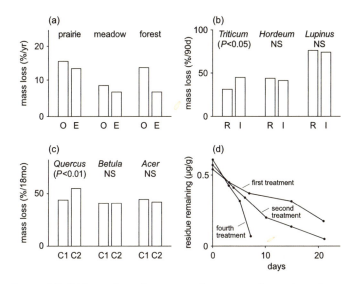

FIGURE 3.5. Evidence that substrates can favor types of organisms that accelerate their decomposition. (a) Results of a litter reciprocal transplant experiment in which litter was taken from the dominant plant species in a meadow (*Agropyron smithii*), a prairie (*Bouteloua gracilis*), and a forest (*Pinus contorta*), and each litter type was then placed in litter bags in all three habitats. For each habitat, O is the decomposition rate observed when litter collected from the plant species dominating that habitat was decomposed in the same habitat and E the decomposition rate expected for that plant species in that habitat based on its decomposition rate in the other two habitats. Overall, the O and E values differ significantly at $P = 0.001$. Derived from contingency table analysis using data extracted from Hunt et al. (1988). (b) Decomposition of residues of three plant species in plots in which *Triticum aestivum* (wheat) residue was either previously incorporated (I) or removed (R). NS indicates not significant at $P = 0.05$. Data from Cookson et al. (1998). (c) Decomposition of added litter of three tree species placed in field plots in 1993 (C1 or litter cohort 1) and litter of the same species placed on the same plots in the following year (C2 or cohort 2). Data from Hansen (1999). (d) Decomposition of residues of the herbicide metolachor in field plots during four successive applications. Data from Sanyal and Kulshrestha (1999).

aestivum litter as a result of previous addition of residues from that species (fig. 3.5b). Hansen (1999) performed an experiment in which forest litter was removed from field plots and replaced with litter from each of three tree species for each of two consecutive years. *Quercus rubra* litter placed on these plots in the second year decomposed significantly faster than that placed on the same plots in the first year (fig. 3.5c). Assessment of invertebrate densities revealed that the initial placement of *Q. rubra* leaves selected for a micro-arthropod guild (endogenous or litter-burrowing oribatid mites) which increased in population over the first year and presumably accelerated decomposition of the second-year leaves. Accelerated decomposition (or "enhanced degrada-tion") is already well known from the pesticide literature, and several studies have shown that repeated pesticide ap-plications to the same soil result in progressively more rapid degradation of the pesticide with each successive applica-tion, as a result of selection for soil organisms adapted for degrading the chemical (e.g., Sanyal and Kulshrestha 1999; fig. 3.5d). In light of the above examples it would appear that there are instances in which mutualisms between the producer and decomposer subsystems can involve selection by the plant for a community of soil organisms that is adapt-ed for decomposing substrates produced by that species.

Differences in the quality of basal resources provided by different plant species, and the consequent effects on soil biota, also affect the level of sequestration of carbon and nitrogen in the soil. This has important consequences for the availability of habitats for soil organisms as well as for the long-term soil fertility. Literature syntheses reveal that plant species which produce basal resources of higher quali-ty, have a greater rate of turnover, and produce tissues that decompose rapidly generally also store smaller pools of car-bon in the soil (Cebrian and Duarte 1995; Cebrian 1999). Tree species which produce low-quality acidic litter with higher phenolic concentrations can cause enhanced accu-mulation of carbon and nitrogen in the soils under them

(Tanner 1977; Binkley and Giardina 1998). For example, Wardle et al. (1997b) found that islands in a lake system that were dominated by *Picea abies* contained over five times as much carbon and nitrogen stored in the soil than did islands dominated by other tree species, and the poor-quality litter (containing high levels of phenolics) produced by *P. abies* may have contributed to this pattern. Similarly, Binkley and Valentine (1991) demonstrated that plots planted with *P. abies* accumulated soil organic matter much more rapidly than did plots planted with other tree species over a 50-year period. Large differences in soil organic matter levels across forests dominated by different tree species (see, e.g., studies by Van Wesemael and Veer 1992; Finzi et al. 1998) are often attributable in large part to differences across forests with regard to rates of carbon and nitrogen mineralization. Although fewer data exist for herbaceous plants, there is some evidence that herbaceous species can also differ with regard to the degree of accumulation of organic matter under them (Fisher et al. 1994; Derner et al. 1997; but see Wardle et al. 1999b).

TEMPORAL AND SPATIAL VARIABILITY

While most studies considering plant species effects on soil biota and ecosystem properties have considered differences among plant species with regard to resource quality, plant species have important, though less well understood, effects on both the spatial patterning and temporal variability of the soil biota, as well as the processes that it regulates. This has important implications for ecosystem functioning at much greater spatial and temporal scales than those considered so far in this chapter.

Spatial Patterning

Soil biota often shows clear spatial patterning, which may result in hotspots of biological activity in the landscape. For example, Saetre and Bååth (2000) used geostatistical analyses to demonstrate that spatial distributions of soil microbial communities (assessed using PLFA analysis) in a Swedish *Picea abies–Betula pubescens* forest was influenced by the spatial positioning of the trees and that these two tree species differed in their effects; these differences appeared to be related to differences in the quality of resources produced by the two species. Further, Ettema et al. (1998) considered the spatial patterning of bacterial-feeding nematodes in soils of riparian wetlands and found that for most taxa a high proportion of sample population variance was spatially dependent over a spatial scale of tens of meters. However, different nematode taxa were aggregated in different hotspots in the landscape. While there are a number of causes of spatial variability of soil biota and processes, there is evidence that vegetation is a key determinant of spatial distribution of belowground properties both across and within plant species.

Effects of plants on the spatial distribution of the soil biota may be expected to predominate in situations in which individual plants are spaced at a sufficient distance to have a dominating influence on the resources that accumulate under them. Such effects are especially apparent in situations in which individual trees or shrubs are scattered throughout the landscape. One example involves "islands of fertility" which are caused by the presence of individual shrubs in arid and semiarid systems (e.g., Schlesinger et al. 1996; Schlesinger and Pilmanis 1998); these shrubs result in hotspots of soil biotic activity and elevated rates of soil processes (e.g., Sarig et al. 1994; Burke et al. 1998). Another example involves mobile "tree islands" caused by krummholtz formations in alpine tundras above treeline in which individual trees migrate across the landsape over the order of decades to centuries. Seastedt and Adams (2001) showed that mobile tree islands in an alpine tundra in the Colorado Rocky Mountains, U.S.A., caused significant reductions of organic matter and concentrations of ammonium from the soil, and that there was no apparent recovery of soil ammonium levels even 250–500 years following the passage of a tree island. Single-plant effects also occur in forests in which litter of differing qualities accumulates under individual trees of different species. For example, in Amazonia individuals of the tree *Dicorynia guianensis* produce litter with a high concentration of phenolics which sequesters over 80% of available nitrogen and which is avoided by endogeic earthworms. Meanwhile, a tree species with which it coexists (*Qualea* sp.) produces litter with low phenolics and which encourages high earthworm activity. The result is a mosaic of patches, some of which support rapid decomposition and others in which litter accumulates (Charpentier in Wardle and Lavelle 1997).

When several plant species co-occur in the landscape but individuals are sufficiently spatially separated to allow each plant to influence the soil directly under it, and when resource transport rates are low, the net result is a spatial mosaic of nutrient availabilities and supply rates. This conclusion is supported by the simulation modeling study of Huston and De Angelis (1994) which predicted that in multispecies communities with low resource transport rates each species should control the resource availability in the patch that it occupies, and this should lead to a greater coexistence of species being possible. If different plant species select for different decomposer communities, and if (as discussed earlier) these communities differ in their effects on soil processes, then spatial variability of plant species should have important effects in maintaining spatial variability of decomposer processes and therefore nutrient supply rates in soil, which may in turn contribute to a greater degree of plant species coexistence. This feedback will be considered further in chapters 5 and 6.

Further, different plant species and vegetation with different compositions and structures are likely to support different levels of heterogeneity of belowground properties, although this issue has seldom been considered. However, Kleb and Wilson (1997) found that heterogeneity (i.e., spatial coefficient of variation) of resources (including soil total and available nitrogen) was greater along transects sampled at 3 m intervals in a forest than along corresponding transects in an adjacent prairie. They found that soils from the prairie that were transplanted and positioned along the transects in the forest attained the higher variability observed for the forest soils, and vice versa. If vegetation composition affects the spatial variability of inputs of soil nutrients, then it is to be expected that this will in turn influence the spatial variability of those components of the soil biota that are bottom-up controlled, as well as the processes that they regulate. This is supported by the results of an experiment in a grassland ecosystem (Wardle et al. 1999b) in which different subsets of the flora were regularly removed from different plots. That experiment showed that spatial variability (coefficient of variation) at the spatial scale of meters for the microbial biomass, some nematode groups, and soil carbon mineralization was influenced by which plant species were removed and therefore by the composition of the plant community.

Temporal Variability, Stability, and Disturbance

Plant species differ in their effects on the temporal dynamics of soil biota and soil processes. There are two mechanisms through which this occurs, i.e., temporal variability of resource input, and determination of disturbance regimes. Many ecosystems are driven by pulsed inputs of resources and this has well-known important consequences for the dynamics of consumer organisms (Ostfeld and Keesing 2000). On the intra-annual time scale, consumers in the decomposer subsystem are driven by seasonal pulses of resources provided by plants, and the nature of these pulses differs across component species of a community. Plant species differ in their seasonal patterns of root death, and in many Northern Hemisphere temperate forests most angiosperm tree species return their leaf litter to the ground during a relatively brief period in the autumn while most gymnosperms do not. These pulses frequently emerge as a major determinant of seasonal variation of the soil microbial biomass (Wardle 1992). On the interannual time scale, plant species vary markedly in their degree of interannual variability of resource addition to the soil. Different plant species and plant functional types show vastly different temporal productivity responses to interannual variations of climate (Grime et al. 2000; see chapter 7), and plant species which are more resistant to climatic variation will therefore provide a more constant input of substrates to the soil over time. In an analysis of multiyear data from eleven field sites across the U.S.A., Knapp and Smith (2001) showed that interannual variability of NPP varied across sites by a factor of 4. The degree of this variability across sites was unrelated to variability in precipitation, but appeared to be related to vegetation type; grasslands and oldfields demonstrated greater variability across years than did forests. This result is important, because it points to the importance of plant functional characteristics in determining interannual variability of resource addition to the soil, which is in turn likely to influence the variability over time of the levels of soil organisms, the rates of decomposer-regulated processes, and ultimately the rates of nutrient supply to the plant community.

Patterns of interannual variability of resource addition to the soil are especially apparent when plant reproductive tissues are considered, for example, as occurs during the phenomenon known as "mast seeding" in forests, in which irregular large synchronized seeding events occur, sometimes over very large spatial scales (see Kelly 1994; Ostfeld and Keesing 2000). These seeding events are well known for their consequences for seed-feeding vertebrates but they almost certainly also have profound consequences for the decomposer subsystem. In the only direct experimental test of this to date, Zackrisson et al. (1999) added sterilized seeds of *Picea abies* to forest field plots at a rate comparable to that which would be expected during a mast seeding event in a Swedish boreal forest, and found that 60–70% of seed nitrogen was made plant available within a year, resulting in a pulse of available nitrogen to the soil of around 5 kg N/ha. This was shown to have important consequences for the nitrogen acquisition and growth of planted *Pinus sylvestris* seedlings in some experimental conditions. Similar sporadic interyear pulses of resources to the soil probably also occur for plants that can produce large amounts of floral material, berries, or pollen. For example, Greenfield (1999) found for 35 plant species that on average 62% of the nitrogen in their pollen could be mobilized in 30 days, and calculated that for stands of some wind-pollinated plant species (e.g. *Pinus radiata*) this may represent a major pulsed addition of nitrogen to the soil.

Variation in the quantity and timing of resource input to soil across plant species is likely to be a major determinant of the stability of the soil biota. The possible effects of vegetation on the stability of the soil biota and soil processes remain little explored other than for the theoretical work described in chapter 2 about how NPP affects soil food web stability. However, a literature meta-analysis of 58 previously published studies demonstrated that while soil microbial biomass varied across three main ecosystem types (forest, arable, grassland), the temporal variability in this biomass across seasons (inversely related to its stability) did not (Wardle 1998a) (fig. 3.6a). There is, however, some evidence from two experiments that plant species can influence the stability of soil biotic components. In the first, monocultures of four grassland plant species were established in a glasshouse, and some monocultures of each species were then artificially droughted while others were not. Resistance to droughting of a range of belowground properties differed across the four plant species (fig. 3.6b) while

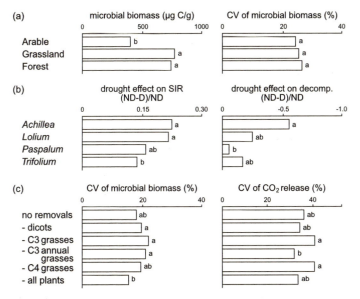

FIGURE 3.6. Effects of vegetation type and plant species on stability of belowground biological properties. (a) Microbial biomass and temporal coefficient of variation (CV) of microbial biomass across seasons for three ecosystem types, derived from a literature meta-analysis of 58 previously published studies (Wardle 1998a). Lower values of CV indicate greater stability. (b) Effect of an experimental drought on substrate-induced respiration (SIR; a relative measure of active microbial biomass) and decomposition rates of added substrates in experimental units planted with monocultures of each of four plant species. ND and D are values of the response variable in nondroughted and corresponding droughted experimental units, respectively. Stability, assessed as resistance, declines as [(ND–D)/ND)] increases. From Wardle et al. (2000). (c) Effect of plant functional group removals in a perennial grassland on temporal CV of microbial biomass and carbon mineralization over a three-year period. From Wardle et al. (1999b). For all panels bars topped with different letters are not significantly different at $P = 0.05$.

resilience of these properties was unresponsive to species effects (Wardle et al. 2000). In the second (Wardle et al. 1999b), treatments involving removal of different plant functional groups from field plots differed in their effects on the temporal variability of soil microbial biomass and activity (fig. 3.6c) and populations of some nematode functional groups over a three-year period. In both these experiments differences between plant species in their effects on belowground stability could not be explained in terms of species differences in productivity or resource input to the soil, meaning that vegetation composition in its own right (and therefore probably plant species effects on resource quality) was a key determinant of stability of the decomposer subsystem.

The second mechanism by which plant species affect temporal dynamics of soil biota and processes, i.e., through determining disturbance regimes in the soil, arises because plant species differ in functional traits that determine their responses to disturbance (Grime 1979; McIntyre et al. 1999) and because disturbance effects are propagated through decomposer food webs (as is apparent from the literature on land use change; see chapter 7). Disturbance regimes play an important role in the dynamics of most, perhaps all, forests, and many tree species are adapted for regeneration associated with tree death and canopy gap formation resulting from disturbance. While gap formation by natural disturbances creates conditions for establishment of new cohorts of trees, the presence of gaps also has important effects on belowground organisms and processes. Studies focused on the ecology of canopy gaps (either naturally occurring or experimentally created) have found gap presence and size to stimulate nitrogen mineralization (Parsons et al. 1994), reduce nitrogen mineralization (Bauhus and Barthel 1995), stimulate litter decomposition (Zhang and Zak 1995; but see Luizão et al. 1998), and reduce soil microbial biomass (Bauhus and Barthel 1995; Zhang and Zak 1995, 1998) (see fig. 3.7). The magnitude of most of the effects depicted in figure 3.7 suggests that plant-mediated disturbances in forests and consequent effects on soil biota and mineralization are likely to be a powerful determinant of nutrient supply for vegetation establishing in the gaps, although the direction of these effects is likely to be context dependent.

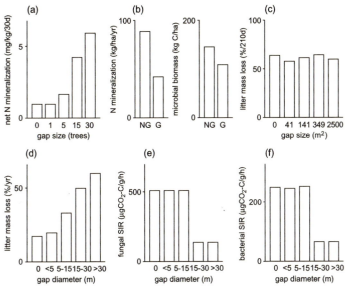

FIGURE 3.7. Effects of forest gaps on microbial biomass and carbon and nitrogen mineralization. (a) *Pinus contorta* forest in Wyoming, United States; data averaged over all sampling dates (from Parsons et al. 1994). (b) Mature *Fagus sylvatica* forest in lower Saxony, Germany; gap size greater than 20 m diameter (from Bauhus and Barthel 1995). NG and G indicate not in gap and in gap, respectively. (c) Rainforest in Roraima, Brazil (from Luizão et al. 1998). (d)–(f) Rainforest in Fukien Province, China; leaf litter data presented only. SIR is substrate-induced respiration (from Zhang and Zak 1995, 1998).

Large catastrophic intermittent disturbances which devastate forests over greater spatial scales (e.g., hurricanes, landslides, earthquakes) would be expected to have major effects on processes driven by the soil biota, at least in part because of the effect of the disturbance in influencing the quality and quantity of the plant residues entering the decomposer subsystem. For example, forest damage caused by Hurricane Hugo in Puerto Rico in 1989 resulted in the return of large amounts of coarse woody debris which led to significant microbial immobilization of nitrogen; experimental removal of woody debris from devastated plots increased soil nitrogen availability and NPP by up to 40% (Zimmerman et al. 1995). Other studies have shown hurricane damage to induce production of phenolics in damaged trees (Hunter and Forkner 1999), and cause large increases in fine litterfall (Herbert et al. 1999b) and significant mortality of fine roots (Silver and Vogt 1993; Herbert et al. 1999b). These effects are all likely to have important consequences for the amounts and quality of resources available to the decomposer subsystem and therefore supply rates of nutrients from the soil.

Another intermittent disturbance regime, which affects a range of ecosystems worldwide, is wildfire. In fire-prone ecosystems many plant species are adapted for both facilitating wildfire and regenerating rapidly after it. Wildfire has adverse short-term effects on soil organisms [i.e., microflora (e.g., Pietikäinen et al. 2000a) and the main soil faunal groups (e.g., Haimi et al. 2000)], due to both soil heating and the loss of soil organic matter (including forest humus). However, there are some taxa of decomposer biota which are adapted for rapid growth after fire, and there is evidence that soon after fire particular species of organisms are favored [e.g., fungi (Widden and Parkinson 1975); arthropods (see McCulloch et al. 1998); earthworms (James 1988)] presumably through the supply of detritus from newly killed plant material, especially belowground. It is apparent that in the longer term fire can have either positive or negative effects on the soil biota depending upon its effects on soil chemistry and the balance of organic matter gains (e.g., through new detritus) and losses (e.g., through the ignition of soil organic matter), and this explains the variety of responses of mineralization of carbon and nitrogen observed in response to burning (e.g., Bauhus et al. 1993; Fenn et al. 1993). However, in some forests naturally adapted to wildfire, burning may have beneficial effects on soil biota and processes through the production of charcoal. In the boreal forests of northern Sweden, prolonged absence of fire results in late successional, highly unproductive systems in which the ground layer becomes dominated by the ericaceous shrub *Empetrum hermaphroditum*, which produces high levels of phenolics that reduce plant growth and nitrogen availability. Zackrisson et al. (1996) demonstrated through experiments that charcoal fragments produced by wildfire act as a form of "activated carbon" source which adsorbs these phenolics and inactivates them, through the fragments serving as foci for microbes that degrade the phenolics. In a complementary study (Wardle et al. 1998c), charcoal addition was found to have a highly positive effect on nitrogen acquisition by *Betula pendula* seedlings grown in humus from under *E. hermaphroditum* (but not in humus from other sources), suggesting that charcoal increased nitrogen release through reducing the activity of soil phenolics from *E. hermaphroditum*. Further, Pietikäinen et al. (2000b) showed through laboratory experiments that charcoal was both an effective adsorbent of dissolved organic carbon, and a determinant of microbial community structure, through selecting for microbes with higher specific growth rates. It is apparent from the literature about how wildfire affects the belowground subsystem that dominant plant species which are adapted for encouraging wildfire result in the ecosystem being subjected to intermittent severe disturbances (usually several years or decades apart), and this introduces an element of long-term temporal variability with regard to decomposer organisms, soil processes, and ultimately nutrient supply rates from the soil.

PLANT TRAITS, STRATEGIES, AND ECOPHYSIOLOGICAL CONSTRAINTS

It is apparent from the material presented so far that plant species differ in their effects on belowground processes, and that this is manifested through plant species effects on belowground organisms. Given that the plant is the ultimate determinant of how the decomposer subsystem functions, it is reasonable to expect that differences in ecophysiological attributes of plants, and differences in plant strategies, are likely to play an important role in influencing the effects that plant species may have on ecosystem properties.

Plant species effects on belowground processes can be clearly demonstrated by comparing species adapted for sites of differing nutrient availability. Plants in nutrient-rich habitats usually grow rapidly, and invest most of their carbon in growth and reproduction. Plants in poorer habitats tend to live longer, retain their leaves for longer, and frequently invest a much higher proportion of their carbon in secondary metabolites such as polyphenolics (Grime 1979; Coley et al. 1985; Horner et al. 1988). These attributes are linked to a number of other traits which will be discussed later in this chapter, but I will first describe examples in which belowground ecological consequences of plant species with vastly differing strategies have been compared.

Possibly the first detailed experimental study to compare plants that differ in their production of secondary metabolites is that of Handley (1961). In that work, *Calluna vulgaris*, which characterizes nutrient-poor sites, was compared with *Epilobium angustifolium* and *Circaea lutetiana*, which both occupy favorable sites. *Calluna vulgaris* produced leaf tannin complexes which were more resistant to breakdown than those of the other two species, which helps to explain the thick moroid humus layer that forms under *C. vulgaris*. Release of nitrogen from both *C. vulgaris* leaf extracts and humus from under *C. vulgaris* plants, as measured by the growth and tissue nitrogen concentration of a phytometer species (*Betula pendula*), was much less than that from leaf extracts from the other species, suggesting that *C. vulgaris* reduces the supply rate of available nitrogen in the soil.

Another example of plants with vastly contrasting strategies affecting supply rates of available nutrients involves the ericaceous dwarf shrub *Erica tetralix* and the grass *Molinia caerulea* in moist heathlands in the Netherlands (Berendse and Aerts 1984; Aerts et al. 1990; Berendse 1990, 1998). In this system, *M. caerulea* is adapted for habitats with high nutrient availability while *E. tetralix* is adapted for poorer habitats. *Molinia caerulea* is capable of a greater productivity than *E. tetralix*, has longer-lived leaves, and produces higher-quality litter with low lignin and high nitrogen concentrations (Berendse et al. 1987). *Erica tetralix* leaf litter decomposes at around 30% of the rate of that from *M. caerulea*, and during decomposition it mobilizes nitrogen at a much slower rate. As a result, *E. tetralix* dominates during early succession when nutrient availability is low, but as organic matter accumulates and the net nitrogen mineralization rates increase, *M. caerulea* begins to replace *E. tetralix* because of its higher potential growth rate. This, and the rapid rate of mineralization of nitrogen from the new litter produced by *M. caerulea*, results in a positive feedback encouraging further growth of *M. caerulea*.

A third example involves boreal forests in northern Sweden. Shortly after wildfire, plant species that produce relatively high-quality litter (e.g., *Pinus sylvestris*, *Vaccinium myrtillus*) dominate, but in the prolonged absence of fire and increasing unavailability of nitrogen, forests frequently become dominated by slower-growing plants which produce a much poorer litter quality with high concentrations of phenolics (e.g., *Picea abies*, *Empetrum hermaphroditum*). The leaves of *E. hermaphroditum* contain extremely high concentrations of the phenolic compound Batatasin III, which greatly inhibits tree seedling growth, nutrient acquisition, and mycorrhizal colonization (Nilsson et al. 1993). Another, otherwise functionally very similar species (*Empetrum nigrum*), which does not produce this phenolic even when coexisting with *E. hermaphroditum*, does not have these strong negative effects (Nilsson et al. 2000). Activated carbon or charcoal added to soils dominated by *E. hermaphroditum* adsorbs these phenolics and significantly reduces their adverse effects on plant growth and nitrogen uptake (Nilsson 1994; Wardle et al. 1998c). Wardle et al. (1997b) demonstrated, through the use of 50 lake islands with differing fire histories, that islands dominated by *E. hermaphroditum* and *P. abies* (as a result of prolonged fire absence) had vastly reduced soil microbial biomass, slower rates of litter decomposition and nitrogen mobilization, lower leaf nitrogen concentrations (indicative of low nutrient acquisition), and much greater soil carbon storage than did more recently burnt islands dominated by *P. sylvestris* and *V. myrtillus*.

It is clear from the above examples that plant species dominating in nutrient-poor sites function very differently from those dominating in nutrient-rich sites. This is consistent with the theory about mull and mor (Muller 1884; Handley 1954; Grubb and Tanner 1976). Mull sites are usually characterized by plant species which produce high-quality litter, encourage high microbial activity and turnover, and promote burrowing earthworms. The result is rapid plant litter decomposition and nutrient mineralization rates and low levels of soil carbon storage. Mor sites are dominated by slower-growing plants which favor a high fungal contribution to the soil microbial biomass, and high populations of microarthropods. In these systems, decomposition rates are much slower and the soils have deep humus layers in which nitrogen is bound in complexes and therefore of low biological availability. Although there is a dearth of studies comparing soil food web structure in contrasting forest types, it would seem likely that food webs in mull systems are more dominated by bacterial-based (versus fungal-based) energy channels and large burrowing invertebrates, and are less dominated by arthropods, than are food webs in mor systems (see chapter 2). These differences should in turn lead to a more rapid turnover of nutrients in mull than in mor systems.

It is important to note that within-species variations in response to site fertility can be important in driving the decomposer subsystem, and that phenotypic plasticity of a given species can therefore be an important component of its effects on belowground properties. In the case of mull and mor theory, some tree species (e.g., *Fagus sylvatica*) seem capable of producing either a mull or a mor humus type depending upon site conditions. Further, in Hawaii, the tree *Metrosideros polymorpha* produces litters of vastly differing quality (and which decompose at very different rates under standardized conditions) in response to variations in site fertility and soil development (Crews et al. 1995; Ostertag and Hobbie 1999), macroclimate (Vitousek et al. 1994; Austin and Vitousek 2000), and subtle genetic differences among geographically isolated populations (Tresender and Vitousek 2001). Abiotic factors can also be important determinants of the production of secondary metabolites such as phenolics by a given species (see Hättenschwiler and Vitousek 2000). For example, Northup et al. (1995, 1998) observed that along a gradient of decreasing fertility and increasing acidity, *Pinus muricata* trees produced litter with increasing concentrations of phenolics which upon decomposition released much lower amounts of mineral nitrogen and greater amounts of dissolved organic nitrogen. Further, Nicolai (1988) found that litter from *F. sylvatica* growing on an infertile site had much greater concentrations of phenolics than that from a fertile site, and that palatability of *F. sylvatica* litter to isopods was inversely related to its phenolic content. Despite the dearth of studies investigating the effects of within-plant-species variation on the decomposer subsystem and soil food web, the available evidence suggests that the phenotypic response of species to soil fertility is likely to have important ecosystem-level consequences.

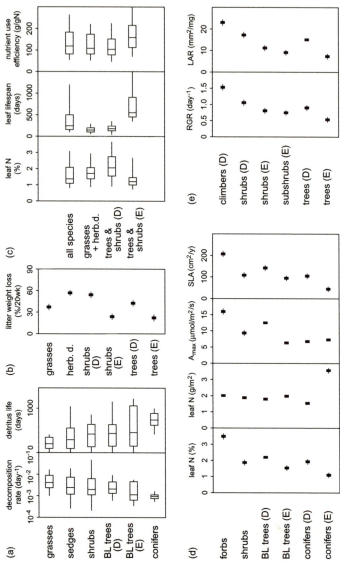

The Role of Suites of Plant Traits

The effects of plant species on ecosystem properties are ultimately determined by their ecophysiological attributes or suites of plant traits. Ecologists have long recognized that plants in habitats with low resource availability have a differing set of traits from those in higher-resource habitats (e.g., Grime 1977; Chapin 1980; Coley et al 1985; Grime et al. 1997). Chapin et al. (1993a) describe the suite of traits required for adaptation to low-nutrient conditions as a "stress resistance syndrome." Those studies which have synthesized data from large numbers of species have found that different plant groups which are favored by habitats of different fertilities also differ in a range of ecophysiological attributes (fig. 3.8). In particular, conifers, which often dominate in low-nutrient conditions and frequently have low rates of tissue production relative to other plants [especially at the seedling stage; see Bond (1989)], also produce longer-lived leaves and have lower leaf nutrient concentrations, lower specific leaf areas and leaf area ratios, and lower photosynthetic rates (see, e.g., Cornelissen et al. 1996; Reich et al. 1998; Aerts and Chapin 2000). Further, comparative studies that test for correlations across large numbers of species provide evidence for consistent relationships among these

---

FIGURE 3.8. Litter decomposition and ecophysiological attributes of a range of different plant life forms, derived from studies that have synthesized data from large numbers of species. Studies are those of (a) Enríquez et al. (1993), (b) Cornelissen (1996), (c) Aerts and Chapin (2000), (d) Reich et al. (1998), and (e) Cornelissen et al. (1996). The study of Cornelissen et al. (1996) involved plants growing in standardized conditions while the others involved plants growing in their natural environment. D indicates deciduous, E evergreen, BL broad leaved, herb. d. herbaceous dicotyledonous plants, A_{max} net photosynthetic capacity; SLA specific leaf area; RGR relative growth rate; and LAR leaf area ratio. For panels (a) and (c) boxes include 50% of the data values and bars include 95% of the data values, and for the other panels horizontal bars indicate standard errors of mean values.

*looking
× refs*

traits (e.g., Poorter and Remkes 1990; Poorter and Berg-kotte 1992; Berendse et al. 1998).

These combinations of traits are important in enabling plants to retain nutrients and preventing their losses in situations of low nutrient supply including losses through herbivory (see chapter 4). As such these traits play a role in regulating the decomposer subsystem. Plants that are attuned by natural selection for low-nutrient conditions often maintain low concentrations in their tissues, and produce smaller, less photosynthetically active leaves with a much higher content of structural carbohydrates (Poorter and Bergkotte 1992). The leaves of such plants also contain much higher concentrations of secondary metabolites such as phenolics. All these properties make the leaf litter produced by these species less favorable as a resource for the microflora and fauna responsible for degrading the litter, and the litter is therefore decomposed much more slowly (see Enríquez et al. 1993; Cornelissen 1996; fig. 3.8). Those plant species that are best suited for growth at low-nutrient conditions and produce litter that is mineralized more slowly are often better capable of taking up nitrogen in non-mineralized forms directly through their mycorrhizas, reducing their dependence on mineralization of plant litter by decomposer biota (see chapter 5). Comparative studies that have considered a broad spectrum of plant species have shown rates of litter mass loss to be correlated with a range of ecophysiological traits, e.g., negatively with plant growth rate, size, and longevity (Wardle et al. 1998a); negatively with leaf tissue strength (Cornelissen and Thompson 1997; Cornelissen et al. 1999); and positively with leaf base content (Cornelissen and Thompson 1997).

The concept of nutrient use efficiency (NUE) (Vitousek 1982), which predicts that plant species that dominate in nutrient-poor environments should have a higher rate of production (e.g., litter fall dry mass) per unit nutrient content, integrates the variety of ecophysiological processes driven by suites of plant traits. However, in understanding species effects on nutrient cycling processes, an approach which also incorporates the significance of storage and re-translocation of nutrients within the plant biomass is desirable. In this light, Berendse and Aerts (1987) refined this concept by breaking NUE into two components, i.e., the mean residence time of the nutrient in the plant and the productivity per unit nutrient uptake (NUE is the product of these components), and Aerts and Van der Peijl (1993) concluded that only one of these components, mean residence time, was likely to be consistently advantageous in nutrient-poor environments, at least when plant species are considered in monoculture. However, when coexisting plants are considered in real communities, competitive interactions among species may select for reduced nutrient loss rates (and hence greater residence times) much more strongly than can be expected in monoculture (Berendse 1994b). The NUE concept has been suggested as a physiological control of litter quality and therefore decomposition rate. Aerts (1997a) found, upon synthesizing data from previous studies, that there was a statistically significant negative relationship between decomposition rate and NUE for nitrogen, suggesting a possible tradeoff between the two (or at least between decomposability and mean residence time). However, the relationship between these variables was relatively weak ($R^2 = 0.24$) and for many low-productivity plant species NUE and decomposition rates were both low. This was interpreted to mean that slowly growing species which produce high levels of secondary metabolites may have both low NUE and low decomposition rates, and that this may lead to low rates of nutrient cycling in the ecosystem. In particular, evergreen tree species, many of which are adapted for nutrient-poor conditions, appear to benefit from long tissue life spans, low nutrient concentrations, low NUE, and low litter decomposition rates (Berendse 1994a).

91

Plant ecophysiological traits must ultimately affect decomposition and mineralization of nutrients through influencing decomposer organisms. The issue of how plant traits affect decomposer organisms has seldom been specifically addressed, although it is clear that litter quality controls of decomposition (e.g., concentrations of nitrogen, lignin, and phenolics) have important effects on the activity, biomass, and composition of the decomposer community. For example, plants that dominate in infertile conditions (e.g., many coniferous species) and have a particular set of traits that result in poor-quality litter being returned to the ground favor strong domination by the fungal-based energy channel and certain groups of saprophagous arthropods. Other plant species, which grow in more favorable conditions, often favor a greater role by the bacterial-based energy channel and endogeic earthworms. The fact that globally coniferous forest soils contain on average a lower amount of microbial biomass per unit carbon in the soil than do other soils (fig. 3.2) is consistent with the traits that characterize coniferous forest species (fig. 3.8). The net result is that plant species attuned by natural selection for growth in infertile conditions often favor the formation of deep mor type humus profiles, high sequestration of carbon and nitrogen in the soil, and low rates of nutrient turnover and supply from litter, in contrast to plant species adapted for fertile conditions. Based on the material presented so far in this chapter, it is clear that as a consequence of their different ecophysiological attributes, different plant species select for different subsets of the decomposer biota (through selecting for those organisms that are best adapted for breaking down litter from that species), and that these plant species effects therefore have profound implications for both decomposer processes and supply rates of plant-available nutrients.

Stress, Disturbance, and Competition

The plant strategy theory of Grime (1977, 1979) proposes that plant species with different suites of traits are adapted for growth in high-stress–low-disturbance and low-stress–high-disturbance habitats. However, within these broad stress-tolerant and disturbance-adapted categories there are a variety of strategies possible at finer levels of resolution; different plant species posess different adaptations for different types of disturbance (Grubb 1985), and Grubb (1998) identifies three different types of plant strategy which can be shown by plant species adapted for different types of stress, and which are only partially consistent with Grime's "stress tolerant" category. However, in general terms, plant species that dominate in high-stress habitats associated with nutrient limitation usually possess specific sets of traits consistent with the theory of Grime (1979), and which have important consequences for the soil biota and soil processes as discussed earlier. Plants that dominate in disturbed conditions (e.g., ruderals) generally possess a separate set of traits, i.e. rapid growth, high tissue nutrient concentrations, high rates of nutrient acquisition, high turnover of tissues, and high specific leaf areas and photosynthetic rates (see Grime 1979; Reich et al. 1998; fig. 3.8). As such, leaf litter from ruderal species is highly favorable to decomposers and breaks down very rapidly (Cornelissen 1996; fig. 3.8b). Disturbance-adapted plant species will therefore be expected to have opposite effects on decomposer biota to those of stress-adapted plants.

Plant species that can be regarded as "competitors" (*sensu* Grime 1979), i.e., those that dominate in low-stress–low-disturbance habitats, could be expected to have effects on decomposer biota via their litter quality which are intermediate between those of ruderals and stress tolerators (see Wardle 1993). However, competitive plant species could potentially affect microbes through resource depletion, and based on the theory of Grime (1979) we could expect these effects to be most important in habitats that encourage the greatest productivity. The concept of plants competing most strongly for nutrients in nonstressed (i.e., productive) habitats has been directly challenged by Tilman (1988), who maintains that competition is important in both fertile and infertile conditions, and that as conditions become less fertile, plants shift from competing for light to competing for nutrients. This is consistent with the predictions of Tilman's R^* model in which the plant species that reduces the limiting resource to a level below that which other plant species can tolerate has the competitive advantage. However, the available evidence is mixed; some experimental studies claim to provide support for the view that competition is greater in unproductive environments (e.g., Wilson and Tilman 1991) while several studies provide support for the opposite view (e.g., Campbell and Grime 1992; Gerdol et al. 2000; Keddy et al. 2000). Part of this inconsistency is undoubtedly due to variation across studies in the way that competition is measured. In the one field study to date for which species effects along an experimental nutrient availability gradient were controlled (Peltzer et al. 1998), neither aboveground nor belowground competition intensity varied with increasing productivity, suggesting that differences in competition which may occur across other productivity gradients probably emerge because different plant species with different traits dominate different parts of the gradient (see Schippers et al. 1999).

Literature syntheses (e.g., Goldberg and Novoplansky 1997) and meta-analyses (Gurevitch et al. 1992; Goldberg et al. 1999) have failed to provide consistent evidence for declining competition with declining productivity; indeed, Goldberg et al. (1999) found some evidence for the reverse trend. However, as stated in chapter 2, the published literature upon which such syntheses depend may select for certain types of results above others. Although some competition clearly does occur in nutrient-poor conditions in both grasslands (Wilson and Tilman 1991; Campbell and Grime 1992) and forests [several examples involving competitive effects of trees on seedlings are provided by Coomes and Grubb (2000)], the position of Tilman (1988) is difficult to reconcile with the frequently observed hump-backed relationship between diversity and productivity; a likely reason for this relationship is because intensity of competition is sufficient to induce competitive exclusion of species in productive but not in unproductive habitats (see chapter 6).

In any case, while the R^* model with regard to nutrients may adequately explain competitive interactions in homogeneous media such as water, it may be of less relevance in soils, especially in infertile conditions. This is because nutrient availability and supply are governed in the first instance by the decomposer biota associated with plants, and only secondarily by soil water availability, which plants influence only in a narrow range of conditions. The R^* model depends upon the assumptions that belowground resources are sufficiently mixed for all species to encounter the same resource concentration, and that the plants themselves regulate the availability of resources in this homogenous environment (see Huston and De Angelis 1994). However, both these assumptions are unrealistic especially in infertile soils. In stressed habitats individual plants are often spatially discrete (several species even show clonal growth) and have direct control of the soil environment under them, minimizing resource transport rates. The ability of plants in stressed environments (and their mycorrhizal symbionts) to utilize organic nitrogen as a primary resource (Northup et al. 1995; Näsholm et al. 1998) means that individuals are able to access a resource which has a much lower transport rate than do mineral nitrogen forms, and each plant occupying its own patch therefore should have preferential access to the nutrients present within it. The same is true for phosphorus, given that most plant species have mycorrhizal symbionts that directly mobilize and take up immobile forms of that nutrient. Further, as discussed through much of this chapter, plants only indirectly regulate their own nutrient availability, because availability and supply are governed in the first instance by their associated decomposer biota. In a modeling study in which both of these assumptions were relaxed, Huston and De Angelis (1994) demonstrated that large numbers of plant species could coexist on a single limiting resource under steady-state conditions. Further, individuals of different plant species growing in a spatially separated manner may have vastly differing effects on the decomposer communities in the soil under them, with each community being adapted for releasing nutrients in the environment provided by that plant (see fig. 3.5). Finally, it is relevant that the mechanism proposed by the R^* theory has rarely actually been tested in soils. Even in the study of Tilman and Wedin (1991), which is presented as evidence that those species which reduce mineral nitrogen concentrations to the lowest levels are also those which can best tolerate the lowest levels, only five species were used, and this is not a statistically adequate number of species for demonstrating a phenomenon that can be tested for convincingly only through the use of a comparative approach in which a large number of species is included. The assumption that must hold for the R^* theory to apply is that competitive effect and competitive response across species are correlated, and this assumption is not supported by the results of comparative studies (e.g., Keddy et al. 1994; Wardle et al. 1998a). It would appear that when adequate consideration is given to the soil biota and soil processes, the assumptions upon which the R^* theory is based are unrealistic especially in unproductive environments. However, much still remains to be resolved about how plant competition varies in relation to productivity, and the likely involvement of soil biota and processes in this.

SOIL BIOTIC RESPONSES TO VEGETATION SUCCESSION

Plant species effects on soil biota and processes are highly apparent from patterns of species replacement during succession. Changes in succession are governed by biotic interactions, i.e., the ability of a plant species to improve the physical environment for subsequent species (facilitation), competitive interactions and other inhibitory mechanisms, and tolerance by a given species of the effects of others (see Connell and Slayter 1977; Huston and Smith 1987; Pickett et al. 1987; Walker and Chapin 1987; Callaway and Walker 1997; Wilson 1999). These interactions are in turn driven by changes in resource availability. It is also clear that during succession there are often three distinct phases: the build-up phase during which plant biomass aggrades; the maximal biomass phase [the "climax community" of Clements (1928)]; and (given sufficient time) the decline phase (e.g., Walker and Syers 1976; Walker et al. 1983; Vitousek et al. 1993). Those plants that dominate at different phases of succession have vastly different suites of ecophysiological traits (Bazzaz 1979; Grime 1979) and as succession proceeds there is a shift from "ruderals" (build-up phase) to "competitors" (maximal biomass phase) to "stress tolerators" (decline phase) (see Grime 1987).

During the build-up phase of succession, ecosystem development is plant driven and is ultimately determined by the functional traits of the dominant plant species present (see Odum 1969). During this phase there is frequently a rapid net accumulation of organic matter; for example, Schlesinger et al. (1998) determined that over the first 110 years following the Krakatau eruption in Indonesia organic carbon accumulated at a rate of 45–112 g $C/m^2/yr$. This accumulation often occurs through the microflora being nitrogen limited during the initial stages of succession, resulting in net photosynthesis exceeding net decomposition. However, as the build-up phase progresses, nitrogen levels increase in the soil (often largely through bacterial nitrogen fixation) and corresponding increases frequently occur for both microbial biomass and microbial activity (e.g., Insam and Haselwandter 1989; Ohtonen et al. 1999). This shift results in the microbial biomass changing from nitrogen limited to carbon limited, and consistent with this a decrease occurs over time for both the microbial carbon-to-nitrogen ratio (e.g., Diquélou et al. 1999) and the rate of carbon loss through respiration per unit microbial biomass (Insam and Haselwandter 1991). These changes in microbial properties are frequently the most dramatic when plant species that are capable of symbiotic nitrogen fixation occur early in the succession. For example, Halvorson et al. (1991) found along the disturbance gradient created by the eruption of Mt. St. Helens (Washington state, U.S.A.) that the patches colonized by *Lupinus* species encouraged greater soil organic matter development, microbial biomass, and microbial activity than did adjacent patches, especially in the areas most devastated by the eruption. Enhancement of microbial biomass and activity during the build-up phase results in increased decomposition rates and supply rates of available nutrients (Zak et al. 1990a). For example, Crews et al. (1995) found that decay rates of *Metrosideros polymorpha* litter increased along the primary stages of a chronosequence created by volcanic activity, and Chapin et al. (1994) determined that rates of nitrogen uptake by *Picea sitchensis* seedlings (indicative of nitrogen supply rates) were greatest when seedlings were planted in those sites along a chronosequence (induced by deglaciation) that were dominated by nitrogen fixing shrubs and of intermediate successional age. Increases in ecosystem nitrogen content during the aggradation phase of succession, and the corresponding enhancement of microbial activity and nutrient mineralization, can be an important means of facilitation of new plant species during succession (see Callaway 1995; Berendse 1998).

Although organic matter levels, microbial biomass, and nitrogen mobilization all follow predictable trends during the successional build-up phase, the development of the soil food web during this phase remains poorly understood. However, microbial community composition does show clear changes during vegetation development. For example, Ohtonen et al. (1999) showed that during succession following deglaciation there was a shift from bacterial to fungal domination as well as community level shifts in the microflora as assessed by soil PLFA profiles. Shifts in community structure during ecosystem development have also been shown for both saprophytic fungi (Brown 1958; Frankland 1998) and mycorrhizal fungi (e.g., Dighton and Mason 1985). Few studies have specifically addressed the issue of how communities of higher-level consumers in the soil food web (soil fauna) change during succession. However, Wasilewska (1994) determined that nematode taxa known to be r selected dominated in younger successional meadows while K-selected taxa dominated in older meadows. In a succession from clear-cut forest areas to old growth *Pseudotsuga menziesii* forests in Vancouver Island (Canada), Setälä and Marshall (1994) found that both community composition and abundance of springtails were closely related to the shifts in the abundance and decay stages of tree stumps along the succession. Successional trends in community composition have also been observed during forest development for larger invertebrates, e.g., earthworms (Scheu 1992) and spiders and carabid beetles (Niemelä et al. 1996). Little is known about how the functional characteristics of decomposer food web structure change during succession, although it would seem likely that as the maximal biomass phase of the succession is approached there should be a shift toward the fungal-based energy channel and an increasing role played by larger invertebrates.

99

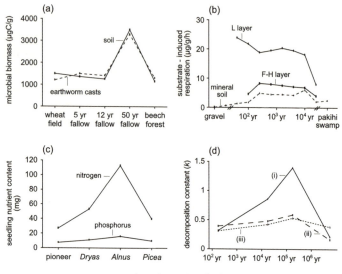

FIGURE 3.9. Changes in soil biotic properties and ecosystem processes during the build-up and decline phases of ecosystem succession. (a) Microbial biomass in a successional sequence following retirement from cultivation of wheat fields in northern Germany. From Scheu (1990). (b) Substrate-induced respiration (a relative measure of active microbial biomass) along a chronosequence created by the retreat of the Franz Josef Glacier, South Westland, New Zealand. From Wardle and Ghani (1995). (c) Total nutrient content of *Picea sitchensis* seedlings planted in each of four stages of a succession caused by deglaciation at Glacier Bay, Alaska, U.S.A. Here, seedling nutrient content can be considered as a phytometer of nutrient availability. From Chapin et al. (1994). (d) Decomposition constants for *Metrosideros polymorpha* litter along a chronosequence created by volcanic activity in Hawaii, U.S.A. (i) In situ decomposition of leaves placed at the sites from which they were sourced; (ii) decomposition of litter from one source placed at all four sites; (iii) decomposition of litter from all four sites placed at a common site. From Crews et al. (1995). No statistical analysis is presented because for each study no true within-stage replication exists.

As the maximal plant biomass phase is approached or surpassed, the decline phase often occurs, in which a marked drop is observed for NPP and subsequently for standing biomass. In some instances this results from domination by plant species that produce litter of poor quality containing high concentrations of secondary metabolites, resulting in diminished nitrogen availability. In the example described earlier, late successional forests in northern Sweden that develop through the prolonged absence of fire become dominated by *Picea abies* and *Empetrum hermaphroditum*, both of which produce phenolics that adversely affect microbial activity, decomposition, nitrogen mineralization, and plant nitrogen uptake (Zackrisson et al. 1996; Wardle et al. 1997b). Similarly, in the floodplains of Alaska, production of tannins by *Populus balsamifera*, which replaces *Alnus tenuifolia* during succession, has adverse effects on symbiotic nitrogen fixation by *A. tenuifolia*, and on microbial activity, litter decomposition, and nitrogen mineralization in *A. tenuifolia* litter (Schimel et al. 1998). Further, in the Glacier Bay chronosequence, domination of late successional stages by *Picea sitchensis* has negative effects on nutrient acquisition by planted seedlings of *P. sitchensis* relative to the earlier successional stages where it is absent (Chapin et al. 1994; fig. 3.9c), indicative of a reduced rate of nutrient supply from the soil.

In other successions, phosphorus limitation becomes important instead in inducing the decline in soil activity. For example, Scheu (1990) found microbial biomass to be maximal in intermediate successional stages of the development of fallow to forest (fig. 3.9a) and used a fertilizer experiment to demonstrate that while microbial activity was limited by nitrogen early in the succession phosphorus became limiting at the final phase. Further, Crews et al. (1995) determined for a chronosequence in Hawaii that litter from an intermediate age stand decomposed more rapidly than that from both older and younger stands, and that standardized

101

litter placed in the intermediate stand also decomposed fastest (fig. 3.9d). Those data, together with the results of fertilization experiments performed at these sites to investigate the nature of nutrient limitation to ecosystem processes (Vitousek and Farrington 1997; Hobbie and Vitousek 2000) showed that while nitrogen was limiting to productivity for ecosystem processes during the build-up phase of succession, phosphorus later became limiting to both productivity and decomposition through being converted to unavailable forms (see Walker and Syers 1976). Further, the increase, plateau, and decrease of microbial biomass which occurs along the Franz Josef Glacier chronosequence (fig. 3.9b), and the corresponding decline, plateau, and rise of the microbial respiration-to-biomass ratio (Wardle and Ghani 1995) reflects microbial nitrogen limitation being followed by carbon limitation and finally by phosphorus limitation along the sequence (see Walker and Syers 1976).

Few studies have considered how soil food web structure responds to the decline phase in late successional systems. However it is likely that since late successional forests develop mor-type soil characteristics, a shift towards a soil food web typical of forests with moroid humus should occur. There is some evidence that this may be the case. For example, Coderre et al. (1995) found that *Acer saccharum* forests which were in a state of decline (probably through acidification) lacked earthworm species that were dominant in vigorous stands. Further, Bernier and Ponge (1994) provided evidence through microscopic analysis of humus profiles in a *Picea abies* forest for a feedback in which burrowing earthworms become scarce when the forests approach maturity, and the humus shifts from a mull type to a moder type. However, as the trees senesce the humus suddenly becomes palatable to earthworms, creating a mull type profile in which organic matter becomes incorporated into the mineral soil. This results in a pulse of mineralization which helps maintain the growth of a new cohort of *P. abies* seedlings.

SYNTHESIS

While it is apparent from chapter 2 that NPP could be an important regulator of soil food webs and processes, this chapter shows that plant species differences have important consequences for decomposer trophic levels that are independent of plant productivity. These effects arise largely through differences in the quality of resources returned to the soil, either through litter characteristics or through the exudation of substrates into the rooting zone. Resource quality effects are fundamental determinants of soil processes and ecosystem properties, including decomposition, supply rates of available nutrients, and sequestration of carbon and nitrogen in the soil. There is some evidence that plant species select for decomposer communities that preferentially break down the litter from that species, suggesting the existence of feedbacks between plant and decomposer communities.

Plant species also affect soil biota through a number of other mechanisms which although less recognized may also be of importance. These include exploitation of nutrients from the soil, physical effects (including ecosystem engineering effects by plants), and alteration of both spatial and temporal variability of decomposer food web components and processes. Plant species that irregularly return large pulses of resources to the ground, or which facilitate periodic large disturbances, may have especially important effects on the dynamics of decomposer organisms. However, few studies have investigated the effects of plant species on the temporal variability or other measures of stability of the soil biota.

The influence of plant species on soil food webs and processes is ultimately determined by the ecophysiological traits of the plant species themselves. Some species allocate carbon to rapid growth while others allocate much of their carbon to secondary metabolites and grow more slowly. These basic differences are in turn related to suites of traits incorporating net photosynthetic rates, nutrient use characteristics, and leaf properties. Plant species adapted for rapid growth dominate in fertile soils and select for decomposer communities that rapidly release nutrients from high-quality litter (e.g., the bacterial-based energy channel and earthworms) while plant species adapted for infertile soils and which grow more slowly select for decomposers that are adapted for poorer-quality litter but which cycle nutrients more slowly (e.g., the fungal-based energy channel and microarthropods). As a result, soils on fertile sites usually have a mull type profile while infertile soils have a mor type profile with a deep layer of undecomposed organic matter. Plant species effects on decomposer organisms and processes are especially apparent during vegetation succession. During the build-up phase, resource quality improves and this favors an increase in decomposer organisms and rates of litter decomposition and cycling of nutrients, while during the decline phase that eventually follows nutrient availability and litter quality both diminish and this adversely affects decomposers and soil process rates.

CHAPTER FOUR

Belowground Consequences of Aboveground Food Web Interactions

It is apparent from the previous two chapters that the soil food web and the processes that it regulates are influenced both by the productivity of the plant community and by the quality of the resources that the plant community produces. However, in addition to driving the belowground subsystem, primary producers also represent the basal trophic level of the aboveground food web. The primary consumers of the aboveground food web, i.e., the foliar herbivores, frequently have important effects on both the ecophysiology of individual plants and the structure of the plant community; these effects may in turn influence the quantity and quality of resources that are returned to the soil. Because the nature of resources entering the soil has important effects on soil organisms, the potential exists for aboveground food web interactions to indirectly affect soil food webs as well as the processes that they regulate.

A better understanding of how plant species affect the performance of soil communities requires specific acknowledgment of the sorts of effects that aboveground consumers exert on plants and plant communities. These effects may be important drivers of ecosystem function, because indirect effects of herbivores on decomposers, as mediated by the plant, should be able to influence nutrient supply rates from the soil. This should in turn have aboveground consequences. In this chapter I will consider the various mechanisms through which aboveground food webs affect soil biota and soil processes, and how this contributes to our understanding about how aboveground and belowground food webs are interlinked.

INDIVIDUAL PLANT EFFECTS

When plant material is consumed by foliar herbivores, there are a number of effects on the physiology of the whole plant which are relevant at a range of temporal scales. In the short term, herbivory can alter the allocation of carbon to the rhizosphere. In the longer term, browsed plants show altered patterns of NPP both aboveground and belowground. Further, plants respond to herbivory through altering the quality of litter derived from both leaves and roots. These effects can strongly influence the soil biota and processes, and each will be discussed in turn.

Rhizosphere Carbon Allocation

As mentioned in chapter 2, many herbaceous plant species allocate a substantial proportion of their fixed carbon to the rhizosphere, and this in turn results in a substantial stimulation of the soil microflora and microfauna. When a plant is defoliated, more resources can be allocated belowground (Dyer et al. 1991; but see Miller and Rose 1992), and as a consequence of this increased exudation of carbon into the rhizosphere usually occurs (Bardgett et al. 1998b). For example, Bokhari and Singh (1974) found that artificially defoliated plants of *Agropyron smithii* grown in hydroponic solution released nearly 10% more carbon into the growth medium than did intact plants, and also had a greater rate of respiration associated with the rooting zone. Further, ^{14}C pulse labeling studies of monocotyledonous herbaceous plants provide evidence that herbivory by grasshoppers causes an allocation shift of carbon from shoots to roots (Dyer et al. 1991) and that this leads to both greater exudation of ^{14}C into the rhizosphere and enhanced rooting zone respiration (Holland et al. 1996).

Studies that have investigated the effects of defoliation on soil organisms often obtain results which are consistent with the stimulation that would be expected due to enhanced exudation of carbon in the rhizosphere. For example, Holland (1995) found that microbial biomass associated with *Zea mays* roots was maximized when plants were subjected to intermediate levels of herbivory by grasshoppers in a no-tillage cropping system, and proposed that increased root exudation due to defoliation was responsible. Similarly, Mawdsley and Bardgett (1997) found greater amounts of microbial biomass, activity, and plate counts of various microbial groups in the rooting zones of *Trifolium repens* and *Lolium perenne* when plants were defoliated, despite defoliation reducing total root biomass (fig. 4.1a). Their results can be explained in terms of defoliation effects on root exudation but not root productivity. It has also been hypothesized by Bardgett et al. (1996, 1998b), based on the results of experiments showing that grazing by sheep (*Ovis aries*) in Welsh upland grasslands enhances soil bacteria relative to fungi, that defoliation causes "fast cycles" of nutrients in the rooting zone due to increased labile substrates in the rhizosphere and selection for bacteria, as opposed to "slow cycles" caused by recalcitrant substrates and domination by fungi.

It would appear likely that increased microbial (and especially bacterial) productivity in the rhizosphere following defoliation should also favor higher-level consumers in the soil microfood-web. Mikola et al. (2001b) performed a glasshouse experiment in which mixtures of three grassland plant species were defoliated at a range of intensities. Although the greatest intensity of defoliation caused a decline in root mass, there was a marked increase in many soil faunal groups, including enchytraeids and microbe-feeding nematodes (fig. 4.1b). Microbial biomass was reduced by de-

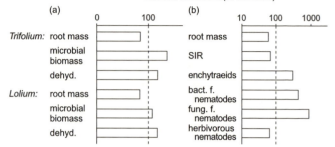

FIGURE 4.1. Examples of stimulation of subsets of the soil biota but reductions of root biomass by experimental defoliation. (a) From Mawdsley and Bardgett (1997). Dehyd indicates dehydrogenase activity. Data presented are the means of the last three sampling dates. (b) From Mikola et al. (2001b). Data obtained using a three-species mixture of *Trifolium repens*, *Lolium perenne*, and *Plantago lanceolata*. SIR is substrate-induced respiration, a relative measure of active microbial biomass; bact. f. indicates bacterial feeding and fung. f. fungal feeding. Data are presented for the most intense defoliation treatment only.

foliation, probably due to top-down regulation resulting from stimulation of microbe-feeding fauna. These results are consistent with multitrophic stimulation in the micro-food web due to defoliation that may have arisen through increased rhizosphere exudation. However, the results did not suggest that defoliation favored the bacterial energy channel since defoliation strongly stimulated both bacterial-feeding and fungal-feeding nematodes.

Plant Productivity

In the longer term, foliar herbivory also has the potential to influence the decomposer biota by inducing shifts in NPP. The issue therefore arises as to how NPP is affected by herbivory. A global literature survey by Milchunas and Lauenroth (1993) of paired grazed and ungrazed plots revealed that herbivore effects on aboveground NPP were usually negative, although many exceptions existed. Indeed, there are several instances in which grazing has been shown to enhance or optimize aboveground NPP, especially in grassland systems (e.g., McNaughton 1985; Holland et al. 1992; Dyer et al. 1993), although the importance, generality, and evolutionary significance of these sorts of effects has been debated (e.g., Belsky 1986; McNaughton 1986; Järemo et al. 1996). Although the extent to which individual plants show overcompensatory responses to herbivory is uncertain (Belsky 1987; Järemo et al. 1996), there is clear experimental (Dyer et al. 1993) and theoretical (Loreau 1995; De Mazancourt et al. 1998) evidence that enhancement of aboveground NPP by herbivores occurs at the level of the plant community especially when effects of grazers on the circulation of nutrients within the ecosystem is considered. This optimization usually occurs at intermediate levels of herbivory (Dyer et al. 1993; Loreau 1995). Although this issue has usually been considered in relation to browsing by mammals, there is also evidence that moderate densities of invertebrates (i.e., grasshoppers) can promote NPP through influencing the cycling of nutrients (Belovsky and Slade 2000).

There is also mixed evidence as to whether foliar herbivory may reduce belowground NPP. In pot trials based on individual plants, experimental defoliation usually reduces root biomass (e.g., Ruess 1988; Bardgett et al. 1998b; Mikola et al. 2001b). However, in field situations the evidence is mixed. McNaughton et al. (1998) found no evidence for greater root productivity by exclusion of grazers from 22–25 year old fenced exclosure plots in the Serengeti plains, Tanzania. However, Ruess et al. (1998) found through the use of minirhizotrons that browsing by moose (*Alces alces*) and snowshoe hares (*Lepus americanus*) in a central Alaskan forest caused a reduction of fine root productivity, an increase in root mortality, and an increase in root turnover rate. Further, the literature survey by Milchunas and Lauenroth

(1993) found that comparable numbers of available data sets reported reductions and enhancements of root mass in response to herbivore exclusion.

Little is known about how herbivore effects on NPP or root mass directly affect soil biota, although several studies have found that defoliation simultaneously reduces root biomass and enhances components of the soil microfood-web (e.g., Mawdsley and Bardgett 1997; Guitian and Bardgett 2000; Mikola et al. 2001b). This suggests that herbivory may often affect decomposers by altering resource quality rather than quantity. However, there are also situations in which severe herbivore-induced defoliation has been shown to reduce soil biota through reducing the amounts of resources entering the soil (e.g., Kristinsen and McCarty 1999; Northup et al. 1999). This would indicate that soil organisms can be optimized by intermediate levels of herbivory (probably through a variety of mechanisms), but are adversely affected when the intensity of herbivory is sufficient for the effects of reduced resource quantity to override other, potentially beneficial, effects of herbivory.

Resource Quality

Herbivores have the potential to influence the soil biota not just by affecting plant productivity but also by altering the quality of resources that the plants return to the soil. These effects can occur through both leaf litter and root litter. In grassland systems, herbivory has frequently been observed to enhance leaf nutrient concentrations (e.g., Ruess and McNaughton 1987; Holland and Detling 1990), and this may result either through reallocation strategies in response to browsing, or through improved nutrient availability due to enhanced activity of soil organisms in the plant rooting zone (Hiernaux and Turner 1986; Bardgett et al. 1998b). This effect should in turn result in higher leaf litter quality being returned to the soil, which could be expected to enhance components of the soil biota and mineralization processes. Browsing of foliage of woody plant species by mammals can also enhance leaf nutrient concentrations (probably through reducing the number of growing points on the plant as well as removing apical dominance) and reduce the levels of secondary metabolites in subsequently produced foliage (Bryant and Reichart 1992). This has obvious implications for litter quality; for example, Kielland et al. (1997) and Kielland and Bryant (1998) found that mammalian browsing of deciduous tree species in an Alaskan forest enhanced the quality and decomposition rates of litter produced by the trees (fig. 4.2a), and that soil carbon mineralization was greater under browsed than under unbrowsed trees.

However, severe defoliation of trees, such as can be caused by periodic invertebrate herbivore outbreaks, often results in reduced concentrations of nitrogen and increased concentrations of secondary metabolites (e.g., phenolics) in subsequently produced foliage (Tuomi et al.1990). Whether this production of secondary compounds is an "induced" response with plants actively producing defenses in response to browsing (Rhoades 1985) or whether secondary metabolite production is simply linked to reduced nutrient availability following defoliation (Bryant et al. 1993) is unclear. However, regardless of the mechanisms involved, the net result is likely to be a reduction of leaf litter quality and production of resources that are less favorable for utilization by decomposer organisms. This aspect has been little investigated, although Findlay et al. (1996) found that damage by spider mites to *Populus deltoides* seedlings caused plants to produce litter with high levels of phenolics and retarded decomposition rates (fig. 4.2b). The direction of this effect is opposite to that observed by Kielland et al. (1997) with regard to mammalian browsing. Further, induced defense responses to invertebrate herbivory have also been shown for herbaceous plants in controlled experiments over much shorter time scales (e.g., Agrawal 1998, 1999), which could

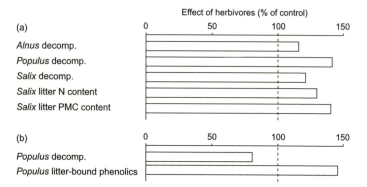

FIGURE 4.2. Quality and decomposition rates of litter from plants subjected or not subjected to herbivory. (a) Litter from plants subjected to browsing by moose (*Alces alces*) and snowshoe hares (*Lepus americanus*), relative to that from unbrowsed plants. PMC indicates potentially mineralizable carbon. From Kielland et al. (1997). (b) Litter from plants infested by herbivorous mites, relative to that from uninfested plants. From Findlay et al. (1996).

conceivably result in diminished quality of litter produced by the plant, although this issue remains unexplored.

Indirect stimulation of root-associated decomposer biota and root herbivores by foliar herbivory (e.g., Ingham and Detling 1984) has been interpreted as evidence that foliar herbivory improves the quality of root material (Seastedt 1985; Seastedt et al. 1988b). One mechanism may involve herbivore-induced enhancement in root nitrogen concentration, either through changes in nitrogen allocation patterns, or through concentration of a given unit amount of nitrogen into a smaller unit of root mass. Seastedt et al. (1988b) found that trimming of shoots of the grass *Agropyron gerardii* caused an increase in root nitrogen concentration of around 25% and concluded that intermediate levels of foliar herbivory may therefore result in optimization of components of the soil biota through improved root resource quality. In contrast, Merrill et al. (1994) found for fenced exclosure plots that while ungulate herbivores increased root densities of microbe-feeding and root-feeding nematodes, root nitrogen concentrations remained unaffected by browsing. They concluded that the enhancement of nematodes by foliar browsers was therefore more likely to be due to either a reduction of secondary chemicals in the roots or changes in root morphology (i.e., an increasing density of root hairs).

Root Mutualists

Although it is apparent that decomposer biota and root herbivores frequently benefit from herbivory as a result of physiological changes in the plant, there is little evidence for this being a common pattern for those soil microorganisms that form intimate mutualisms with plant roots. With regard to mutualisms involving nitrogen fixing bacteria, Butler et al. (1956) demonstrated that experimental defoliation of *Lotus uliginosus* prevented the formation of new rhizobial nodules entirely and reduced nodule formation on *Trifolium pratense* by over 90%. Effects of defoliation on *Trifolium repens* nodulation were less consistent. Similarly, herbivores usually adversely affect mycorrhizal infection of plant roots. Colonization by ectomycorrhizal fungi of *Pinus edulis* roots was found to be about a third less for seedlings that were susceptible to herbivory by stem and cone boring larvae of the moth *Dioryctria albovitella* relative to nonsusceptible seedlings or seedlings from which larvae were excluded (Gehring and Whitham 1991). Similarly, Rossow et al. (1997) used exclosure plots to demonstrate consistent adverse effects of browsing by moose (*Alces alces*) and snowshoe hares (*Lepus americanus*) in Alaska on mycorrhizal colonization of *Salix* spp and *Populus balsamifera*, and proposed that browsing-induced reduction of mycorrhizal infections of these species plays a role in their decline under browsing. In a literature survey, Gehring and Whitham (1994) synthesized data for 37 plant species for which herbivore effects on mycorrhizal infection had been assessed. They found that herbivory had adverse effects on mycorrhizae for 23 of these species and positive effects for only two. This suggests that the frequently observed pattern of optimization of decomposer and root-feeding biota by intermediate intensities of herbivory does not usually apply to mutualists intimately associated with plant roots. It would appear that the mechanism here involves herbivores removing fixed carbon, meaning that the plants have less carbon to allocate to their mutualists (see Smith and Read 1997).

DUNG AND URINE RETURN

Browsing mammals return undigested material and non-assimilated nutrients to the soil as dung and urine. In some ecosystems this can represent a major redistribution of nutrients; for example, it has been estimated that in managed pastures 60–90% of herbage ingested by livestock is returned to the pasture in these forms (Haynes and Williams 1993). This material contains labile forms of carbon and nitrogen which can serve as powerful stimulants of the soil microbial biomass, microbial activity, net carbon and nitrogen mineralization, and ultimately plant nutrient acquisition and growth. For example, Tracy and Frank (1998) found that chronic grazing of grasslands in Yellowstone National Park, U.S.A., did not reduce soil carbon, microbial biomass, or nitrogen mineralization, and concluded that this was largely because of the inputs of labile forms of carbon from dung. Similarly, large increases in the rates of certain soil processes (notably nitrogen mineralization) caused by moose in an Alaskan *Salix planifera* forest were attributed to the return to the soil of fecal and urinary nitrogen (Molvar et al. 1993). However, Pastor et al. (1993) found that while the mixing of moose fecal pellets with soil stimulated rates of carbon and nitrogen mineralization, these effects were insufficient to offset other, adverse effects of browsing on these processes. The elevated rates of nitrogen mineralization caused by returns of dung and urine can be expected to enhance nitrogen loss rates from the ecosystem in the longer-term perspective. For example, Frank and Evans (1997) used $\delta^{15}N$ enrichment patterns in soils inside and outside ungulate exclosures in grasslands in Yellowstone National Park, U.S.A., to demonstrate that dung and urine patches induced nitrogen loss through leaching and losses of gaseous forms of nitrogen.

The impact of return of fecal material on soil processes is apparent from studies that have experimentally added this material to soils. For example, Bardgett et al. (1998a) added sheep (*Ovis aries*) dung to microcosms containing grassland soil, and found that dung had multitrophic effects through stimulating microbial biomass and inducing a severalfold increase in the populations of bacterial-feeding nematodes. With regard to invertebrate herbivores, Lovett and Ruesink (1995) found that frass from larvae of the gypsy moth (*Lymantria dispar*), which had been fed on foliage of *Quercus velutina*, caused large increases in microbial growth and net nitrogen immobilization when incubated in microcosms containing soil. This was due to the frass containing high concentrations of labile carbon. They concluded that following large defoliation events by *L. dispar*, microbial immobilization of nitrogen due to return of frass to the soil may be important in conserving nitrogen within the forested ecosystem.

Urine addition to the soil is associated with the return of very high amounts of nitrogen, potassium, and sulfur, at least in pastoral agroecosystems. The high amounts of ammonium and urea in urine select for nitrifier and denitrifier bacteria, resulting in significant nitrogen loss from the soil (see the review by Haynes and Williams 1993). Williams et al. (2000) found, upon adding synthetic urine to field plots, that although microbial biomass was unaffected there were clear shifts in microbial community composition (with bacteria being favored) and idiosyncratic effects on carbon mineralization. Meanwhile, Haynes and Williams (1999) found that addition of dairy shed effluent (including urine) to a pasture increased soil organic matter content, nutrient accumulation, microbial activity, and a range of soil enzymes; this was consistent with the patterns that were observed in livestock camp areas relative to noncamp areas.

Honeydew Production

Many hemipterous insects, such as aphids and scale insects, suck phloem sap from within the plant, and this resource typically contains high concentrations of soluble carbon and low amounts of nitrogen. In order to gain enough nitrogen to satisfy their requirements, these insects secrete significant amounts of excess sugars in droplets known as "honeydew." A high proportion of this honeydew ends up in the soil through falling or being washed from the vegetation canopy, where it acts as an important labile source of carbon. Dighton (1978) added synthetic honeydew to field plots at rates produced on an areal basis by lime aphids (*Eucalipteris tilidae*) on trees of *Tilia* spp., in both a woodland and a grassland. Honeydew addition stimulated fungal populations by 30% and bacterial populations by 300% in the woodland soil, although microarthropod populations were unaffected. There was no detectable effect of honeydew addition on microbial populations in the grassland soil, although there was a brief pulse of carbon mineralization following honeydew addition. In forests of central Europe, honeydew-producing aphids on *Picea abies* have been shown to cause substantial increases in dissolved organic carbon in the soil and to reduce soil concentrations of mineral and dissolved organic nitrogen, indicative of enhanced immobilization of nitrogen by the soil microbial biomass (Stadler and Michalzik 1998; Michalzik et al. 1999). A similar pattern probably occurs in *Nothofagus* forests in New Zealand, in which the bark-inhabiting scale insect *Ultracoelostoma* spp. produces substantial quantities of honeydew (up to 1200 l/ha/yr), much of which ends up being washed to the soil surface (Beggs et al. 2001).

EFFECTS OF PALATABILITY DIFFERENCES
AMONG PLANT SPECIES

Herbivores also have the potential to exert important effects on soil biota and processes through influencing the composition of the entire plant community. This is because different plant species growing in the same community may differ considerably in their palatability, which in turn causes selection for certain plant functional types above others. Those plant species that are preferentially browsed have different ecophysiological characteristics from those that are not; they generally grow more rapidly, accumulate higher concentrations of nitrogen in their tissues, and contain lower amounts of secondary metabolites (Grime 1979; Coley et al. 1985). Palatable species tend to dominate earlier in succession. For example, Cates and Orians (1975) found, for 100 plant species, that earlier successional annual species were more palatable to generalist herbivores (slugs) than were early successional perennial species, which in turn were more palatable than later successional species. Similarly, Fraser and Grime (1999) used microcosms planted with mixtures of 24 grassland plant species to demonstrate that four generalist herbivore species preferentially consumed earlier successional ruderal species under higher-fertility conditions.

For selective herbivory in plant communities to influence the soil biota and processes requires that herbivores induce significant changes in the functional composition in the vegetation. This has often been observed to be the case, and the net result can be the acceleration of succession through herbivory promoting the dominance of less palatable plant species (e.g., Bryant 1987; Brown and Gange 1992). In the microcosm study of Fraser and Grime (1999), generalist

117

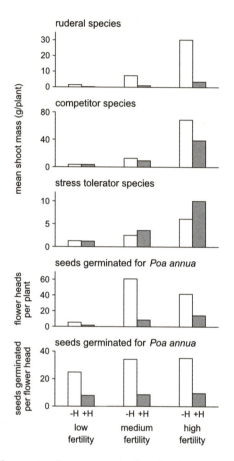

FIGURE 4.3. Changes in plant community functional composition (using the strategy classification of Grime 1979), and reproductive properties of a ruderal species (*Poa annua*) in outdoor mesocosms either amended with four species of generalist herbivores (+H) or without herbivores (−H) under three levels of soil fertility. Data from Fraser and Grime (1999).

herbivores resulted in earlier successional plant species (i.e., those classified as "ruderals" and "competitors") being suppressed by herbivory in higher-fertility treatments, resulting in a net increase in later successional ("stress tolerator") species through competitive release (fig. 4.3). This was not, however, observed in the lowest-fertility treatment. Further, reproductive output of the early successional species *Poa annua* was consistently reduced by herbivory, which helped contribute to its decline relative to later successional species (fig. 4.3). The differential responses of plant species to herbivory in synthesized communities are also apparent in the microcosm study of Buckland and Grime (2000). Here, herbivory by invertebrates, notably the slug *Deroceras reticulatum,* suppressed the most palatable and fastest-growing plant species in low- and intermediate-fertility conditions, resulting in slower-growing plant species dominating. Meanwhile, in high-fertility conditions, faster-growing plant species dominated in the presence of herbivory, presumably because their high growth rate under favorable conditions enabled them to compensate for tissue loss by foliar consumption. In this system, herbivory contributed to domination by slower-growing uncompetitive plant species in infertile conditions, while competitive exclusion of subordinate species coupled with tolerance of herbivory by faster-growing species resulted in corresponding dominance of these faster-growing species in fertile conditions. Further, herbivory can also retard succession, especially when those plants that dominate at early or intermediate successional stages (notably grasses) are adapted to regular herbivory, have a high capacity for compensatory growth, or are optimized by browsing (Brown et al. 1988, Augustine and McNaughton 1998). Indeed, several published studies have presented evidence of vegetation succession being both promoted and arrested by herbivory depending upon the mechanisms involved (see the review by Davidson 1993).

When the above information is compared with that in

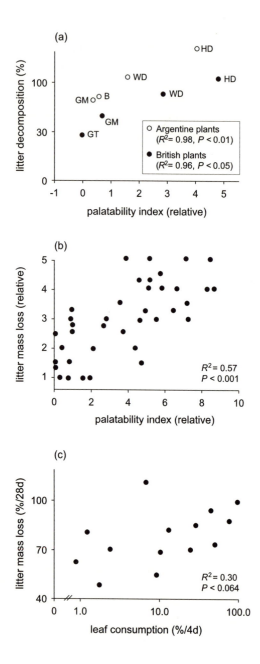

chapter 3 (and especially fig. 3.8) we would expect that herbivores should preferentially feed upon those plant species which have suites of ecophysiological traits associated with higher-quality litter and which therefore promote biomass and activity of the decomposer subsystem. Coniferous tree species are less palatable to generalist herbivores than are dicotyledonous tree species, which in turn are less palatable than dicotyledonous herbaceous species (Cornelissen et al. 1999). This ordering is generally consistent with the trends for these groups with regard to leaf lifespan, leaf nitrogen content, relative growth rate, leaf area characteristics, photosynthetic rates (fig. 3.8), and litter decomposition rate (Cornelissen et al. 1999) (figs. 3.8 and 4.4a). Indeed, the results of comparative studies both across plant functional types and across species provide some evidence that leaf palatability and litter decomposability are generally governed by the same ecophysiological traits (fig. 4.4). Grime et al. (1996) found a strong relationship between palatability to two generalist herbivores of foliage for 43 vascular plant species and litter mass loss for the same species. Similarly, Cornelissen et al. (1999) determined that plant life forms which were more palatable also decomposed more rapidly, for both British and Argentine floras. Wardle et al. (1998a) found an almost statistically significant relationship between

FIGURE 4.4. Relationships between litter decomposability and palatability to generalist herbivores across plant species. (a) Different plant life forms based on a data set involving 48 Argentinian and 72 British plant species. HD are herbaceous dicots, WD woody dicots, GM graminoid monocots, B bromeliads, and GT gymnosperm trees. The herbivores used were generalist snails. From Cornelissen et al. (1999). (b) Forty-three plant species (each represented by a different point) representing a range of plant life forms. The herbivores used were *Helix aspersa* and *Acheta domestica*. From Grime et al. (1996). (c) Twenty dicotyledonous herbaceous species (each represented by a different point). The herbivore used was *Deroceros reticulatum*. From Wardle et al. (1998a).

palatability to slugs (*Deroceras reticulatum*) and litter decomposition for 20 dicotyledonous herbaceous species, although there was no relationship between decomposability and palatability of these species when a generalist herbivorous weevil was used. The weaker relationship for this latter study is probably because the plant species considered were all from one plant functional group rather than several. Based on these examples, it would appear likely that herbivore-plant interactions that lead to domination by unpalatable species, as well as those that lead to domination by palatable species, could therefore be expected to have profound implications for the functioning of the decomposer subsystem.

Although many studies have demonstrated effects of herbivory on vegetation succession and functional composition, few studies have explicitly addressed the effects of herbivore-induced vegetation changes with regard to belowground properties. Probably the strongest evidence emerges from the "moose, microbes, and boreal forest" study of Pastor et al. (1998, 1993) in Isle Royale, Michigan, U.S.A. Here, there is clear evidence from three long-term fenced exclosure plots that moose (*Alces alces*) preferentially browse deciduous tree species that have high litter quality, resulting in a shift to increasing dominance by *Picea* spp., which are of low palatability, are less productive, have high concentrations of foliar secondary metabolites, and produce litter of much poorer quality. In this study, browsing by moose and consequent domination by *Picea* spp. generally caused reductions in the soil microbial biomass, rates of soil carbon and nitrogen mineralization, and concentrations of soil carbon and nitrogen. While this pattern is consistent with what we would expect when browsing accelerates vegetation change, the reverse effect may also be expected in cases where herbivory stalls successional development, such as when the dominant plant species are those that are optimized by browsing (i.e., in many grasslands) (Augustine and McNaughton 1998). Here, herbivory should select for plant communities that are more favorable to soil biota than those communities that would develop in the absence of browsing.

Selective herbivory of some plant species above others can have important effects on decomposers and soil processes even within plant communities. For example, Mikola et al. (2001a) established experimental units in a glasshouse each of which consisted of a mixture of three grassland plant species; treatments consisted of selective defoliation of each of the three species singly, as well as all the possible multiple-species combinations. These treatments differed with regard to their effects on soil carbon mineralization, and on the populations of enchytraeid worms and bacterial-feeding and fungal-feeding nematodes, indicative of multitrophic effects. While some of these effects could be explained simply in terms of the amounts of plant material removed, others could not, meaning that plant-species-specific properties over and above mass removal effects were also important in determining how soil food webs respond to selective foliar herbivory. This pattern was also supported by the results of Guitian and Bardgett (2000) in which defoliation was performed on each of three grass species grown in monoculture. The effects of clipping intensity on soil microbial biomass, respiration, and metabolic quotient all differed across species.

Cross-Ecosystem Effects

Given that plant functional groups differ in the extent to which they can support herbivory, it is perhaps to be expected that the magnitude of herbivore effects that occur will differ across ecosystems dominated by different functional types. Infertile and unproductive habitats tend to be dominated by plant species that have adaptive mechanisms for preventing nutrient losses from their tissues by herbivory, and have long-lived leaves with low concentrations of nutrients and high concentrations of structural carbohydrates and secondary defense compounds, relative to plants from fertile and productive habitats (see chapter 3). McNaughton et al. (1989) estimated, through the use of a global data set, that the proportion of NPP that is not consumed by herbivores (and is therefore available for entry to the decomposer subsystem) varies considerably according to ecosystem productivity, and ranges from below 50% in highly productive systems to above 99% in highly unproductive systems. Further, even within a given plant species the degree of consumption by herbivory can vary significantly across primary productivity gradients (Van der Wal et al. 2000). Cross-ecosystem differences in terms of the proportion of foliage consumed by herbivores are of immense importance in determining the amounts of resources available for the soil biota. Based on an extensive literature synthesis that incorporated data from a range of aquatic and terrestrial ecosystems, Cebrian et al. (1998) and Cebrian (1999) determined that those ecosystems in which plants had higher tissue concentrations of nitrogen and phosphorus, and which were more productive, supported greater levels of consumption by herbivores (which presumably diverted carbon away from the decomposer subsystem), and also greater rates of litter decomposition. The net result of this is greater accumulation of carbon and nutrients in the soils of unproductive versus productive environments. Reconciling these results with the framework developed in chapter 3, we would expect that in productive ecosystems carbon and nutrients would be returned to the soil in labile forms by herbivores (e.g., as dung and urine), and as high-quality litter, leading to low net accumulation of soil organic matter, and probable formation of mull type soil profiles dominated by bacterial energy channels and earthworms. In contrast, in unproductive environments most carbon and nutrients entering the soil would be in the form of poor-quality litter, and with little herbivore intervention, leading to the development of thick moroid humus profiles dominated by the fungal energy channel and microarthropods.

Trends Observed in Real Plant Communities

In this chapter I have indicated a number of mechanisms by which aboveground herbivores may indirectly affect decomposer biota through influencing the quality and quantity of resources entering the soil subsystem. These mechanisms operate both through altering the physiology of the plant at the level of the individual, and by influencing the structure of the plant community. At each of these two levels I have presented plausible explanations of how herbivores may have positive effects on the quality and quantity of resources entering the soil, and also how they may have negative effects. Therefore, in real ecosystems we would expect soil biota to show either positive or negative responses to browsing depending upon the relative balance of these various mechanisms. The net effect of these mechanisms in real ecosystems can be assessed through the use of fenced exclosure plots from which certain herbivores (usually browsing mammals) are excluded, and which can be compared with corresponding unfenced areas. Such approaches have been widely used for assessing browsing mammal impacts on vegetation composition (see the review by Milchunas and Lauenroth 1993) but have been used less commonly for evaluating the influence of browsers on other biotic components. However, there is evidence from the literature for positive, negative, and neutral effects of browsing mammals both on the soil microbial biomass and on those soil invertebrates that occupy higher trophic levels in the decomposer food web (fig. 4.5; see also Wardle et al. 2001). This means that while browsers can be powerful determinants of decomposer biota, the mechanisms by which these effects occur clearly differs for different situations. For example, in the studies of Pastor et al. (1988) and Suominen (1999) in bore-

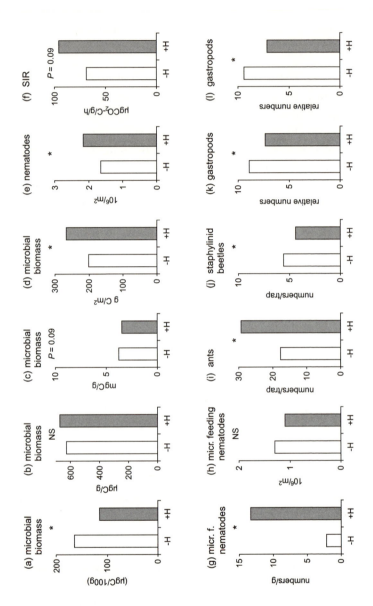

al forests, reduction of soil biota was probably in part due to changes in vegetation composition resulting in domination by plant species that produce poorer-quality litter. In contrast, positive effects of browsers on soil biota (e.g., Bardgett et al. 1997; Stark et al. 2000) may be due in part to the beneficial effects of the return of labile forms of substrates as dung and urine. Further, herbivory by mammals has been shown to significantly alter the community structure of various decomposer invertebrate groups in exclosure studies, including soil arthropods (Suominen et al. 1999) and gastropods (Suominen 1999) in boreal forests browsed by cervids, nematodes, disploods, and gastropods in temperate forests browsed by deer and goats (Wardle et al. 2001), and macroarthropods in grasslands grazed by livestock (Milchunas et al. 1998).

Given that decomposer organisms show variable responses to browsing in field situations, it is perhaps inevitable that the processes that they regulate would also show a

---

FIGURE 4.5. Biomasses or populations of soil biotic components of the decomposer food web, inside ($-$H) and outside ($+$H) exclosure plots designed to exclude browsing mammals. (a) Moose (*Alces alces*) browsing in boreal forest, Michigan, U.S.A. (Pastor et al. 1988); (b) ungulate browsing in grasslands in Yellowstone National Park, U.S.A. (Tracy and Frank 1998); (c) reindeer (*Rangifer tarandus*) grazing in boreal forest, northern Finland and Russia (Väre et al. 1996); (d) and (e) sheep (*Ovis aries*) grazing in upland grasslands in Wales (Bardgett et al. 1997); (f) and (g) reindeer grazing in boreal forest, Muonio, Finland (Stark et al. 2000; only litter layer data presented); (h) cattle (*Bos taurus*) grazing in shortgrass steppe, Colorado, U.S.A. (Wall-Freckman and Huang 1998; only long-term grazing treatment data presented); (i) and (j) moose browsing in boreal forest, eastern Sweden (Suominen et al. 1999; only data from the Sunnäs site presented); (k) moose browsing in boreal forest, central Sweden and southern Finland (Suominen 1999); (l) reindeer browsing in boreal forest, Finnish Lappland (Suominen 1999). SIR, substrate-induced respiration, a relative measure of active microbial biomass; micr. f., microbe feeders; * or NS means that herbivore effect is significant or nonsignificant, respectively, at $P = 0.05$.

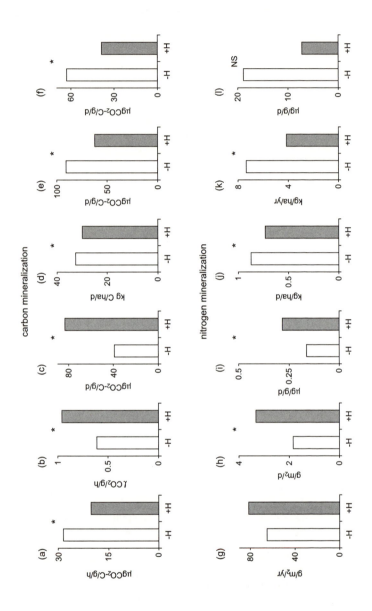

carbon mineralization

(a) µgCO₂-C/g/h
(b) ℓCO₂/g/h
(c) µgCO₂-C/g/d
(d) kg C/ha/d
(e) µgCO₂-C/g/d
(f) µgO₂-C/g/d

nitrogen mineralization

(g) g/m₂/yr
(h) g/m₂/d
(i) µg/g/d
(j) kg/ha/d
(k) kg/ha/yr
(l) µg/g/d

variety of responses. This is indeed the case. With regard to soil carbon mineralization rates, both positive (e.g., Bardgett et al. 1997; Kielland et al. 1997) and negative (e.g., Pastor et al. 1988; Väre et al. 1996; Burke in Milchunas et al. 1998; Stark et al. 2000) responses to browsing have been observed in exclosure studies (fig. 4.6a–f). A similar pattern emerges with regard to nitrogen mineralization rates, with browsing mammals enhancing (McNaughton et al. 1988; Frank and Groffman 1998; Tracy and Frank 1998), reducing (Pastor et al. 1988; Van Wijnen et al. 1999), or not affecting (Burke in Milchunas et al. 1998; Stark et al. 2000) mineral nitrogen release (fig. 4.6g–l). This is critical to determining ecosystem functioning in the longer-term perspective, because browser effects on nitrogen mineralization influence the supply rate of available nitrogen for plants, which in turn impacts upon plant productivity (De Mazancourt and Loreau 2000; discussed further in chapter 5). Browser effects on mineralization rates are also likely determinants of soil carbon and nutrient sequestration, and there are exam-

---

FIGURE 4.6. Soil carbon and nitrogen mineralization rates inside ($-$H) and outside ($+$H) exclosure plots designed to exclude browsing mammals. (a) Reindeer (*Rangifer tarandus*) grazing in boreal forest, northern Finland and Russia (Väre et al. 1996); (b) sheep (*Ovis aries*) grazing in upland grasslands in Wales (Bardgett et al. 1997); (c) moose (*Alces alces*) browsing in boreal forest, Alaska, U.S.A. (Kielland et al. 1997; data averaged for both *Salix* and *Alnus* sites); (d) livestock grazing on native grassland, Colorado, U.S.A. (Burke in Milchunas et al. 1998); (e) moose browsing in boreal forest, Michigan, U.S.A. (Pastor et al. 1988); (f) reindeer grazing in boreal forest, Muonio, Finland (Stark et al. 2000); (g) simulated estimates of mineralization for the Serengeti grazing system, Tanzania (McNaughton et al. 1988; only data for the no-litter simulations included); (h) ungulate browsing in grasslands in Yellowstone National Park, U.S.A. (Frank and Groffman 1998); (i) ungulate browsing in grasslands in Yellowstone National Park, U.S.A. (Tracy and Frank 1998); (j) as for (d); (k) as for (e); (l) as for (f). * or NS means that herbivore effect is significant or nonsignificant, respectively, at $P = 0.05$. Statistical analysis is not possible for panel (g).

ples from exclosure studies of browsing mammals inducing negative (e.g., Pastor et al. 1988), positive (e.g., Derner et al. 1997), and neutral (e.g., Tracy and Frank 1998; Stark et al. 2000) effects on net soil carbon and nitrogen accumulation. Indeed, in a study of 30 long-term exclosure plots located in rainforests throughout New Zealand, Wardle et al. (2001) found evidence of significant net soil carbon and nitrogen sequestration in some locations and net loss of soil carbon and nitrogen in others.

SPATIAL AND TEMPORAL VARIABILITY

Herbivores may also be important determinants of spatial and temporal variability of the decomposer biota and soil processes, although this issue has seldom been addressed. There is, however, some evidence that browsing mammals can increase the spatial heterogeneity of resources entering the decomposer subsystem. For example, Knapp et al. (1999) demonstrated landscape-level increases in spatial variation of plant biomass following reintroduction of bison (*Bos taurus*) to the Konza prairie (Kansas, U.S.A.). Similarly, Schlesinger et al. (1990) provide evidence that long-term grazing of a semiarid grassland in New Mexico (U.S.A.) caused greater spatial heterogeneity of soil resources, which led to invasion by desert shrubs, inducing further spatial heterogeneity. The net result is the development of "islands of fertility" associated with these shrubs (Schlesinger and Pilmaris 1998).

Those herbivores that themselves show highly temporally variable population dynamics may be expected to enhance temporal variability of soil biota and decomposer processes. This is probably the case at least for invertebrates that undergo periodic population outbreaks resulting in large-scale defoliation. In forests, such defoliation can occasionally result in widespread plant mortality, and plants that survive severe defoliation events often produce new foliage with greatly elevated concentrations of secondary metabolites (Tuomi et al. 1990). These effects should in turn reduce both the quantity and quality of litter entering the soil for a period of time following defoliation. Further, large amounts of frass produced during invertebrate herbivore outbreaks may have significant effects on nutrient availability, as has been proposed for outbreaks of the gypsy moth (*Lymantria dispar*) in North American forests (Lovett and Ruesink 1995). Mammals that undergo significant population cycles may also be important drivers of temporal variability of soil biota and processes. Sirotnak and Huntly (2000) used fenced exclosure plots to investigate the effects of voles (*Microtonus* spp.), which show very large interyear population fluctuations, on nitrogen cycling in grasslands in Yellowstone National Park, U.S.A. They found that during and after population peaks, voles increased the availability of soil nitrogen, but this was offset by a decline in nitrogen mineralization in the longer term. This decline was probably a result of vole-induced reductions of palatable, high-litter quality plant species such as legumes. This result suggests that temporal variability of voles governs the temporal variability of soil nitrogen mineralization (and, presumably, the organisms that cause this), and ultimately the nitrogen supply rate from the soil. In the longer term perspective, herbivores can also influence disturbance regimes which may in turn impact upon soil biota, although this issue remains largely unexplored. However, grazing reduces the buildup of fuel (undecomposed detritus) for wildfire in many grassland grazing systems and therefore significantly influences the frequency and intensity of fire (Hobbs et al. 1991; Knapp et al. 1999); based on the discussion in chapter 3 it is apparent that this will in turn affect long-term temporal dynamics of soil biota as well as decomposition and mineralization processes.

The above studies suggest that herbivores influence (and often enhance) temporal variability of decomposer organisms and processes, and therefore have destabilizing effects on the functioning of the belowground subsystem. However, the issue of how herbivores affect soil biotic stability has seldom been addressed. In a mesocosm experiment, Wardle et al. (2000) found that addition of two invertebrate herbivore species to pots sown with grassland plant species reduced the stability (resistance to an experimentally imposed drought) of the soil microbial biomass but did not exert consistent effects on decomposer processes. Bardgett et al. (1997) measured soil microbial biomass and nematode populations at each of four sampling times throughout a year in each of four Welsh pastures, in plots with and without sheep grazing. When measures of temporal variability are calculated from the data presented by Bardgett et al., it is apparent that grazing consistently increased the temporal variability of soil nematode populations (and thus had destabilizing effects), but for three of the four pastures grazing tended to reduce temporal variation for the soil microbial biomass (fig. 4.7).

CONSEQUENCES OF PREDATION OF HERBIVORES

Aboveground food web interactions that involve secondary or higher-level consumers have the potential to affect the decomposer subsystem whenever these consumers induce trophic cascades that are of sufficient strength to influence the quantity and quality of resources that the autotrophs produce. Despite the likely significance of these effects, the issue of how aboveground cascades affect soil organisms and decomposer processes is poorly understood. While there has been debate as to how generally important trophic cascades are in terrestrial ecosystems (see Polis 1994; Pace et al. 1999; also chapter 2) there are clear examples in the literature of trophic cascades which are of sufficient strength to be relevant with regard to the soil biota. For example, Letourneau and Dyer (1998) experimentally

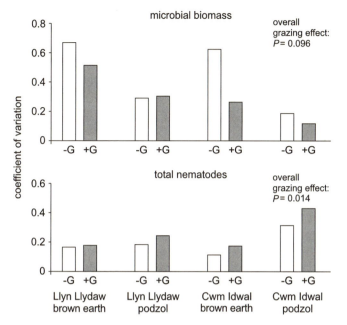

FIGURE 4.7. Temporal variability of soil microbial biomass and nematodes in four Welsh pasture sites with sheep (*Ovis aries*) grazing (+G) and without grazing (−G). Calculated using time course data presented in fig. 1 of Bardgett et al. (1997).

investigated a trophic cascade that involved four trophic levels [i.e., in order of ascending trophic position, the tree *Piper cenocladum*, herbivores, ants (*Pheidole bicornis*), and beetles (*Tarsobaenus letourneauae*)] in a Costa Rican rain forest. They found that the addition of the top predator induced a trophic cascade that resulted in total plant leaf area being reduced by almost 50%. A second example involves a trophic cascade in which wolves (*Canis lupus*) in Isle Royale (U.S.A.) limit the abundance of moose (*Alces alces*), which in turn significantly increases the radial increment of *Abies balsamea* trees (Post et al. 1999). Predation by wolves on moose is driven by macroclimate at an interyear scale (high winter snow levels associated with the North Atlantic oscillation induced wolves to hunt in larger packs, trebling the kill rates of moose), which in turn influences *A. balsamea* growth. Finally, trophic cascades have the potential to affect the amounts of organic matter entering the soil not just through shifts in NPP, but also by causing changes in plant litter chemistry. Evidence in support of this arises from the study of Stamp and Bouwers (1996) which found that the production of defense compounds (iridoid glycosides) by the leaves of *Plantago lanceolata*, in response to invertebrate herbivory, was reduced by addition of a third trophic level.

TRANSPORT OF RESOURCES BY ABOVEGROUND CONSUMERS

Aboveground consumers can affect soil food webs over greater spatial scales than those I have considered so far in this chapter. This involves herbivores or predators causing the spatial redistribution of resources by consuming them in one location and depositing them in another. For example, large mammalian herbivores in the Serengeti plains, Tanzania, cause large-scale spatial transport of nutrients across the landscape through their migration patterns (McNaughton 1985). Spatial transport of resources is especially important when habitats of high and low NPP are in close proximity to one another; in these cases aboveground consumers ingest resources in the productive habitat and egest them in the unproductive habitat, enabling the unproductive habitat to support a much greater biomass and productivity of consumers (including decomposers) than is otherwise possible (see the review by Polis et al. 1997a). For example, a range of vertebrates and invertebrates are important vectors of resource transport into caves, maintaining a diverse assemblage of microbial and faunal decomposers in the absence of autotrophs (Howarth 1983; Polis et al. 1997a).

Transport by seabirds of resources from the sea to the land can be of major significance at an ecosystem-level scale whenever an unproductive landmass is adjacent to a productive water body. This is especially apparent for coastal ecosystems in Antarctica. Here the main terrestrial autotrophs are lichens and algae, and NPP is extremely low. In these systems, Adélie penguins (*Pygosecelis adeliae*) form ornithogenic soils from guano originating from the sea. These soils support a diverse and active assemblage of soil bacteria (Ramsay 1983; Ramsay and Stannard 1986); bacterial density and biomass were found to be nearly 50 times greater in areas with penguin rookeries relative to a site outside the rookeries (Ramsay 1983), and soil microbial biomass was measured as being 100 times greater in active colonies than in sites remote from colonies (Roser et al. 1993).

Another example that points to the role of seabirds in affecting terrestrial decomposers involves islands in the Gulf of California, which receive very little rainfall and are extremely unproductive. On these islands, energy derived from marine sources exceeds that resulting from terrestrial NPP in the majority of cases, and this has major implications for terrestrial consumers; arthropods are 2.2 times more abundant on islands with seabird colonies than on those without (Polis and Hurd 1996). The abundance of the dominant group of detritivorous arthropods on these islands, i.e., tenebrionid beetles, varies over three orders of magnitude across these islands and they are most abundant on those islands in which seabirds roost (Sanchez-Piñero and Polis 2000). Here, guano stimulates plant productivity and increases availability of detritus for the beetles. These beetles also serve as a food source for higher trophic level consumers, indicative of multitrophic consequences of resource subsidies from the sea. Guano addition also directly increases concentrations of soil nitrogen and phosphorus by up to six times, and plants on islands colonized by seabirds have tissue concentrations of these nutrients that are at least double those of plants on noncolonized islands (Anderson and Polis 1999). There is also evidence that fairy prions

(*Pachyptila turtur*) inhabiting Stephens Island, New Zealand, substantially increase soil phosphorus levels, and to a lesser extent leaf tissue nitrogen concentrations (Mulder and Keall 2001). Such studies suggest that removal of nutrients by seabirds from sea to land may have important effects on the quality of litter that plants produce, and therefore the abundance of decomposer organisms and the processes that they regulate.

SYNTHESIS

It is apparent from the preceding chapters that both the quantity and quality of plant material returned to the soil are important in influencing soil biota and processes. Given that plants represent the trophic level that links the aboveground and belowground food webs, aboveground trophic interactions that affect the quantity and quality of resources produced by the plant can exert important effects belowground. At the level of the individual plant, herbivores can increase and change patterns of root carbon exudation, and either increase or decrease both NPP and the quality of resources entering the soil; as a result both positive and negative effects of herbivory on the decomposer subsystem are possible.

At the across-species level, suites of plant traits that confer greater palatability of foliage to herbivores also confer higher quality of plant litter. Herbivores that selectively remove palatable species from a plant community therefore induce domination by less palatable species that produce poorer quality litter, retarding soil biota and processes. Conversely, when herbivores cause perpetuation of early or mid successional plant species, successional development is stalled and domination of the community by later successional species with poorer litter quality is prevented. Differences among plant species in terms of palatability also results in across-ecosystem differences. Ecosystems that are more productive and dominated by palatable plant species also have a higher proportion of NPP entering the herbivore (versus the decomposer) food chain, which contributes to higher-quality organic matter being returned to the soil. This has important consequences for the levels of carbon and nutrient sequestration in the soil, the type of humus that is formed, and ultimately the composition and activity of the decomposer biota.

There are further mechanisms by which aboveground food webs can have important effects on soil biota but which have attracted little attention to date. First, herbivores have the potential to influence both the spatial and temporal variability of soil biota and soil processes. In particular, those herbivores that show highly temporally variable population dynamics have the potential to cause large temporal variation in the addition of resources to the soil. Secondly, the issue of belowground consequences of predation of herbivores remains little explored, although these predators have the potential to influence the soil biota and processes whenever they cause trophic cascades of sufficient strength to affect the quality and quantity of resources produced by autotrophs. Finally, aboveground consumers may have important, although largely unrealized, effects on resource redistribution across habitats. In conclusion, aboveground consumers can influence the soil subsystem through a number of mechanisms and over a range of spatial and temporal scales, leading to a variety of possible consequences for both soil organisms and the processes that they regulate.

Completing the Circle

How Soil Food Web Effects Are Manifested Aboveground

It is clear from the material that I have presented so far that the soil biota is responsive to the amounts of plant material returned to the soil (chapter 2), the quality of resource input and the composition of the plant community (chapter 3), and the composition of the aboveground food web and nature of aboveground trophic interactions (chapter 4). For feedbacks to exist between the aboveground and belowground biota it is required that soil organisms also influence what is observed aboveground.

The interaction between plants and the decomposer food web is ultimately mutualistic because the two components are usually obligately dependent upon one another; the plants add sources of carbon to the soil which fuel microbial activity, and this in turn increases nutrient supply for the plants. As discussed in chapter 2, the soil invertebrate community also participates in this association through catalyzing the release of nutrients by microorganisms, by a variety of mechanisms that operate at a range of spatial and temporal scales. Those soil organisms which are more intimately associated with plant roots also have the potential to induce important aboveground effects, and these include both mutualists such as mycorrhizal fungi and nitrogen fixing bacteria, and antagonists such as root pathogens and root herbivores. As a result of these types of feedback, soils previously conditioned by different plant species vary in the

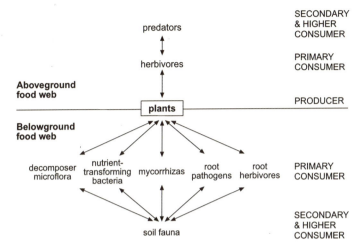

FIGURE 5.1. Routes by which belowground organisms may influence aboveground organisms, as outlined throughout chapter 5. Arrows indicate direction of possible effects, and all groups of organisms identified in this figure are capable of either directly or indirectly influencing all other groups of organisms.

soil communities that they support, and this has important effects on the relative performance of different plant species subsequently grown on the same soil (Bever et al. 1997).

The types of interaction that may exist between plants and soil organisms, and therefore the routes by which the soil biota may influence both plants and aboveground consumers, are shown in figure 5.1. In this chapter I will consider the ways in which aboveground biota are affected by belowground biota. This will be done by considering how plants and their consumers are influenced by communities of decomposer biota, nutrient-transforming organisms, and those organisms that form close positive and negative associations with plant roots. These considerations will in turn be used to evaluate the nature of the linkage that exists between the aboveground and belowground food webs as mediated by the plant.

139

THE DECOMPOSER FOOD WEB

The primary consumers of the decomposer food web, i.e., the bacteria and fungi, are directly responsible for most of the mineralization of nutrients that occurs in the soil, and are therefore the primary biotic regulators of nutrient supply for plants. As discussed in chapters 2 and 3, soil microbial biomass often increases with elevated NPP and improved quality of plant-derived resources, and rates of nutrient mineralization are often observed to increase with increasing microbial biomass. Microbial activity results in localized, short-term pulses of nutrient availability (Clarholm 1990; Wardle 1992), and many plant species are well adapted to take advantage of nutrient pulses of the duration of a few hours (see Campbell and Grime 1989). Nutrients immobilized in the microbial biomass represent a labile source for plant growth as a result of microbial tissue turnover; this is especially apparent during wetting-drying or freezing-thawing cycles. For example, Singh et al. (1989) used temporal dynamic data for microbial biomass and nutrient availability in a dry tropical forest and a savanna in northern India to suggest that microbial biomass conserves nutrients during the dry summer period (i.e., high biomass, low turnover) when plant activity is low, and releases these nutrients during the monsoon period (i.e., low biomass, high turnover) to stimulate plant growth when plant nutrient demand is high. Studies involving isotopic labeling of the microbial biomass also find that a significant proportion of nutrients present in microbial tissues are taken up by plants over the time scale of several weeks (e.g., Lethbridge and Davidson 1983; Schnürer and Rosswall 1987). Further, in the Negev desert in Israel, Sarig et al. (1994) found that the annual grass *Stipa capensis* was more productive under the halophyte shrub *Hammada scoparia* but only when microbial biomass was elevated, and concluded that high microbial biomass (resulting from carbon input from *H. scoparia*) was

140

required to ensure access of *S. capensis* to sufficient amounts of available nitrogen.

Although the microflora and autotrophs are clearly mutually dependent upon one another, they both also require nutrients from the same nutrient pool. Therefore when the microbial biomass is nutrient limited rather than carbon limited, microbes can compete with plants for nutrients, negatively affecting plant nutrient acquisition and therefore plant growth. This effect can be demonstrated through the experimental addition of simple sugars (e.g., glucose) to field plots. When this is done, the soil microbial biomass increases and immobilizes a greater amount of available nutrients; glucose addition has been shown to reduce growth and tissue nitrogen concentrations of *Hordeum vulgare* in cropping systems (Rutherford and Juma 1992; fig. 5.2a), leaf nutrient concentrations of forest tree species (Yarie and Van Cleve 1996; fig. 5.2b), and the growth and nitrogen concentrations of shoot (but not root) material from grassland plants (Jonasson et al. 1996b; fig. 5.2c). Negative interactions between soil microflora and plants are also apparent from studies in which negative relationships have been detected between the biomass of roots and soil microbes (e.g., Okano et al. 1991; Raghubanshi 1991). Further, Zak et al. (1990b) showed that for soil cores that had been isolated in PVC cylinders in a hardwood forest, microbial immobilization of added ^{15}N was 10–20 times greater than that taken up by the ephemeral herb *Allium tricoccum*, indicative of a strong competitive ability on the part of the microflora for uptake of nitrogen. This was interpreted to mean that microbial immobilization was the most important mechanism for retaining nitrogen within the ecosystem prior to canopy development. On balance it therefore appears that microbial biomass can act as a reservoir of nutrients through net immobilization, particularly when plants are inactive, and can directly compete with plants for nutrients. However, microbial activity and turnover of microbial tissues result

141

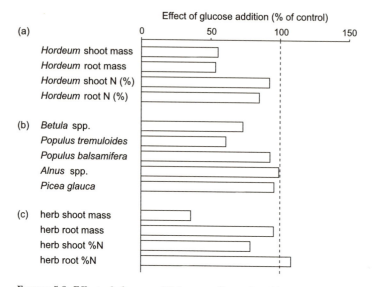

FIGURE 5.2. Effect of glucose addition to soil on plant biomass and tissue nitrogen concentrations. (a) *Hordeum vulgare* in a cropping system, Alberta, Canada (from Rutherford and Juma 1992); (b) leaf nitrogen concentrations one year after glucose addition in interior Alaska, U.S.A. (from Yarie and Van Cleve 1996; data averaged for all sites); (c) herbaceous vegetation in mixed grassland near Copenhagen, Denmark (from Jonasson et al. 1996b).

in release of nutrients, making them available for plant growth.

Involvement of Soil Fauna

As discussed in chapter 2, the nutrient mineralization processes driven by the soil microflora are in turn influenced by the soil fauna through the microfood-web system in which direct predator-prey relationships between microbes and soil fauna affect soil processes, and through the litter transformer and ecosystem engineering systems in which fauna influence microbial activity either by altering their habitats or through mutualistic associations with microbes. Increased rates of nutrient supply due to soil faunal activity could in turn be expected to stimulate plant growth. Those studies that have involved amending experimental microcosms containing plants with soil fauna from the microfood web system have frequently found positive effects of faunal activity on plant shoot and root growth (e.g., Ingham et al. 1985; Jentschke et al. 1995; Setälä et al. 1999; Bonkowski et al. 2000), shoot-to-root ratios (e.g., Kuikman and van Veen 1989; Bonkowski et al. 2000), and tissue nutrient concentrations (e.g., Kuikman and van Veen 1989) (fig. 5.3). However, several studies have also found variable, neutral, or even strongly adverse effects of soil fauna on these plant properties (e.g., Ingham et al. 1985; Alphei et al. 1996; Bardgett and Chan 1999) (fig. 5.3).

Experimental microcosm studies have also shown that larger soil fauna can exert strong positive effects on plant growth and nutrient acquisition, presumably through promoting microbial activity and therefore nutrient mineralization (e.g., Setälä and Huhta 1991; Haimi et al. 1992; Alphei et al. 1996). In addition, Spain et al. (1992) showed that ^{15}N transfer from the microbial biomass to plants was enhanced by the addition of earthworms, and that ^{15}N incorporated into both microbial biomass and live earthworms was able to serve as a source of nutrients for plants. Further, some larger soil animals are of sufficient body size to enable manipulation of their populations in field plot experiments, and there are several examples of earthworm manipulation influencing plant production in the field largely through increasing nutrient availability. For example, Lavelle (1994) found that for 27 pairs of plots in three sites in the humid tropics, in which one of each pair was inoculated with earthworms and the other was not, the proportional increase in the yield of sown arable crops caused by earthworm inoculation was significantly correlated with the enhancement of earthworm biomass due to inoculation (fig. 5.4). However, there is also evidence that stimulation of crop yield by earthworms may be partly for reasons other than nutrient miner-

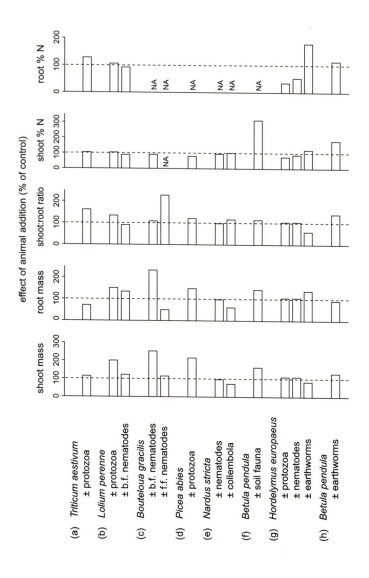

effect of animal addition (% of control)

(a) *Triticum aestivum*
± protozoa

(b) *Lolium perenne*
± protozoa
± b.f. nematodes

(c) *Bouteloua gracilis*
± b.f. nematodes
± f.f. nematodes

(d) *Picea abies*
± protozoa

(e) *Nardus stricta*
± nematodes
± collembola

(f) *Betula pendula*
± soil fauna

(g) *Hordelymus europaeus*
± protozoa
± nematodes
± earthworms

(h) *Betula pendula*
± earthworms

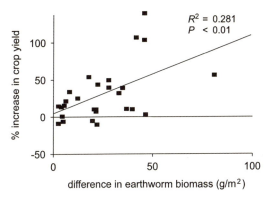

FIGURE 5.4. Relationship between the difference in earthworm biomass in a plot inoculated with earthworms and a plot not inoculated with earthworms, and the increase in crop yield caused by earthworm inoculation. Each point represents a pair of plots. From Lavelle (1994). (Reproduced with permission from the International Society of Soil Science)

alization, given that stimulation of plant productivity by earthworms occurs even when optimal amounts of fertilizers are applied (Lavelle et al. 1994; discussed later in this chapter).

The positive effect of soil fauna on plant growth and tissue nutrient content is largely caused by soil fauna enhancing nutrient mineralization rates. This is also consistent with the mostly positive effects of soil fauna on plant shoot-to-root ratios (fig. 5.3)—the plants have sufficient nutrients to produce more shoot material per unit root mass. In the case of the microfood-web at least, this is likely to be due in part

---

FIGURE 5.3. Effects of decomposer food web fauna on plant growth, root-to-shoot ratios, and tissue nitrogen concentrations, in experimental microcosms. (a) Kuikman and van Veen (1989); (b) Bonkowski et al. (2000) (data from fully mixed treatments only included); (c) Ingham et al. (1985); (d) Jentschke et al. (1995) (nitrogen concentration for needle tissue only); (e) Bardgett and Chan (1999); (f) Setälä and Huhta (1991) (nitrogen concentration for leaf tissue only); (g) Alphei et al. (1996); (h) Haimi et al. (1992). NA, data not available.

to the "microbial loop" (Clarholm 1985; Coleman 1994). This loop is suggestive of a possible mutualistic association in the plant rooting zone, in which plants exude large amounts of soluble carbon into the rhizosphere, stimulating the productivity of various microfood-web components, notably bacteria and protozoa. Protozoan grazing of the bacteria mineralizes nutrients within the rooting zone, which in turn stimulates plant growth.

The influence of soil fauna on nutrient supply for plant growth is constrained both temporally and spatially. An example of temporal constraints of enhancement of soil nutrient supply by soil fauna is presented by Clarholm (1990). In a field study in which daily measurements were made, individual rainfall events stimulated bacterial biomass in the rooting zone of individual plants of *Hordeum vulgare*, and this was followed by a peak in populations of bacterial-feeding naked amoebae. These peaks coincided with rapid in creases in the uptake of nitrogen by the plants, suggesting that these predator-prey relationships directly contributed to plant nutrition through pulsed release of mineral nitrogen. Many plant species are probably adapted to benefiting from nutrient pulses that operate over this time scale (see Campbell and Grime 1989). As an example of spatial constraints of soil faunal enhancement of nutrient supply, Bonkowski et al. (2000) established microcosms containing plants of *Lolium perenne*, across which habitat heterogeneity was varied by creating different levels of patchiness of organic matter hotspots in the soil. These microcosms were then either amended with microfauna (protozoa and microbes, either singly or in combination) or left unamended. Plant biomass was greatest in those treatments in which both soil fauna was present and organic matter aggregated into patches. This type of response would be expected for plant species that show precision in resource foraging (see Campbell et al. 1991; Grime 1994, Hutchings and De Kroon 1994), and demonstrates that soil fauna can enhance nutrient supply from spatially isolated organic matter patches in such a way as to benefit those plant species that have such a foraging strategy. Indeed, there is evidence that increasing patchiness of resources can promote plant growth (Wijesinghe and Hutchings 1997; Korsaeth et al. 2001; but see Hodge et al. 2000) and can change the competitive balance between plants and microbes for nutrients in favor of plants (Korsaeth et al. 2001).

It is clear from figure. 5.3 that soil faunal activity does not always benefit plant growth and nutrient acquisition. Alphei et al. (1996) found that while addition of microfauna to microcosms stimulated growth of *Hordelymus europaeus*, these fauna had generally negative effects on plant tissue nutrient concentrations. Their results suggest that whatever effect these fauna had on plant growth, the mechanism involved did not involve increased rates of nutrient supply by faunal activity. Similarly, Bardgett and Chan (1999) found that soil springtails and nematodes could either enhance or reduce the growth of the grassland plant *Nardus stricta* depending upon the time of harvest, and that plant growth was sometimes suppressed by springtails even when nitrogen mineralization was simultaneously enhanced. It was proposed that increased concentrations of ammonium were actually inhibitory to this plant species. This issue will be addressed later in this chapter.

Effects of Soil Food Web Structure

Although several studies have investigated the effects of decomposer fauna on plant nutrient uptake and growth, the effects of soil food web structure on plant growth remains less well investigated. However, those studies that have compared the effects of different subsets of the soil fauna on plant growth frequently find that different trophically equivalent groups have dissimilar influences (Ingham et al.; 1985; Alphei et al. 1996; Bonkowski et al. 2000). Given that components of the soil fauna that have different body sizes often differ in their effects on nitrogen mineralization (fig. 2.9) it appears plausible that plant growth would be sensitive to these differences. Indeed, Alphei et al. (1996) found that tissue nitrogen concentrations of the grass *Hordelymus europaeus* were greater in microcosms containing earthworms than in those containing protozoa or nematodes. Meanwhile, Setälä et al. (1996) established a greenhouse mesocosm experiment with litter baskets containing varying mesh sizes to allow entry by fauna with different body sizes, and found that plant biomass and nitrogen concentrations of *Populus trichocarpa* seedlings planted in these mesocosms were unresponsive to these mesh size effects. This was despite mesh size influencing the decomposition rate and nitrogen content of litter in these baskets, and also clearly controlling the structure of the soil food web.

Based on the material presented in chapter 2, it could also be expected that the across-trophic-level structure of decomposer food webs should influence plant growth and nutrient uptake. This issue has seldom been addressed. However, Laakso and Setälä (1999a) set up microcosms planted with *Betula pendula* seedlings and containing soil food webs of different structures; treatments consisted of adding bacterial-feeding and fungal-feeding nematodes (either singly or in combination) either with or without specialist and nonspecialist predatory mites that consume nematodes. Although the specialist mite reduced nematode populations by over 50%, microbial biomass was unaffected, and there were only weak effects of top predators on microbial respiration and soil ammonium concentrations. No response of plant growth or nitrogen acquisition to predator additions was detected. Further, Laakso and Setälä (1999b) set up microcosms in which food web structure was varied by including different combinations of faunal species belonging to the saprophagous and fungal-feeding trophic groups, with or without addition of top predators. Soil food web structure generally had small effects on growth and nitrogen uptake of *B. pendula* seedlings, although the saprophagous enchytraeid *Cognettia sphagnetorum* had strongly beneficial effects on plant growth and nitrogen acquisition that none of the other species present in the food web were capable of. In their study they also demonstrated that the functional importance of faunal groups on plant productivity was inversely related to the trophic position of the fauna, with the highest-level consumers having the weakest effect. This suggests that in this example at least, trophic cascades that may occur in the soil food web are unlikely to be functionally important with regard to ecosystem NPP. Based on the limited amount of data available, it would appear that decomposer food web structure may have detectable, though often weak, effects on plant productivity.

The Significance of Plant Litter

The material discussed so far in this chapter relates to the responses of plants to material mineralized from plant litter following soil biological activity. However, the activity of the soil biota also determines what proportion of the litter produced by the plant remains undecomposed on the soil surface, and the residence time of this litter before it is broken down. Accumulated litter can influence plant establishment, growth, and community composition via a range of mechanisms including alteration of light regimes, microclimatic effects, loadings of pathogens and herbivores, phytotoxin release, and nutrient immobilization (reviewed by Facelli and Pickett 1991). Foster and Gross (1997) used a manipulative experimental approach to show that litter negatively affected seedling establishment of the grass *Andropogon gerardi*, and that these effects were greater for communities of greater standing biomass, consistent with the predictions of Carson and Peterson (1990) that litter effects become increasingly negative with greater productivity. Further, Foster and Gross (1998) performed an experiment in which fertil-

149

izer addition and litter removals were both performed on field plots; fertilization increased plant biomass and litter production, which in turn reduced the establishment of subordinate forbs. Removal of this litter improved establishment of these forb species. Indeed, in a meta-analysis of 35 previously published studies, Xiong and Nilsson (1999) found that the effects of litter presence on plant germination and establishment were usually negative, although effects on plant biomass were more variable. They also found that litter effects on plants differed markedly in relation to both ecosystem type and vegetation type; litter sourced from forbs and trees had more negative effects on germination and establishment than did that sourced from grasses (fig. 5.5).

The physical effects of plant litter are not, however, always negative. For example, Suding and Goldberg (1999) found that the presence of plant litter in an old field had mostly facilitative effects on seedling establishment across a plant productivity gradient (probably as a result of improved soil moisture retention), although litter in combination with intact vegetation had adverse effects, presumably through the reduction of available light levels. Further, Facelli and Ladd (1996) found in a glasshouse experiment that addition of litter had differential (including positive) effects on seedling emergence and growth of four species of *Eucalyptus*, and concluded that both positive microclimatic effects and negative shading effects of litter were responsible for these varied responses. A variety of plant responses to accumulated litter appears possible (see Facelli and Pickett 1991) although the meta-analysis of Xiong and Nilsson (1999) suggests that effects are increasingly negative as the amount (and in particular the depth) of the litter present increases. Ultimately, the effects of litter presence on plant community development and composition are likely to be most important when litter production rates exceed the rates at which litter is processed by soil organisms.

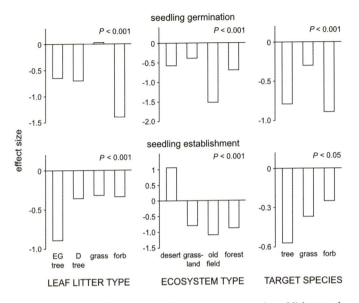

FIGURE 5.5. Effects of litter on seedling germination and establishment determined by a literature meta-analysis of 35 previously published studies performed by Xiong and Nilsson (1999). Effect size is a relative measure (Hedge's *d*) which becomes increasingly positive or negative as mean effect of litter becomes increasingly positive or negative, respectively. EG and D, evergreen and deciduous. Each *P* value relates to the overall comparison of all bars within each panel.

Decomposer-Plant Feedbacks

It is apparent from the material in this chapter that decomposer organisms and the activity of the soil biota have important effects on plant growth and tissue nutrient concentrations. In particular, effects of soil fauna on NPP and leaf nitrogen contents are frequently positive and sometimes very strong [Setälä and Huhta (1991) found that soil fauna increased leaf nitrogen concentrations of *Betula pendula* seedlings by a factor of more than three] and this should lead to both greater quantity and greater quality of the litter that plants return to the soil. When this is consid-

151

ered alongside the material presented in chapters 2 and 3, showing that decomposers and their activity are often favored by both enhanced NPP and enriched litter quality, it is clear that the decomposer subsystem frequently affects the quality and quantity of resources returned by plants to the soil in a manner that benefits decomposer organisms. This is evidence of a strong positive feedback between the autotrophic and decomposer components of terrestrial ecosystems.

NITROGEN TRANSFORMATIONS

While soil saprophytic microorganisms release mineral nitrogen for plant growth as a consequence of their activities, soils also host microflora that are specifically adapted for carrying out certain nitrogen transformations such as atmospheric nitrogen fixation, nitrification, and denitrification. These processes all have the potential to influence plant growth through determining both the amounts and forms of nitrogen present in the soil. Fixation of atmospheric nitrogen is carried out by only a tiny subset of the total soil bacterial community, but nitrogen fixing microbes are of profound importance through providing the major source of nitrogen input in many, perhaps most, ecosystems. These include bacteria that live freely in the soil and are frequently associated with organic substrates, bacteria that live in the root zones of certain plant species (associative nitrogen fixers), and bacteria that form intimate associations with leguminous and various other plant species through inhabiting root nodules. Biological nitrogen fixation has long been recognized as a major source of nitrogen input to agroecosystems, and as an alternative source of nitrogen input to fertilizers (see the reviews by Giller and Cadisch 1995, Peoples et al. 1995). Further, in natural ecosystems, biological nitrogen fixation in the early phases of primary succession is well recognized as a major contributor to ecosystem development.

The importance of the contribution of biological nitrogen fixation to ecosystem development is especially apparent for those seres for which plants capable of biological nitrogen fixation occur at the early stages. One clear example (discussed further in chapter 7) involves early successional systems dominated by *Metrosideros polymorpha* in Hawaii (Vitousek et al. 1987; Vitousek and Walker 1989). In the absence of symbiotic nitrogen fixers, the main nitrogen input occurs via rainfall. However, when the system is invaded by the actinorhizal alien shrub *Myrica faya*, the total nitrogen input to the system increases approximately fivefold. This increases nitrogen availability and facilitates the ingress of other alien species. Symbiotic nitrogen fixation by early successional native plants has also been shown to have a fundamental role in enhancing ecosystem nitrogen accumulation and facilitating later successional plant species for a variety of primary successions. Examples include *Lupinus* spp. on newly created surfaces following volcanic activity on Mt. St. Helens (U.S.A.) (Morris and Wood 1989), and *Alnus* spp. following the deposition of fresh gravels on floodplains (e.g., Luken and Fonda 1983; Walker 1989) and formation of fresh surfaces during eglaciation (Chapin et al. 1994).

For symbiotically fixed nitrogen to benefit associated non-fixing plant species it is required that the fixed nitrogen be released from the host plant in a form that other species can take up. This has frequently been demonstrated to occur for pastoral plant species, in which ^{15}N incorporated by legume species (e.g., *Trifolium* spp., *Medicago* spp.) becomes detected over time in tissues of associated grass species (e.g., Ledgard et al. 1985; Burity et al. 1989), promoting grass growth. Mechanisms involve both exudation of nitrogen as nitrate from root nodules (Wacquant et al. 1989) and mineralization of nitrogen resulting from nodule turnover (Wardle and Greenfield 1991) and residue return (e.g., Vallis 1983; Jensen 1994). These mechanisms form the basis behind much agricultural practice, and also explain why experimental studies investigating the effects of plant functional group diversity on NPP find the group consisting of nitrogen fixing plants to have a disproportionately positive effect (e.g., Hooper and Vitousek 1997; Hector et al. 1999).

Available Forms of Soil Nitrogen and Plant-Microbe Feedbacks

Specific soil microorganisms can also influence plant communities by determining the relative abundance of different forms of nitrogen (dissolved organic, ammonium, nitrate) present in the soil during and following decomposition. A minute proportion of the soil microflora is well adapted for oxidizing ammonium; this includes specific chemoautotrophic nitrifiers (notably *Nitrobacter* and *Nitrosomonas*) as well as certain bacteria capable of heterotrophic nitrification. The degree of nitrification carried out by these organisms is likely to have important ecosystem-level consequences, both through governing the degree of ecosystem nitrogen loss (nitrate is usually more easily lost from the ecosystem than are other forms of nitrogen, e.g., through leaching, denitrification), and because different plant species respond differentially to nitrate.

A feedback occurs between the aboveground and belowground subsystems because those plant species that promote nitrifying bacteria are almost certainly those that are most adapted for the uptake of nitrate nitrogen. Plant species differ in their effects on net nitrification, with plants that produce superior litter quality, improve soil pH, and dominate earlier in succession frequently being observed to promote net rates of nitrification (Iseman et al. 1999; Lovett and Rueth 1999). For example, Steltzer and Bowman (1998) found nitrification rates to be four times greater in alpine meadows dominated by *Deschampsia caespitosa* (with high litter quality) than in meadows dominated by *Acomastylis rossii* (with poor litter quality), and Lovett and Rueth (1999) detected significant nitrification under stands of *Acer saccharum* but no detectable nitrification under several *Fagus grandifolia* stands. Further, disturbance of stands dominated by later successional plant communities (including those dominated by plant species that produce poor litter quality) can cause substantial net nitrification and nitrate losses, as has been demonstrated through studies involving forest clear-cutting (e.g., Likens et al. 1968) and experimental trenching (Vitousek et al. 1979).

While apparent rates of net nitrification are often lower in systems dominated by later successional plant species, the mechanistic basis of this remains unclear. In a classical study, Rice and Pancholy (1973) found both soil nitrate levels and nitrifier bacteria populations to decrease markedly during the build-up phase of an old field succession in Oklahoma (U.S.A.), and proposed that tannins and tannin-derived phenolics present in leaves of the dominant later successional flora (*Quercus* spp. and *Pinus echinata*) were responsible for the virtual cessation of net nitrification at the maximal vegetation biomass phase. Further, White (1986) proposed that vegetation in forests dominated by *Pinus ponderosa* inhibited nitrification through production of volatile terpenoids that inhibit nitrifier activity. However, it has been disputed as to whether secondary metabolites produced by such plants reduce apparent net nitrification through allelopathy (White 1990) or whether rapid microbial immobilization of nitrate following its production (e.g., through increased carbon availability) is responsible (Bremner and McCarty 1988, 1990). In any case, Stark and Hart (1997) used a ^{15}N isotope dilution technique in intact soil to demonstrate that gross nitrification was high in several mature coniferous forests and that net nitrification was consistently low simply because of rapid rates of nitrate assimilation by the soil microflora. This is consistent with the expectation that the microflora in late successional systems can be severely nitrogen limited (chapter 3).

There is much evidence that different plant species vary in their responses to nitrate and ammonium. Olsson and Falkengren-Grerup (2000) provided evidence for five understory herb species in *Quercus* forests in southern Sweden that nitrification rates were greatest in soils under those species that were less acid-tolerant and presumably preferentially took up nitrate, and were low for acid tolerant plant species that favored ammonium. Further evidence for such feedback is presented by Kronzucker et al. (1997). Here, ammonium uptake by a late successional conifer (*Picea glauca*) was 20 times greater than that for nitrate, and *P. glauca* seedlings were far more effective at incorporating ammonium nitrogen than nitrate nitrogen into cytoplasmic material. It was proposed that the inability of late successional tree species to use nitrate may be responsible for their poor establishment on disturbed sites where nitrification, and therefore plant species capable of effective nitrate utilization, would be favored.

There is increasing recognition that plant species may also take up simple forms of organic nitrogen and thus bypass the nitrogen mineralization step entirely. This has been shown for plant species that dominate in nitrogen-limited boreal forests (Näsholm et al. 1998), but there is recent evidence that herbaceous plant species characteristic of fertile agricultural systems can also take up organic nitrogen (Näsholm et al. 2000). It is perhaps unsurprising that plants should be able to take up organic nitrogen when associated with mycorrhizal symbionts with known abilities to degrade organic matter, and indeed it has been shown using $\delta^{15}N$ measurements of leaf tissues that plant species with different types of mycorrhizal associations probably access different pools of nitrogen (Michelsen et al. 1996). However, preferential organic nitrogen uptake by the nonmycorrhizal arctic sedge species *Eriophorum vaginatum* has also been demonstrated (Chapin et al. 1993b). Organic nitrogen uptake may be more important in later successional systems with low nitrogen availability; Northup et al. (1995) demonstrated that *Pinus muricata* growing in infertile acidic sites produced litter with very high phenolic concentrations which resulted in decomposing litter releasing dissolved organic nitrogen rather than mineral nitrogen during breakdown. This mechanism has adaptive significance since it enables plants growing in infertile sites to access nitrogen via their mycorrhizal fungal symbionts in conditions where net microbial mobilization of nitrogen is unlikely to occur.

The material presented so far indicates that ecosystems that differ in fertility will differ in terms of the relative importance of those mechanisms that affect nitrogen availability, and this will in turn affect the functional composition of the plant community. Fertile and relatively early successional ecosystems should favor rapid nitrogen mineralization, high rates of net nitrification and nitrogen fixation, and high nitrogen availability. This will in turn lead to domination by plant species that take up nitrate, which often have rapid relative growth rates, short leaf retention times, high specific leaf areas, high palatability of foliage to herbivores, and high litter quality, leading to rapid decomposition rates. As systems become less fertile and the rates of nitrogen cycling processes decline, plant communities should become increasingly dominated by species that show preferential uptake of ammonium and finally dissolved organic nitrogen, grow more slowly, retain their leaves for longer, and produce foliage and litter with increasing amounts of secondary metabolites to reduce loss rates of nitrogen.

MICROBIAL ASSOCIATES OF PLANT ROOTS

The plant root zone, including the rhizosphere, supports a diverse array of microflora which are promoted by resources released from plant roots. Since many of these microorganisms can in turn either inhibit or stimulate plant growth, both negative and positive feedbacks between plants and soil microbes can occur. Negative effects that root zone microbes exert on plants include competition for nutrients, production of inhibitory metabolites (e.g., hormones, antibiotics, organic acids), and pathogenic activity. Positive effects include mineralization of nutrients (including the "microbial loop" described earlier), nitrogen fixation, suppression of pathogens, production of plant-growth-promoting compounds (e.g., hormones, siderophores that increase iron availability), and increased acquisition of nutrients through mycorrhizal associations. There is a vast agronomic and biotechnological literature about how several of these interactions may affect the growth of crop plants, as well as how the rhizosphere microflora can be manipulated to enhance plant growth (see the reviews by Whipps and Lynch 1986; Schippers et al. 1987; Arshad and Frankenberger 1998; Bowen and Rovira 1999). However, with the exception of those interactions involving nutrient transformations, pathogens, and mycorrhizas, the ecological role of these mechanisms in nonmanipulated systems remains largely unexplored. The nutrient transformation aspects have been discussed earlier in this chapter. Here I will discuss the ecological significance of root pathogens and mycorrhizas in the rooting zone.

Root Pathogens

Despite the extensive literature about the effects of plant root pathogens on agricultural crop production, relatively few studies have considered the role of pathogens in influencing plant communities or vegetation succession. However, it is apparent that root pathogens can be an agent contributing to forest decline (e.g., Florence 1965; Podger and Newhook 1971; Otrosina et al. 1999) and may serve as an agent structuring plant communities through affecting relative plant fitness (Jarosz and Davelos 1995). For example, Van der Putten et al. (1993) showed that the buildup of soilborne diseases under *Ammophila arenaria* colonizing coastal foredunes in the Netherlands resulted in the decline of that species and increased colonization of later successional species. Growth of *A. arenaria* seedlings planted in unsterilized soil from under declining *A. arenaria* stands, relative to their growth in the same soil following sterilization (to kill pathogens), was strongly adversely affected. In contrast, seedling growth of three later successional plant species was less affected by nonsterilized soil sourced from under stands of *A. arenaria* (fig. 5.6a). Meanwhile, the later successional species performed less well in nonsterilized (vs. sterilized) soil collected from under later successional species than in soil from under *A. arenaria*. Further, competition experiments demonstrated that *A. arenaria* seedlings competed poorly with the later successional species *Festuca rubra* when nonsterilized soil collected from under *A. arenaria* was used, but less poorly when this soil was sterilized (Van der Putten and Peters 1997), pointing to a likely role of pathogens favored by *A. arenaria* in inducing successional replacement. Similarly, Bever (1994) found for old-field plant species in North Carolina, U.S.A., that seedlings of *Krigia dandelion* subjected to soil inocula from its own species showed much greater mortality than did seedlings of other species subjected to *K. dandelion* soil inocula (fig. 5.6b). Generally, the other species also performed less well when subjected to inocula sourced from their own species than to those sourced from other species. These effects may have resulted from plant species favoring buildup of fungal pathogens that eventually led to their decline, given that similar (though smaller) effects were also obtained when washed live root segments rather than soil inocula were used. Evidence also emerges from forested systems for plant species favoring the buildup of pathogens that hasten their decline; Packer and Clay (2000) found that seedlings of *Prunus serotina* showed a very high mortality when planted in soil collected from under *P. serotina* trees, but not in soil collected 25–30 m away from the trees. This mortality was apparently due to buildup of *Pythium* spp. under the trees, and seedling mortality was almost eliminated by sterilizing the soil.

159

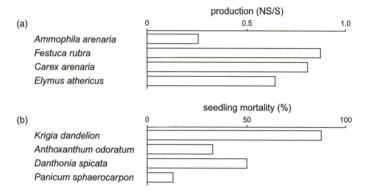

FIGURE 5.6. Evidence that soil sourced from under a particular plant species can be more suppressive of that plant species than of other plant species from the same habitat, probably through buildup of pathogens. (a) Growth of seedlings of four foredune plant species in nonsterilized soil (NS) relative to that in sterilized soil (S) collected from root zones of degenerated *Ammophila arenaria* plants. From Van der Putten et al. (1993). (b) Mortality of seedlings of four herbaceous plant species when grown in soil inocula derived from *Krigia dandelion* plants. From Bever (1994).

Fungi are regulated mainly by bottom-up control (see chapter 2), and for this reason the competitive effects of saprophytic fungi should be able to keep many potential fungal pathogens (including those that can also exist saprophytically) in check, and therefore potentially below the level of inoculum potential required for pathogens to influence vegetation composition. Although the ecological role of supression of pathogens by saprophytic microflora in natural ecosystems remains unexplored, this issue has attracted much attention in the agronomic plant pathology literature, especially in relation to the issue of suppressive soils (i.e., where certain soils support microbial communities that reduce pathogen loadings) (see Baker and Cook 1974; Knusden et al. 1999), and biological control of pathogenic fungi by specific antagonistic microbial species (Papavizas and Lumsden 1980; Hoitink and Boehm 1999). In particular, species of *Trichoderma* and *Gliocladium* have often been shown to reduce the inoculum potential for possible fungal pathogens in soil (e.g., Rai and Upadhyay 1983; Papavizas 1985). In natural soils the resident microflora may therefore keep many fungal pathogen populations in check in the majority of instances simply as a result of the intensity of competition for resources that normally occurs in fungal communities, and this may help explain why documented cases of fungal pathogens affecting vegetation dynamics in indigenous communities are few.

Mycorrhizal Associations

While plants can enter loose mutualisms with free-living saprophytic microbes by providing carbon and receiving mineralized nutrients in return, most plant species also form closer linkages with fungi through mycorrhizal associations. In exchange for providing carbon the fungus enables the plant more direct access to nutrients (notably phosphorus and often nitrogen) than is otherwise possible. Indeed, the literature is replete with examples of plants benefiting from improved nutrition provided by mycorrhizas (reviewed by Smith and Read 1997) although neutral or even negative effects may be possible [Gange (1999) proposes a curvilinear relationship between extent of mycorrhizal colonization and plant benefit with negative effects possible at high colonization densities]. Mycorrhizal fungi can directly access nutrients in plant litter and translocate these to the plant. For example, Perez-Moreno and Read (2000) grew seedlings of *Betula pendula* inoculated with the ectomycorrhizal fungus *Paxillus involutus*, with or without discrete patches of leaf litter of each of three tree species buried in the soil. Mycorrhizal mycelia exploited both nitrogen and phosphorus from the litter, translocating it to the seedlings and improving seedling growth. There is also recent evidence that mycorrhizal fungi can also access phosphorus from sources other than organic matter. For example, ectomycorrhizal fungal hyphae appear capable of penetrating weatherable rock minerals through production of organic acids, possibly facilitating phosphorus release and uptake from them (Jongmans et al. 1997; Van Breemen et al. 2000). Another possible source of phosphorus derives from antagonistic interactions between mycorrhizal and saprophytic fungi. Lindahl et al. (1999) demonstrated that ectomycorrhizal fungi hosted by *Pinus sylvestris* could lyse hyphae of the wood-decomposing saprotroph *Hypholoma fasciculare*, and transfer ^{32}P from the saprotroph to the host plant.

The mechanisms of phosphorus acquisition by mycorrhizal fungi can in turn be important in regulating ecosystem phosphorus cycling. For example, an intersite comparison of ectomycorrhizal leguminous tree stands in the tropical rain forests of Cameroun (Newbery et al. 1997) provided evidence that mycorrhizas promoted the utilization of a greater range of phosphorus compounds than was otherwise possible, and were effective at continually reaccessing and retaining phosphorus in the soil organic layer. This mechanism served both to increase the availability of phosphorus in the soil organic layer and to prevent its loss from the system through leaching and surface water movement. Ultimately, this retention mechanism was proposed as being responsible for maintaining the dominance over time by ectomycorrhizal tree species.

In addition to improved nutrition, mycorrhizal fungi can also confer several other benefits upon plants (Brundrett 1991; Newsham et al. 1995; Smith and Read 1997), including reduced uptake of heavy metals, improved water uptake, production of plant-growth-stimulating compounds, and resistance to herbivores (discussed later in this chapter) and pathogens. The interaction between mycorrhizal fungi and root pathogens in particular may be an important determinant of plant growth and fitness. Carey et al. (1992) and Newsham et al. (1994) proposed, based on experiments in which fungicides had been applied to grassland field plots, that while root pathogenic fungi reduced plant fecundity, arbuscular fungi increased fecundity, probably in part because they were able to protect the host plant against pathogenic attack.

Mycorrhizal fungal community structure is likely to be important in influencing plant growth and therefore ecosystem productivity, given that different species of mycorrhizal fungi capable of colonizing the same plant roots can differ markedly in their effects on nutrient acquisition by the host plant. For example, Jakobsen (1992) found that *Trifolium subterraneum* inoculated with each of three different arbuscular mycorrhizal species differed by a factor of nearly 100 in terms of the amount of ^{32}P taken up into plant tissue. With regard to ectomycorrhizal associations, Le Tacon et al. (1992) demonstrated that growth rates of nursery seedlings of both *Pinus sylvestris* and *Pseudotsuga menziesii* were highly responsive to which fungal species they were inoculated with, and Jonsson et al. (2001) found that eight fungal species differed markedly in their effects on productivity and shoot-to-root ratios of *Pinus sylvestris* and *Betula pendula* seedlings. Different growth forms of mycorrhizae may also differ in their effects on plant growth, and this is apparent for plant species that are capable of forming both arbuscular and ectomycorrhizal associations. For example, M. D. Jones et al. (1998) demonstrated that *Eucalyptus coccifera* growth and phosphorus uptake were stimulated by inoculation with either type of mycorrhizal fungus, although ectomycorrhizal fungal species had greater positive effects than did arbuscular mycorrhizal fungi.

Associations between mycorrhizal fungal species and plant species are likely to be important determinants of plant community structure. This is clearly demonstrated by the experiment of Van der Heijden et al. (1998a) in which each of three grassland plant species was inoculated with each of four arbuscular mycorrhizal fungal species. There were significant effects of fungal species on the relative performance of the plant species in terms of biomass production,

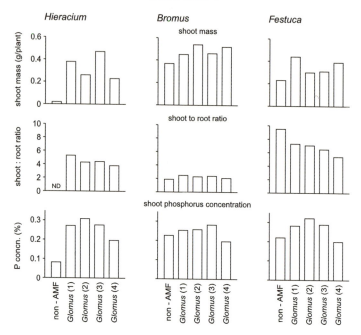

FIGURE 5.7. Interactive effects of arbuscular mycorrhizal species and plant species on plant shoot mass, shoot-to-root ratios, and shoot phosphorus concentrations. Non-AMF, no arbuscular mycorrhizal fungi; mycorrhizal species are as follows: *Glomus* (1), *Glomus* sp. (BEG 19); *Glomus* (2), *Glomus geosporum*; *Glomus* (3), *Glomus* sp. (Basle Pi); and *Glomus* (4), *Glomus* sp. (BEG 21). ND, not determined. Derived from Van der Heijden et al. (1998a).

allocation patterns, and tissue phosphorus concentrations (fig. 5.7). Similarly, Kiens et al. (2000) provided evidence that three tropical tree species differed in their relative growth response (total leaf area) to arbuscular mycorrhizal inoculum sourced from different tree species. Both studies suggest that mycorrhizal fungal species composition may be an important determinant of vegetation community composition. Evidence that this is indeed the case emerges from one of the experiments reported by Van der Heijden et al.

(1998b), in which different experimental units (each consisting of synthetic plant communities with 11 species) were inoculated with each of four different arbuscular mycorrhizal fungal species. Their results showed that while different mycorrhizal fungal species did not differ much in their effects on total plant biomass, there were very large fungal species effects on the relative performance of the different plant species.

The effects of mycorrhizal fungi on plant community structure probably arise from fungal effects on both plant competition intensity and the competitive balances between plant species. The intensity of plant interspecific competition is often reduced by the presence of mycorrhizae [e.g., Fitter (1977) and Hetrick et al. (1989) for arbuscular mycorrhizae; Perry et al. (1989) for ectomycorrhizae], although this pattern is not universal (e.g., Eissenstat and Newman 1990). Mycorrhizal colonization can also cause greater coexistence (and therefore diversity) of plant species, apparently as a consequence of reducing plant competition intensity (Grime et al. 1987; Van der Heijden et al. 1998b; but see Connell and Lowman 1989; Hartnett and Wilson 1999). One possible mechanism by which mycorrhizae may reduce plant competition involves hyphal links between coexisting plants, and it has been shown in several studies that shared hyphae among different plants facilitate transfer of carbon and nutrients between them, both within and across species (e.g., Francis and Read 1984; Read et al. 1985; Bethenfalvy et al 1991). Further, Simard et al. (1997) used reciprocal isotope labeling (in which one plant is labeled with ^{13}C and the other with ^{14}C) to demonstrate two-way transfer of carbon between ectomycorrhizal seedlings of *Betula papyrifera* and *Pseudotsuga menziesii*. It has also been suggested that hyphal networks are likely to be important in assisting the recruitment of new seedlings into plant communities (Read et al. 1985; but see Eissenstat and Newman 1990) and maintaining forest understory plants in low light conditions through net carbon flow from canopy trees (Högberg et al. 1999). However, there are difficulties in demonstrating that interplant hyphal linkages, and the existence of hyphal transport mechanisms between plants, are in themselves important in influencing plant fitness or are major determinants of plant community structure (see Newman 1988; Fitter et al. 1998; Newbery et al. 2000).

If mycorrhizal hyphal links facilitate greater coexistence of plant species through reduced competition, then fungal-regulated positive interactions between plant species might not be uncommon in plant communities. This further indicates that mutualisms may be more important in influencing community and ecosystem properties than is often appreciated when appropriate consideration is given to the belowground biota (see chapter 2). However, mycorrhizal fungi can also induce some plant species to more effectively compete against other plant species, particularly those species not connected within the same hyphal network. For example, Hartnett and Wilson (1999) used fungicide applications to field plots in a tallgrass prairie in Kansas, U.S.A., to reduce mycorrhizal effects, and this in turn reduced obligately mycorrhizal C4 grasses, inducing corresponding increases in nonobligately mycorrhizal C3 grasses and forbs which were favored by competitive release. Further, it has been proposed for tropical rainforests that ectomycorrhizal associations with high specicifity may encourage domination by a small number of ectomycorrhizal tree species and displacement of arbuscular mycorrhizal tree species, resulting in a reduced level of coexistence among tree species (Connell and Lowman 1989). This arises because in rainforests ectomycorrhizal associations can probably afford greater protection of trees from natural enemies and enable improved acquisition of a wider range of nutrients than is the case for arbuscular mycorrhizal associations (Connell and Lowman 1989).

Mycorrhizae, Food Webs, and Plant Growth

As well as directly contributing to plant nutrition, mycorrhizal fungi also comprise part of the decomposer food web and are therefore consumed by soil animals. Faunal feeding activity may therefore affect plant growth and nutrient acquisition not just by mineralizing nutrients (as discussed earlier in this chapter) but also by grazing on mycorrhizal hyphae, although relatively few studies have considered this interaction. In a microcosm study, Setälä et al. (1997) grew ectomycorrhizal *Pinus sylvestris* seedlings in nitrogen-rich and nitrogen-poor humus, and with or without a complex soil faunal community. In both substrates the complex community reduced mycorrhizal hyphal density, but plant mass was reduced only in the nutrient-poor substrate (Fig. 5.8a). However, Setälä (1995) found that while soil faunal addition to microcosms reduced ectomycorrhizal density of roots of both *P. sylvestris* and *Betula pendula*, there were substantial increases in both shoot mass and foliar nutrient concentrations (Fig. 5.8b). This suggests that reduction of mycorrhizal hyphae may not be detrimental to plant growth so long as the fauna also induce sufficient mineralization of nutrients to maintain plant growth. Soil fauna may also have important effects on endomycorrhizal associations [reviewed by Gange (2000) with regard to Collembola]. For example, Klironomos and Kendrick (1995) investigated the effects of litter addition and microarthropods on endomycorrhizal colonization and growth of seedlings of *Acer saccharum*. Faunal addition stimulated plant growth only when litter was also added, and this was accompanied by a substantial faunal stimulation of arbuscular and vesicular colonization of plant roots (fig. 5.8c). The mechanistic basis of this stimulation is unclear, but probably involves arthropods preferentially consuming saprophytic fungi growing on the litter (Klironomos and Kendrick 1995; Klironomos et al. 1999).

The effects of the structure of the soil food web on mycor-

167

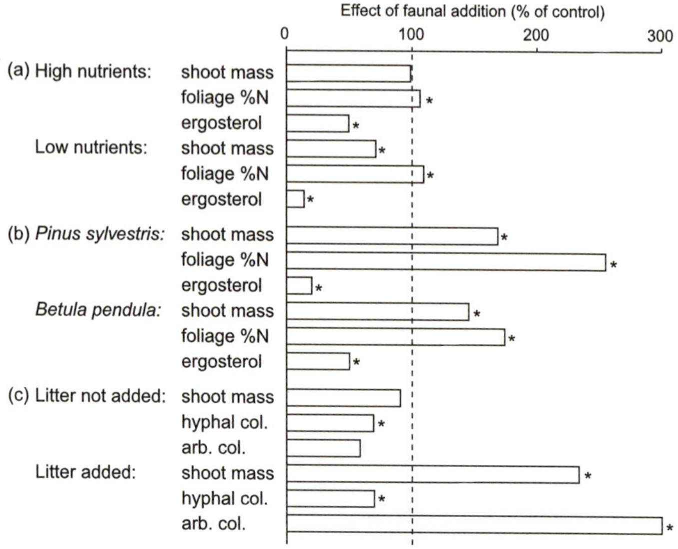

FIGURE 5.8. Effects of addition of soil fauna on mycorrhizal fungi, plant growth, and nutrient acquisition. (a) Effect of complex forest floor fauna relative to effect of microfauna only, on ectomycorrhizal *Pinus sylvestris* seedlings, in both a nutrient-rich and a nutrient-poor substrate. Soil ergosterol concentration is used as a relative measure of the amount of ectomycorrhizal fungal hyphae. From Setälä et al. (1997). (b) Effect of complex forest floor fauna on ectomycorrhizal seedlings. Ergosterol concentrations relate to mycorrhizal fungal density on plant roots. From Setälä (1995). (c) Effect of microarthropod addition on endomycorrhizal *Acer saccharum* seedlings, with and without the addition of *A. saccharum* litter. Col, colonization of roots; arb, arbuscular. From Klironomos and Kendrick (1995). * indicates faunal effect significant at $P = 0.05$.

rhizal performance and plant growth have been little explored. However, the results of Setälä et al. (1997) provided evidence that adding a complex soil fauna to microcosms induced effects that a food web consisting of only microflora and microfauna could not. In a subsequent experiment, Setälä et al. (1999) planted *P. sylvestris* seedlings into microcosms containing different soil microfood-web structures (i.e., bacterial-feeding and fungal-feeding nematodes

either singly or in combination, with and without top predatory nematodes). Food web structure had no detectable effects on ectomycorrhizal fungal colonization or root growth,
except that when fungal feeding nematodes were present in
the food web, mycorrhizal hyphal density showed a relatively
small reduction. In total, while little is known about how soil
food webs interact with the plant-fungal mycorrhizal association, the nature of these effects appears to be context dependent, and such interactions might be of considerable
(though largely unrealized) importance in affecting ecosystem productivity.

Hypogeous fruiting bodies produced by ectomycorrhizal
fungi are also consumed by a wide range of fauna, and can
comprise most of the diet of several forest dwelling mammal
species (Johnson 1996; McIlwee and Johnson 1998). These
mammals can be important in transporting spores of ectomycorrhizal fungi from forests to nearby unforested areas
(Cazares and Trappe 1994) and the possibility exists that
they can be powerful agents of dispersal of mycorrhizal fungal spores (Johnson 1996; Terwilliger and Pastor 1999). If
this is the case then the trophic interactions between mycorrhizal fungi and the consumers of their fruiting bodies may
have an important role in facilitating the establishment of
ectomycorrhizal tree species during successional development (Terwilliger and Pastor 1999).

ROOT HERBIVORES

Belowground herbivores consist of a diverse range of fauna, and include representatives of the Gastropoda, Acarina,
Symphyla, three orders of Nematoda, and ten orders of In-
secta, as well as various vertebrate taxa. While the commu-
nity- and ecosystem-level consequences of aboveground herbivory have been extensively studied, the ecological
consequences of belowground herbivory remain much less
well understood. This is despite belowground NPP fre-

quently exceeding that which occurs aboveground (Eissenstat and Yanai 1997).

Several studies have shown that severe densities of root herbivores can reduce aboveground plant growth, especially that of agriculturally important plant species (reviewed by Andersen 1987; Brown and Gange 1990; Mortimer et al. 1999; Wardle et al. 1999a). However, modest levels of root feeding may not have adverse effects on plant biomass because of the capacity of many plant species to show compensatory belowground growth responses to root herbivory (Andersen 1987). For example, Detling et al. (1980) found that while simulated root herbivory of hydroponically grown *Bouteloua gracilis* plants temporarily reduced net photosynthetic rates, compensatory root growth (assessed as production per unit mass) occurred. Further, Riedell et al. (1989) observed that *Zea mays* plants supporting modest levels of the corn rootworm (*Diabotrica* spp.) had larger root systems and higher yields than uninfested plants. It may be that the growth of some plant species can actually be optimized by low levels of root herbivory (comparable to what has been proposed for aboveground herbivory; see chapter 4), although there is insufficient information at present to evaluate the generality of this (see Brown and Gange 1990).

Root herbivory may influence plant growth not only directly but also indirectly by affecting the release of materials from the plant root zone, and this can occur at both the within-plant and between-plant levels. At the within-plant level, root herbivory can influence the rate of carbon release by roots, increasing soil microbial activity and ultimately the supply rates of nutrients in the soil. For example, Yeates et al. (1999a) grew plants of *Trifolium repens* in pots inoculated with each of five species of plant parasitic nematodes, and found that all nematode species increased the rate at which ^{14}C (sourced from $^{14}CO_2$) was transferred from the roots to the soil microbial biomass. However, Holland and Detling (1990) determined that root herbivory by

black-tailed prairie dogs (*Cynomys ludovicianus*) in a grassland in North Dakota, U.S.A., reduced microbial immobilization of nitrogen as a result of diminished availability of carbon (through reduced root productivity) to decomposer
organisms. It may be that root herbivory induces increased
microbial activity and biomass at low but not high densities;
Denton et al. (1999) found that low levels of clover cyst
nematode (*Heterodera trifolii*) on roots of *T. repens* enhanced
soil microbial biomass while higher levels of root herbivory
had neutral or even negative effects.

At the across-plant level, root herbivory may affect the release of nutrients from a target plant to its neighbors. This
aspect has received very little attention, but two studies
point to its probable importance in grasslands. In the first,
Hatch and Murray (1994) showed that significant transfer of
nitrogen from roots of *T. repens* to tissues of *Lolium perenne*
occurred only when part of the root system of *T. repens* was
damaged in order to simulate the effects of belowground
herbivores such as *Sitona* spp. In the second, Bardgett et al.
(1999b) demonstrated that low levels of infestation of the
root herbivorous nematode *Heterodera trifolii* (which feeds on
the roots of *T. repens* but not *L. perenne*) increased shoot
growth of *T. repens* by 141% and that of adjacent *L. perenne*
plants by 219%. Root herbivory also caused a significant increase in the soil microbial biomass and a fourfold increase
in the transfer of ^{15}N from *T. repens* to *L. perenne*; it is therefore apparent that low levels of herbivory can greatly accelerate the rate at which fixed nitrogen is transferred from
legumes to associated grasses.

The differential effects of root herbivores on different
plant species can in turn serve as an important determinant
of plant community structure as well as the relative abundance of different plant functional types. Field experimentation at Silwood Park, U.K., which involved manipulation of
aboveground and belowground hervivores through the addition of insecticides to an early successional grassland com-

munity (Brown and Gange 1989a, 1990, 1992), demonstrated that reduction of root herbivory encouraged the persistence of annual forbs and increased the colonization of
perennial forbs. It was proposed that root herbivores therefore promoted succession by reducing the success of early
successional forb species. In contrast, foliar-feeding insects
served to retard successional development by reducing colonization of grasses. In a mesocosm experiment investigating
effects of invertebrate herbivores on plant community development, Wardle and Barker (1997) also found that addition
of root herbivores affected forbs more than grasses, although aboveground herbivores had effects that were similar to those of root herbivores. It is therefore apparent that
belowground herbivores have the potential to induce shifts
in the functional composition of the vegetation, which
could in turn induce longer-term feedbacks involving decomposers and nutrient mineralization, comparable to
those effects described for aboveground herbivores in chapter 4.

One issue that remains little studied with regard to root
herbivory is whether consumers of these herbivores induce
trophic cascades affecting plant growth in the way that may
operate for aboveground food webs (see chapter 4). However, one example has been presented by Strong (1999) and
Strong et al. (1999) in which the root-feeding larvae of the
ghost moth *Hepialis californiacus* can serve as a major source
of mortality of *Lupinus arboreus,* and in which the entomopathogenic nematode *Heterorhabditis hepialus* can cause
high mortality of the *H. californiacus* larvae. Although this
system may serve as a potential trophic cascade through
which the nematodes contribute to the success of *L. arboreus,* Strong (1999) proposes that other interactions may induce an element of complexity which may work to reduce
the degree of direct causation of *H. hepialis* in improving
the success of *L. arboreus* (as predicted by the green world
hypotheses described in chapter 2). One possibility is that
promotion of nematode-trapping fungi through increased

detrital addition resulting from high productivity of *L. arboreus* may in turn reduce populations of *H. hepialis* (Strong 1999).

PHYSICAL EFFECTS OF SOIL BIOTA

As discussed in chapter 2, the subset of the soil fauna which operate as ecosystem engineers can exert important ecosystem-level effects through altering the physical structure of the soil. This in turn has significant effects on plant growth and community composition, and can be an important determinant of vegetation succession. For example, Derourad et al. (1997) found that three tropical earthworm species present in cropping systems in Côte d'Ivoire differed in their effects on soil infiltration, with *Millsonia anomala* compacting the soil and *Chuniodrilus zielae* and *Hyperiodrilus africanus* having the opposite effect. This in turn resulted in the earthworm species exerting differential effects on the growth of three crop species grown in buckets. Further, Chauvel et al. (1999) describe an example in which conversion of forest to pasture in Amazonia encouraged domination by the earthworm *Pontoscolex corethurus* and reduction of the indigenous macrofaunal community. The casts produced by this earthworm have very low porosity, resulting in severe compaction comparable to that caused by mechanical compression. The net result is the development of conditions that are unfavorable for plant growth. In contrast, in the study of Bernier and Ponge (1994) in *Picea abies* forests described in chapter 3, the physical effects of anecic earthworms in senescent stands (including mixing of organic matter with humus) create conditions that are conducive to the development of new cohorts of *P. abies* seedlings.

Soil fauna that produce mounds can also induce the formation of patches with different vegetation from that of the surrounding landscape. For example, Brown and Human (1997) found that mounds produced by the ant *Messor andrei* in a serpentine grassland in northern California, U.S.A., promoted grass species at the expense of forbs. In the Chihuahuan Desert (New Mexico, U.S.A.), Whitford and Di-Marco (1995) observed that mounds produced by the ant *Pogonomyrmex rugosus* had greater nitrogen availability than corresponding non-nest areas, and this in turn promoted plant species that rapidly respond to available nitrogen such as spring-annual grasses and forbs. Further, Blomqvist et al. (2000) provided evidence from a grassland in the Netherlands that mounds of the ant *Lasius flavus* favored domination by the clonal sedge *Carex arenaria* while *Festuca rubra* dominated the surrounding vegetation. This appears to be because properties of the soil under the mounds are more favorable for *C. arenaria* growth, probably in part because this soil supports lower levels of harmful organisms (pathogenic nematodes and fungi) than does the surrounding soil. With regard to termites, Spain and McIvor (1988) investigated the vegetation composition of termitaria in four sites in northeast Australia, and found that annual grasses and forbs dominated on the mounds while perennial grasses and sedges were prevalent further away from the mounds. Mounds produced by pocket gophers (*Thomomys bottae*) in northern Californian serpentine grasslands result in "microhabitat islands" which support development of plant communities that differ from the surrounding vegetation, through influencing germination, survival, seed production, and competitive interactions of individual species (Hobbs and Mooney 1985; Wu and Levin 1994). In each of the above examples, structures created by soil animals result in an altered functional composition of the vegetation, frequently promoting plant species that are adapted to disturbance and high nutrient availability, and which may be expected to produce litter of superior quality. This is in turn likely to result in feedback mechanisms with the vegetation also affecting belowground organisms and processes.

Larger soil animals may also influence vegetation composition and growth through dispersal of plant and microbial propagules. Earthworms in particular can have an important role in the transport of seeds through the soil profile (Grant 1983). Thompson et al. (1994) evaluated the species composition of the seed bank of both earthworm casts and non-cast soil in a sown grassland in Devon, U.K. Some plant species that were common in the soil seed bank were rare in worm casts, while seeds of *Cerastium fontanum*, a species not dominant in the soil seed bank or resident vegetation, comprised 85% of the total seeds present in the casts. Seeds in casts were on average much smaller than those in the soil seed bank, suggesting that earthworm ingestion and transport to the soil surface of buried seeds operates as a filter which may encourage the abundance of smaller-seeded plant species in the longer term.

Larger soil fauna such as earthworms may also influence plant growth and community structure through affecting the transport of propagules of microbial symbionts of particular plant species. For example, it has been shown that earthworm casting favors the nodulation of *Trifolium subterraneum* by encouraging the redistribution of *Rhizobium leguminosarium* (Doube et al. 1994), improves nodulation by *Frankia* and growth of seedlings of *Casuarina equisteifolia* (Reddell and Spain 1991), increases the dissemination of arbuscular mycorrhizal propagules of crop plants (Harinikumar and Bagyaraj 1994), and promotes the dissemination of the soilborne pathogen *Fusarium oxysporum* f. sp. *raphani* (Toyota and Kimura 1994). Structures created by earthworms may therefore have a range of possible effects on both plant growth and plant communities, depending upon the types of propagules that are concentrated within the casts.

SOIL BIOTIC EFFECTS ON ABOVEGROUND FOOD WEBS

A recurrent theme of this chapter is the importance of soil organisms in determining plant nutrient acquisition,

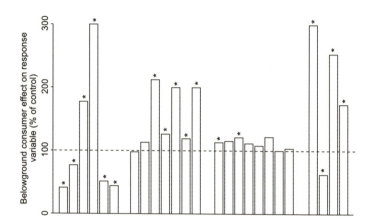

Belowground consumer	Plant species	Aboveground consumer (response variable)
MYCORRHIZAL FUNGI		
(a) Arbuscular mycorrhizae	Plantago lanceolata	Arctia caja larval growth rate
		A. caja food consumption
		Myzus persicae
		M. persicae embryo prodn.
(b) Glomus mosseae	Fragaria x ananassa	Otiorychnus sulcatus survival
Glomus intraradices	F. x ananassa	O. sulcatus survival
Glomus mosseae		
+ Glomus intraradices	F. x ananassa	O. sulcatus survival
(c) Glomus intraradices	P. lanceolata	Myzus ascalonicus growth rate
		M. ascalonicus fecundity
		M. persicae growth rate
		M. persicae fecundity
(d) Glomus spp.	Lotus corniculatus	Polyommatus icarus survival
		P. icarus larval weight
ROOT HERBIVORES		
(e) Phyllopertha horticola	Sonchus oleraceus	Chromatomyia synergesiae pupae
		C. synergesiae mine intensity
(f) P. horticola	S. oleraceus	M. persicae weight
		M. persicae growth rate
		M. persicae fecundity
(g) All root herbivores	Tripleurospermum perforatum	Olibrus aeneus emergence
		Gall midge emergence
		Apion hookeri emergence
DECOMPOSER FAUNA		
(h) Collembola	Poa annua	M. persicae juvenile production
	Trifolum repens	M. persicae juvenile production
Earthworms	P. annua	M. persicae juvenile production
	T. repens	M. persicae juvenile production

growth, and community structure. It is therefore reasonable to expect that aboveground (foliar) consumers should frequently be indirectly responsive to interactions involving belowground consumers. Only a handful of studies have explicitly addressed this issue, but they provide some evidence of plant-mediated linkages between aboveground and belowground trophic interactions.

Arbuscular mycorrhizae are known to exert both positive and negative effects on aboveground invertebrate herbivores. Foliar aphids (*Myzus* spp.) have been shown to respond positively to mycorrhizal infection of the host plant (Gange and West 1994; Gange et al. 1999; fig. 5.9), although the occurrence of these effects appears to be dependent upon soil nutrient availability (Gange et al. 1999). Meanwhile, mycorrhizal infection has been shown to have effects on foliage-chewing insects that are either positive [e.g., Goverde et al. (2000) for the butterfly *Polyommatus icarus*] or negative [e.g., Gange and Bower (1997) for the bee-

---

FIGURE 5.9. Effects of belowground consumers on aboveground invertebrate consumers as mediated by plants. (a) Gange and West (1994); *A. caja* and *M. persicae* belong to the Arctidae (Lepidoptera) and Aphididae (Homoptera) respectively. (b) Gange and Bower (1997). Data for *O. sulcatus* survival are for larvae; *O. sulcatus* belongs to the Curculionidae (Coleoptera). (c) Gange et al. (1999). (d) Goverde et al. (2000). All data are for *P. icarus* larvae; *P. icarus* belongs to the Lycaenidae (Lepidoptera). Data are averaged for all mycorrhizal treatments. (e) Masters and Brown (1992). Pupae, pupal weight; mine intensity, intensity of leaf mining. *P. horticola* and *C. synergesiae* belong to the Scaraedidae (Coleoptera) and Agromyzidae (Diptera), respectively. (f) Masters (1995). (g) Müller-Schärer and Brown (1995). All data are for emergence of fauna from flower heads. *O. aeneus* and *A. hookeri* belong to the Phalacridae (Coleoptera), and Circulionidae (Coleoptera) respectively. (h) Scheu et al. (1999). Data presented for the 16-week measurement only; juveniles produced, numbers of juvenile aphids produced per 8 days. Collembola species used are *Heteromurus nitidus* and *Onychiurus scotarius*; earthworm species are *Aporrectodea caliginosa* and *Octolasion tyrtaeum*. For all bars, * indicates that treatment and control values are significantly different at $P = 0.05$.

tle *Otiorychnus sulcatus*; Gange and West (1994) for the moth *Arctia caja*] (fig. 5.9). The mechanism behind these effects remains unclear, but positive effects of mycorrhizae may arise through their effects on host plant nutrition while negative effects may occur as a result of mycorrhizal induction of increased allocation to carbon-induced defenses (Gange and West 1994; Gange et al. 1999). Mycorrhizal infection can also influence pathogenic aboveground consumers. Examples exist of both positive and negative effects of arbuscular mycorrhizae on the degree of leaf infection by fungal foliar pathogens (West 1995, 1997) and the nature of these effects is probably linked to aspects of soil fertility including soil phosphorus availability (West 1995).

Individual plants also mediate the associations between aboveground and belowground herbivores. Masters et al. (1993) and Masters and Brown (1997) proposed a mechanism where root herbivores stimulate aboveground herbivores (through causing a plant stress through removal of roots, resulting in greater concentrations of soluble amino sugars and carbohydrates in the foliage), but where aboveground herbivores have adverse effects on root herbivores (by reducing carbon flow to roots and reducing root growth). This latter prediction is only partially consistent with much of the material presented in chapter 4, given that aboveground herbivory can also improve root substrate quality and probably optimize plant productivity at low intensities. Nevertheless, experimental support for this theory is apparent in the study of Masters and Brown (1992) in which the root herbivore *Phyllopertha horticola* exerted positive effects on the foliar leaf miner *Chromatomyia synergesiae* (fig. 5.9e), while the foliar miner had adverse effects on the growth rate of the root herbivore. In addition, Masters (1995) demonstrated beneficial indirect effects of *P. horticola* on the growth and fecundity of the aphid *M. persicae* (fig. 5.9f). However, in a field experiment, Müller-Schärer and Brown (1995) found no consistent effect of soil insecticide addition (applied to reduce root herbivores) on phytophagous insect emergence from flower heads of *Tripleurospermum perforatum* (fig. 5.9g). Further, in a mixture of glasshouse and field experiments, Moran and Whitham (1990) found that the presence of root-feeding aphids on plants of *Chenopodium album* had no detectable effect on leaf colonization by gall-forming aphids, although the gall-forming aphids often had strongly negative effects on the root feeders. These studies suggest that the effects of aboveground and belowground herbivores on one another may be able to work in both positive and negative directions, given the varied mechanisms by which herbivory influences both plant productivity and tissue chemistry (see chapter 4 and earlier in this chapter).

Given the role of soil decomposer fauna on nutrient cycling, it is intuitive that these fauna should therefore confer indirect benefits on aboveground consumers. Indeed, earthworms are often inoculated into managed pastures with the intention of increasing the production of herbivorous livestock. However, the potential effects of decomposer biota on herbivore performance remain largely uninvestigated. In what appears to be the only study of its type performed to date, Scheu et al. (1999) added collembolans and earthworms to microcosms containing plants of *Trifolium repens* and *Poa annua*, and used small cages to confine parthenogenic females of the aphid *M. persicae* to the leaves of these plants. Collembolans significantly enhanced aphid reproduction on *P. annua* but reduced aphid reproduction on *T. repens* (fig. 5.9h); it was concluded that Collembola promote aphid reproduction only on host plants of low palatability (i.e., *P. annua*). Earthworm addition had generally positive effects on aphid reproduction.

All of those studies that have found significant effects of belowground biota on foliar invertebrate herbivores have considered relatively palatable herbaceous plant species, and have involved microcosm or glasshouse experiments in

179

which predators of the herbivores are not present. For these reasons, the results of such experiments may overemphasize the importance of these types of effect in nature. First, the effect of soil biota on invertebrate herbivores of less palatable plant species (e.g., woody plants) remains unknown. Secondly, if aboveground invertebrate herbivores are actually regulated primarily by top-down rather than by bottom-up control (as proposed by various trophic theories discussed in chapter 2), then the effects of soil biota on foliar quality and quantity may not benefit the herbivores themselves but rather their predators. However, the effects of soil biota on predators of foliar herbivores, and the effects of these predators on the responses of foliar herbivores to soil biotic effects remain unexplored.

Linking Aboveground and Belowground Food Webs

It should be apparent from both this chapter and the last one that aboveground and belowground food webs are likely to influence one another, and that trophic interactions that occur on one side of the soil/surface interface would therefore be able to indirectly affect trophic levels on the other side of the interface. This issue remains largely unexplored at least for trophic levels higher than herbivores and microbe feeders. However, foliar feeders influence the soil biota through a variety of mechanisms, and this should in turn affect the rate of nutrient supply for plant growth and ultimately determine the palatability of the foliage for the herbivore. Indeed, most situations of optimization of NPP by herbivory probably involve the herbivores promoting the cycling of nutrients (see McNaughton 1979; Dyer et al. 1986; Belovsky and Slade 2000; De Mazancourt and Loreau 2000), and it is inevitable that indirect herbivore effects on components of the decomposer food web will have a key involvement in this. Similarly, the role of soil food web biotic interactions in determining nutrient availability for plants should impact upon both aboveground NPP and foliar quality; this is likely to influence the extent to which herbivory (and probably higher-level trophic interactions) occurs aboveground, which should in turn affect the quantity and quality of resources entering the decomposer subsystem.

SYNTHESIS

It is apparent from chapters 2–4 that the aboveground biota can drive the belowground biota through a diverse range of mechanisms. This chapter completes the circle by also outlining several mechanisms by which soil communities can affect aboveground communities. The most universally important of these mechanisms is the decomposer pathway, since it is responsible for regulating the supply of nutrients available for plant growth. While the soil microflora is directly responsible for mineralizing plant-available nutrients, soil faunas which consume microbes and each other can have very strong, usually positive, effects on plant growth, shoot-to-root ratios, and tissue nitrogen concentrations, through improving nutrient turnover and availability. Some experiments also point to a possible role of soil food web structure in affecting plant growth. Soil biotic effects on plant nutrient supply may induce important feedbacks by affecting the quantity and quality of plant litter that subsequently enters the decomposer subsystem.

A range of other soil biota also affect plant growth, community structure, and plant succession through a variety of positive and negative interactions. These include nitrogen fixing bacteria, which often play a role in ecosystem development, and nitrogen transforming bacteria that determine the relative abundance of nitrate, ammonium, and dissolved organic nitrogen in the soil (different plant species respond differentially to these nitrogen forms). Root pathogenic and mycorrhizal fungi can have a key role in influencing plant community structure, and some examples specifically point to the role of individual fungal species in affecting aboveground responses. Interactions involving root herbivores may alter plant growth and plant communities, not only through direct effects of root removal, but through promoting the release of carbon from the root zone (stimulating the microflora and therefore nutrient turnover) and the transfer of nitrogen from host plants (notably legumes) to their neighbors. Soil organisms that create physical structures (ecosystem engineers) can operate as powerful determinants of plant community composition, because these structures favor establishment of certain plant species and functional types above others.

The effects of belowground biota on foliage production and quality can in turn influence aboveground consumers; examples exist for aboveground herbivore responses to the effects of arbuscular mycorrhizae, root herbivores, and decomposer fauna. A recurrent theme through this chapter is that of aboveground-belowground feedbacks, and it is apparent that feedbacks occur between foliage-based and soil-based food webs that are mediated by the plant, although much remains unknown in this area. In conclusion, linkages between aboveground and belowground biota are extremely important determinants of long-term ecosystem function, and in particular the associations between the photosynthetic and decomposer subsystems are essential for the sustained maintenance of the ecosystem.

CHAPTER SIX

The Regulation and Function of Biological Diversity

The previous four chapters of this book have focused on biotic interactions among organisms, with an emphasis on associations between aboveground and belowground organisms, and the consequences of these for community and ecosystem properties. However, the conceptual basis of these chapters is also relevant to understanding what regulates biological diversity and the ecological consequences of this. Traditionally the study of mechanisms that determine diversity has had a predominantly aboveground focus, although the issue of belowground diversity or "soil biodiversity" has emerged rather suddenly as an area of considerable topicality. That notwithstanding, there have been few serious attempts to develop general principles about what regulates soil biodiversity or how aboveground and belowground diversity feed back to one another (but see Wardle 1999a; Hooper et al. 2000).

A major emphasis of this book so far has been on the biotic controls of ecosystem function, and I have provided numerous examples at a range of temporal and spatial scales on how aboveground and belowground biota, and their interactions, influence key ecosystem properties and processes. However, there is also a school of thought that regards certain components of biological diversity (notably, the number of species present within a given trophic level or grouping) as a fundamental ecological axis that determines ecosystem functioning in its own right. The whole issue of how biodiversity affects ecosystem function has snowballed since the publication of the synthesis volume edited by Schulze and Mooney (1993). A recent development in this topic is the so-called "diversity debate," with some researchers presenting the results of experimental studies as evidence that species diversity determines key ecosystem functions, and others questioning that evidence or emphasizing other evidence providing little support for such a causation (Kaiser 2000). Indeed, this book arrives at a very different conclusion based on the available evidence about the significance of diversity in regulating ecosystem function from that of another recently published book in this series (Kinzig et al. 2001). An approach that considers aboveground-belowground feedbacks is necessary for better understanding how diversity affects ecosystem function, because of the mutual dependence of the producer and decomposer subsystems on one another.

In this chapter, I will first consider the issue of how diversity is regulated in belowground communities and assess the linkages between aboveground and belowground diversity within this context. This will provide the background for the second issue I will consider, i.e., the effect of biological diversity on ecosystem properties and processes, including ecosystem stability, with special emphasis on the significance of aboveground-belowground linkages.

ASSESSMENT OF SOIL DIVERSITY

Before evaluating the regulation and functional significance of soil biodiversity, it is first necessary to briefly discuss the practicalities of assessing it. As outlined in chapter 1, soils contain an immensely greater diversity of organisms than that occurring aboveground, on both local and global scales. It is therefore apparent that assessment of species diversity of many groups of soil organisms should be much more difficult than assessing diversity of most aboveground groups. These problems are compounded by the fact that probably around only about 1% of bacteria and 3% of nematode species have been described (Klopatek et al. 1992), making measurement of species richness of these two groups effectively impossible. Further, only a small subset of soil bacterial species are potentially culturable, and analysis of 16S rRNA sequences in natural communities points to immense numbers of species of microbes that have never been cultured (Ward et al. 1990). Even for those groups of organisms for which identification of all species is possible, the sampling effort required to obtain a reliable estimate of total species richness is immense and frequently impractical. Christensen (1981) provides examples which demonstrate that the total number of microfungal species identified in a given soil sample increases continually with the number of isolates cultured from that sample, at least up to 500 isolates. Further, Linkel et al. (1992) used random simulations to demonstrate that in a community with 100 microbial species organized according to a lognormal frequency distribution, even assessment of thousands of isolates would probably not achieve an adequate enumeration of the species richness of that community.

Adequate assessment of species richness is feasible for soil macrofaunal groups, and with access to adequate taxonomic expertise for many mesofaunal groups. However, for many groups of smaller soil organisms, attempts at quantifying species richness are unlikely to yield a realistic picture. This therefore limits the types of questions that can be addressed for such organisms, as well as the kinds of experiments that can be used to evaluate the regulation and functional significance of their diversity. For soil microfauna, these issues can probably be adequately addressed for higher levels of taxonomic resolution, emphasizing diversity of genera or families. However, for the microflora, in particular bacteria, approaches with a very different conceptual basis are required. One approach is the analysis of composition of microbial (principally bacterial) phospholipid fatty acids (PLFAs) extracted from soil; different PLFAs originate from different subsets of the soil microflora (see Tunlid and White 1992). However, the majority of PLFAs used for evaluating microbial community structure are of bacterial, not fungal, origin. Other approaches involve the use of molecular techniques for assessing DNA or RNA directly extracted from soils; these have the advantage of enabling assessment of nonculturable micro-organisms (Tiedje et al. 1999; reviewed by Øvreås 2000). A third type of approach is based on evaluating microbial "functional diversity" through assessing the abilities of suspensions from soil containing microbes to utilize suites of substrates, e.g., through commercially available kits such as BIOLOG (see J. Zak et al. 1994) and ECOLOG. Despite its relative ease of use, this approach has limitations because it uses relatively simple substrates while many soil organisms use much more complex ones, and more importantly only a small subset of the microbial community can be assessed using this technique, i.e., those that grow most rapidly on simple substrates. For this reason, Øvreås (2000) concludes that this approach has limited value for soil ecological studies. A more realistic approach for assessing microbial functional diversity which goes part of the way to overcoming this latter problem is the in situ technique proposed by Degens and Harris (1997), in which the relative respiratory response of the soil microflora to a range of added substrates is assessed in the soil itself. However, this approach is still likely to be disproportionately influenced by faster-growing microbes that utilize low-molecular-weight substrates. It would appear that there is no method available that on its own yields an adequate representation of soil microbial community composition, and that a combination of methods is probably necessary for assessing the community structure of the soil microflora.

STRESS AND DISTURBANCE AS CONTROLS
OF SOIL DIVERSITY

The species diversity of a given group of organisms on a local scale can be influenced by a range of abiotic factors, and these can be broadly classified in terms of degree of stress (e.g., resource availability, temperature, pH, moisture status, levels of heavy metals) and disturbance (e.g., wildfire, cultivation, rapid climatic events) (see chapter 3). With regard to plants, diversity is characteristically low in extremely unproductive environments, and here amelioration of adverse conditions causes an increase in both productivity and diversity. However, as productivity continues to increase diversity frequently begins to decline, so that in highly productive conditions diversity is again low. The net result is the classic hump-backed relationship between productivity and diversity proposed by Grime (1973a,b) and subsequently validated by large numbers of observational and experimental studies (see Grace 1998; Waide et al. 1999). In this context, diversity is also often maximized by intermediate levels of disturbance, with reduction of disturbance beyond this intermediate level leading to domination by fewer species and declining diversity (Connell 1978; Huston 1979). While diversity reduction by extreme stress or disturbance is clearly because few species have the traits required for survival in very harsh conditions, several explanations have been proposed for the diversity reduction frequently observed under conditions that lead to high productivity (i.e., low stress, low disturbance) (Grace 1998); indeed Rosenzweig (1995) provides nine possible explanations for this decline. One plausible explanation for this (see chapter 3) is based upon the assumption that competition intensity is least under highly productive conditions, and that as productivity is reduced, greater coexistence of species is possible through reduced competitive exclusion. While Rosenzweig (1995) suggests that a drawback of this explanation is that it "needs a coherent theoretical treatment to formalize it and its set of predictions," no mention is made of the strategy theory of Grime (1977, 1979), which is based on the role of functional plant traits in conferring greater competitive ability upon plant species that dominate in productive environments. An alternative mechanism proposed for the decline in diversity, i.e., that lower spatial heterogeneity of resources in more productive (e.g., fertilized) habitats reduces organism diversity through reducing habitat variability (Tilman 1982), is not supported by the available empirical evidence [see the critique by Abrams (1995) of this concept and of the resource ratio theory used in support of it]. A further explanation involves "noninteraction" mechanisms (e.g., Abrams 1995; Oksanen 1996; Stevens and Carson 1999a,b) through increased productivity resulting in a reduction of plant diversity by reducing the number of plants present per unit area. Although experimental evidence has been presented in support of this explanation (Stevens and Carson 1999b), noninteraction mechanisms appear to be unlikely as explanations for several other unimodal diversity-productivity relationships documented in the literature (see Grime 1997; Rapson et al. 1997).

In any case, the unimodal pattern between diversity and productivity (and drivers of productivity such as stress and disturbance) is generally more likely to be important for those organisms for which bottom-up (competitive) rather than top-down forces are dominant. If this is indeed the case then some trophic groupings in the soil food web will show a unimodal diversity response to such factors as resource availability and quality, nutrient availability, pH, etc., while others will not. There are few belowground data sets that lend themselves to testing such a hypothesis, no doubt in part because of problems in assessing soil organism diversity. However, nine studies that do present relevant data are depicted in figure 6.1. Two studies have shown greater mi-

crobial functional diversity with increasing availability of belowground resources (Derry et al. 1999; Degens et al. 2000;
fig. 6.1a,b). This is inconsistent with the prediction that
high resource availability should reduce diversity, but both
studies employed substrate utilization approaches for measuring diversity, and these are likely to reflect mainly the
activity of faster-growing microbial groups (i.e., bacteria)
which are less likely to be regulated by competition than are
slower-growing groups (i.e., fungi) (see chapter 2). There is,
however, strong evidence from figure 6.1 that soil faunal
diversity does not decline as conditions become more favorable. Diversity of nematodes (a group hypothesized in chapter 2 not to be regulated by competition) declines monotonically with increasing lead concentration (Yeates and
Bongers 1999; fig. 6.1e), is not reduced by lime addition
(De Goede and Dekker 1993; fig. 6.1f), and drops monotonically with resource depletion during soil storage (Wright
and Coleman 1993; fig. 6.1g). In the last study, resource depletion actually caused the extinction of nematode taxa.
Further, in the study of Schaefer and Schauermann (1990)
the diversity of most groups of soil fauna was greater in a
mull *Fagus sylvatica* soil with relatively favorable conditions
for soil activity (i.e., pH of 4.3–6.8 and a high cation exchange capacity) than in a *F. sylvatica* moder soil with less
favorable conditions (i.e., pH of 3–4 and a low cation exchange capacity) (fig. 6.1h). In addition, data presented by
Paoletti (1988) (fig. 6.1i) point to a monotonic increase in
the species richness of chilopods and isopods with increasing soil nitrogen status. Of the nine studies depicted in fig,
6.1, all seven that involve decomposer organisms provide no
support for diversity decreasing along the most favorable
portions of stress gradients, and these organisms therefore
do not conform to the hump-backed model frequently demonstrated for aboveground biota, notably plants (see Grime
1973a; Rosenzweig 1995; Grace 1998; Waide et al. 1999). In
the case of soil faunal groups, these findings are more con-

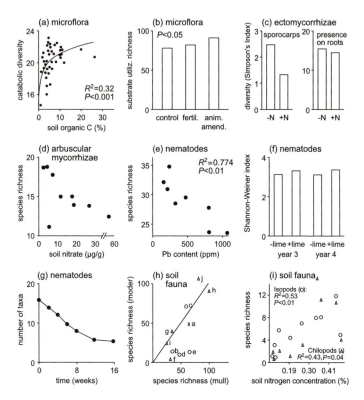

FIGURE 6.1. Diversity of soil organisms along stress gradients. (a) Relationship between catabolic diversity (calculated using the Simpson-Yule index) of the soil microbial biomass and soil organic carbon content, for a range of New Zealand soils. From Degens et al. (2000). (b) Soil substrate utilization richness (measured using BIOLOG) for soils from arctic Canada that are unamended, or amended by nutrient addition or herbivorous mammal activity. Only highest-temperature incubation data used. From Derry et al. (1999). (c) Diversity of mycorrhizae in a *Picea abies–Pinus sylvestris* forest in southwest Sweden, in nitrogen-amended ($+N$) and control ($-N$) plots. From Jonsson et al. (2000). (d) Arbuscular mycorrhizal fungal diversity across a nitrogen deposition gradient in sage scrub in California, U.S.A. Derived from tables 1 and 2 of Egerton-Warburton and Allen (2000). (e) Nematode species richness response to aerial lead pollution in Italy. From Zulinni and Peretti (1986) as adapted by Yeates and Bongers (1999). (f) Diversity of nematode genera three or four years after addition of lime in a *P. sylvestris* stand in the Netherlands. Determined from frequency data

sistent with the hypothesis outlined in chapter 2 that many groups of soil fauna are not strongly regulated by competition, and competitive exclusion is therefore less likely to occur when resource supply is increased.

Some groups of soil biota are clearly regulated by competition, however, and these include fungi (see chapter 2). The microbial substrate utilization measurements made by Degens et al. (2000) and Derry et al. (1999) (fig. 6.1a,b) are unlikely to have reflected fungal responses. The two fungal examples shown in fig. 6.1 both involve mycorrhizal fungi (Jonsson et al. 2000; Egerton-Warburton and Allen 2000; fig. 6.1c,d); in both cases nitrogen addition reduced their diversity. These results are consistent with increasing resource supply promoting competitive exclusion. Further, Giller et al. (1998) provide evidence from a study in Braunchweig, Germany, that genetic diversity of *Rhizobium leguminiosarum* bv. *trifolii* populations shows a hump-backed response to a heavy metal gradient, consistent with competitive exclusion operating at the most favorable end of the gradient.

Application of the "intermediate disturbance hypothesis" (*sensu* Connell 1978; see also Grime 1973a; Huston 1979) to belowground communities would predict that there should be conditions in which increasing disturbance should increase diversity through reducing competitive exclusion. Al-

---

presented by De Goede and Dekker (1993). (g) Response of diversity of nematode taxa to soil storage and depletion of food resources over time. Data averaged for all treatments. Derived from Wright and Coleman (1993). (h) Species richness for moder (Solling) vs. mull (Göttingen Wald) soils in German *Fagus sylvatica* forests, for each of ten taxa. Crosses, circles, and triangles represent microfaunal, mesofaunal, and macrofaunal groups, respectively. Codes: a, Testacea; b, Enchytraeidae; c, Cryptostigmata; d, Collembola; e, Gamasida; f, Gastropoda; g, Diptera larvae; h, Araenida; i, Carabidae; j, Staphylinidae. From Table 1 of Schaefer and Schauermann (1990). (i) Species richness of isopods and chilopods in relation to soil nitrogen concentration in woodlands and fields in northeast Italy. From Paoletti (1988).

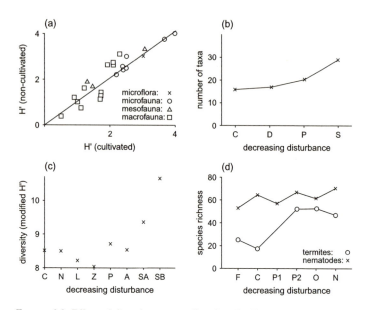

FIGURE 6.2. Effect of disturbances on diversity of soil organisms. (a) Shannon-Weiner diversity index (H′) for species or genera of groups of soil organisms in cultivated vs. noncultivated soil for each of 24 comparisons from 19 different published studies. Derived from data tabulated by Wardle (1995). (b) Numbers of nematode taxa identified along a disturbance gradient in South Australia. C indicates conventional tillage; D, direct drill (no tillage); P, pasture; S, native shrubland. From Yeates and Bird (1994). (c) Diversity (modified Shannon-Weiner diversity index) for nematode taxa along a disturbance gradient in Michigan, U.S.A. C indicates conventional tillage; N, no tillage; L, low input; Z, zero input; P, poplar planting; A, alfalfa planting; SA, natural succession "A"; SB, natural succession "B." From Freckman and Ettema (1993). (d) Numbers of species of nematodes and termites identified along a disturbance gradient in the Mbalmayo Forest Reserve, Cameroun. F indicates manually cleared farm fallow; C, complete clearance with young plantation; P1, partial mechanical clearance with plantation; P2, partial manual clearance with plantation; O, old-growth secondary forest; N, near-primary forest. From Lawton et al. (1998a).

though relatively few belowground studies have presented relevant data, little support for this hypothesis arises from those that have. For example, a synthesis by Wardle (1995) of 19 previous studies that have provided data on the effects of soil cultivation on taxonomic diversity of various groups (or community data from which diversity responses could be calculated) found that diversity was frequently adversely affected by disturbance, especially when larger soil fauna were considered (fig. 6.2a). However, there were several exceptions to this pattern, and no significant overall effect of cultivation on diversity across the 19 studies. Those studies that have assessed soil faunal diversity across gradients of land disturbance provide little evidence for optimization of diversity along intermediate portions of these gradients. Yeates and Bird (1994) and Freckman and Ettema (1993) both found diversity of soil nematodes to increase monotonically along gradients of decreasing disturbance ranging from intensive agricultural systems to natural ecosystems (fig. 6.2b,c). Further, studies in Cameroun involving gradients of disturbance ranging from intensive management to near-natural forest show either neutral or negative effects of increasing disturbance on species richness of soil termites (Lawton et al. 1998a; Davies et al. 1999) and soil nematodes (Bloemers et al. 1997; Lawton et al. 1998a) (fig. 6.2d). These patterns are consistent with trends identified along stress gradients (see above) in providing little evidence of reduction of diversity of soil biota by favorable conditions, and are again explicable in terms of the unimportance of competition in structuring communities of many of the groups of soil biota. The available belowground data are also consistent with the predictions of Huston (1994) and the results of the theoretical modeling study of Wootton (1998) in suggesting that diversity within higher trophic levels is less likely than plant diversity to show predictable optimization by intermediate levels of disturbance.

193

BIOTIC CONTROLS OF DIVERSITY

The diversity of a given group of organisms can be influenced by the activities of other organisms, and this has often been demonstrated in aboveground and aquatic systems. The mechanisms involved can include predation, competition, and habitat modification. The diversity within a given trophic level can be strongly modified by predation by organisms in the next higher trophic level. For example, in a literature survey of 44 previous studies, Proulx and Mazumder (1998) found that herbivores generally increased plant richness under fertile conditions and reduced it under infertile conditions. Species richness within consumer groups can either increase or decrease in response to the presence of their predators, and a variety of mechanisms can be involved, including effects of predators on competitive interactions among species in the lower trophic level (see Paine 1966; Holt and Lawton 1994; Schoener and Spiller 1996; Gurevitch et al. 2000). Competition in its own right can also be a powerful determinant of diversity. For example, removal experiments demonstrate that the loss of dominant species within a trophic level can increase the diversity (including species richness) of the remainder of that trophic level; this has been shown to occur for both plants (Goldberg and Barton 1992; Wardle et al. 1999b) and animals (Valone and Brown 1995). It has also been proposed through a modeling approach that mutualistic interactions can theoretically enhance the diversity within each of the guilds of organisms that participate in the mutualism (Bever 1999), although there is a dearth of available data about whether and how mutualism influences diversity. Finally, certain organisms can affect the diversity of other organisms by determining the structure of their habitat. There is abundant evidence that habitat diversity is an important determinant of organism diversity (MacArthur 1964; Huston 1994; Rosenzweig 1995), and organisms that affect habitat diversity therefore affect the diversity of organisms that occupy that habitat. For example, the diversity of birds in forests is well known to reflect the diversity of habitat provided by vegetation structure (MacArthur 1964; Recher 1969). I will now discuss the biotic controls of soil organism diversity, and the effects of soil biota on plant diversity, against the background of these mechanisms.

Biotic Controls of Soil Diversity

Despite the large number of studies investigating how biotic interactions influence diversity in aboveground and aquatic food webs, there have been relatively few attempts to investigate how diversity of soil organisms is affected by belowground biotic interactions. However, McLean et al. (1996) found no effect of addition of an oribatid mite and a collembolan, either singly or in combination, on the fungal diversity of pine needles incubated in microcosms. It appears inevitable, however, that there should be instances in which fungal-feeding fauna influence fungal diversity, given the role of competition in structuring fungal communities coupled with the selectivity of fungal feeders for certain fungal species (Newell 1984; see chapter 2). A handful of studies have also investigated the effects of belowground ecosystem engineers on diversity of other belowground consumers through habitat modification. For example, Tiwari and Mishra (1993) found that fungal species diversity was greater in earthworm casts than in surrounding soil (fig. 6.3a), and Loranger et al. (1998) detected greater species diversity of oribatid mites in patches with high levels of earthworms (fig. 6.3b). A further example involves the invasion of the alien epigeic earthworm *Dendrobaena octaedra* into *Pinus contorta* forests of southwest Alberta, Canada (discussed further in chapter 7). In a mesocosm study, McLean and Parkinson (1998a) showed that *D. octaedra* could alter fungal community structure and in some instances promote fungal species diversity, apparently through reducing competition intensity

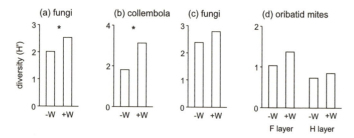

FIGURE 6.3. Effect of earthworm presence (+W) vs. absence (−W) on diversity of other soil organisms, expressed as the Shannon-Weiner diversity index. (a) Fungal species diversity in earthworm casts vs. mineral soil near Shillong, India. From Tiwari and Mishra (1993). (b) Collembolan species diversity in zones of high vs. low density of the earthworm *Polypheretima elongata* at St. Anne, Martinique. From Loranger et al. (1998). (c) and (d) Species diversity of fungi and oribatid mites in mesocosms amended vs. nonamended with the earthworm *Dendrobaena octaedra*. From McLean and Parkinson (1998a,b); only data for six-month measurements shown. * indicates earthworm effect significant at $P = 0.05$.

among fungal species, although the effects on diversity were rarely statistically significant (fig. 6.3c). In a corresponding field experiment, McLean and Parkinson (2000a) determined that addition of *D. octradea* could also exert some negative effects against fungal species diversity, and appeared to promote domination by faster-growing fungal species. A further mesocosm experiment showed that *D. octradea* addition could increase diversity of oribatid mites in some instances and influence the relative dominance of different mite species (McLean and Parkinson 1998b) (fig. 6.3d).

There is a number of examples demonstrating that the species identity of vegetation can influence the diversity of components of the soil biota. For example, Badejo and Tian (1999) found that in agroforestry plots in southwest Nigeria, monospecific stands of *Leucaena leucocephala* supported a higher diversity of mite genera than did stands of three other tree species. Hansen (1999) determined that litter of *Quercus rubra* in a deciduous forest in North Carolina, U.S.A., contained a greater species richness of oribatid mites than did litter of *Betula alleganiensis* or *Acer saccharinum.* In southern France, David et al. (1999) observed that shrub-dominated sites had a nearly twofold higher species richness of macrofauna than did forested sites. Arbuscular mycorrhizal fungal species richness on plant roots was found by Eom et al. (2000) to differ across five different host plant species in a tallgrass prairie in Kansas, U.S.A. Further, Wardle et al. (1999b) found, using a removal experiment in a grassland near Hamilton, New Zealand, that both the abundance and diversity of decomposer organisms of each of three trophic levels of a soil food web (microbes, microbe-feeding nematodes, top predatory nematodes) were sometimes responsive to which plant functional group was removed. The role of vegetation composition in affecting diversity of soil organisms is also apparent from studies that have investigated soil biological diversity along successional gradients (in which different stages are dominated by different plant functional types) and this has been demonstrated for a range of groups including microbes (e.g., Skujins and Klubek 1982; Frankland 1998), nematodes (e.g., Wasilewska 1994), and arthropods (e.g., Paquin and Coderre 1997).

The mechanistic basis by which plant species affects diversity of soil organisms has been little studied. However, plant species differ with regard to their effects on both the stress regime and the degree of disturbance in the soil environment (chapter 3), and given that diversity of soil organisms can be determined by factors known to be influenced by plants [e.g., organic matter content (fig. 6.1a,b,g), nitrogen concentration (fig. 6.1c,d,i), pH (fig. 6.1f), and disturbance regime (fig. 6.2)], it seems reasonable that plant species effects should serve as important drivers of soil biodiversity. Another likely explanation is through plant species differing in their effects on soil habitat heterogeneity. Few studies have explicitly investigated the effects of habitat diversity on soil biotic diversity, but some evidence points clearly to the importance of this. Anderson (1978) quantified the diversity

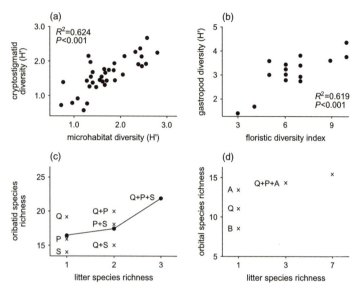

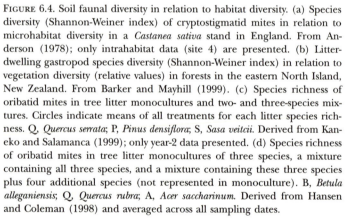

FIGURE 6.4. Soil faunal diversity in relation to habitat diversity. (a) Species diversity (Shannon-Weiner index) of cryptostigmatid mites in relation to microhabitat diversity in a *Castanea sativa* stand in England. From Anderson (1978); only intrahabitat data (site 4) are presented. (b) Litter-dwelling gastropod species diversity (Shannon-Weiner index) in relation to vegetation diversity (relative values) in forests in the eastern North Island, New Zealand. From Barker and Mayhill (1999). (c) Species richness of oribatid mites in tree litter monocultures and two- and three-species mixtures. Circles indicate means of all treatments for each litter species richness. Q, *Quercus serrata*; P, *Pinus densiflora*; S, *Sasa veitcii*. Derived from Kaneko and Salamanca (1999); only year-2 data presented. (d) Species richness of oribatid mites in tree litter monocultures of three species, a mixture containing all three species, and a mixture containing these three species plus four additional species (not represented in monoculture). B, *Betula alleganiensis*; Q, *Quercus rubra*; A, *Acer saccharinum*. Derived from Hansen and Coleman (1998) and averaged across all sampling dates.

of microhabitat categories in *Castanea sativa* woodlands in England, and found that cryptostigmatid mite diversity reflected this habitat diversity (fig. 6.4a). In New Zealand forests, Barker and Mayhill (1999) showed that species diversity of litter-dwelling gastropods increased with increasing floristic diversity (fig. 6.4b), probably through plant diversity determining habitat diversity in the litter layer. Two litter mixing experiments also provide evidence that increasing habitat diversity can increase soil biological diversity. In the first (Kaneko and Salamanca 1999), litter bags were prepared that contained all the possible one-, two- and three-way combinations of litter from three tree species. Increasing litter diversity generally increased the diversity of oribatid mites in the litter bags (fig. 6.4c). In the second (Hansen and Coleman 1998; Hansen 2000), plots were established in a forest in North Carolina, U.S.A., in which resident litter on the forest floor was removed and replaced with each of five litter treatments, i.e., monoculture litters of each of three tree species, a mixture of all three litter types, and a seven-species mixture treatment including these three species plus four additional species. Diversity of oribatid mites was mites was greater in the three-species mixture than in the corresponding monocultures (Hansen and Coleman 1998) (fig. 6.4d). The seven-species mixture sometimes supported a higher mite diversity still, although the possibility remains that this diversity increase may have been due to litter characteristics of one or more of the four species not present in monoculture, rather than because of a habitat diversity effect.

Given that effects of higher aboveground trophic levels (notably herbivores) can influence vegetation properties, resource quality, and soil biota (chapter 4), it is also reasonable to expect that diversity of soil biota could be affected by aboveground consumers. Few studies have provided relevant data, but Suominen (1999) showed for 23 exclosure plots (designed to exclude cervids) in Sweden and Finland that browsing mammals reduced species richness of ground-dwelling gastropods by on average 12.8%. However, Stark et al. (2000) found that grazing by reindeer (*Rangifer tarandus*) increased the diversity of soil nematodes. Further, in grasslands in Colorado, U.S.A., grazing by cattle (*Bos taurus*) was found to have little effect on the diversity of soil microarthropods (Milchunas et al. 1998) or nematodes (Wall-

Freckman and Huang 1998). Wardle et al. (2001) investigated the effects of browsing mammals on aboveground and belowground communities in each of 30 rainforest locations throughout New Zealand, and found that, while browsing generally reduced vegetation diversity, there was a range of response to browsing for the diversity of each of several soil faunal groups depending upon location. It is to be expected that herbivores would have a variety of effects on the diversity of soil organisms which may be positive, negative, or neutral, given the variety of mechanisms by which herbivores can either increase or reduce the quantity and quality of organic matter entering the decomposer subsystem (see chapter 4).

Soil Biotic Controls of Plant Diversity

Belowground organisms that influence plant growth and community structure have the potential to also influence plant diversity; indeed most of the examples of belowground biotic effects on plant communities described in chapter 5 could conceivably alter plant diversity. Regulation of nutrient supply by decomposer microflora and fauna would almost certainly affect the competitive balance between plant species and thus influence plant diversity, especially given the hump-backed relationship between soil nutrient availability and plant diversity that has been demonstrated in a large number of studies. However, to date there have been no direct experimental tests of the effects of decomposer communities on plant diversity. Larger soil organisms (including "ecosystem engineers") promote vegetation on the structures that they create which differs in functional composition from the surrounding vegetation [e.g., Whitford and DiMarco (1995) for ants; Spain and McIvor (1988) for termites; Hobbs and Mooney (1985) for burrowing mammals; see also chapter 5] and in this light a response of plant diversity to these organisms seems inevitable. In particular, organisms that create structures clearly promote greater habitat heterogeniety in the landscape, presumably leading to a greater diversity of plant species.

Organisms associated with plant roots can also influence plant diversity. In the Silwood Park grassland experiment described in chapter 5, Brown and Gange (1989b) found root feeding by insect herbivores to depress plant species richness largely through reducing establishment of forbs. Meanwhile, Wardle and Barker (1997) determined that adding root herbivores (larvae of beetles of the Scarabidae) to mesocosms had little effect on the diversity of grassland vegetation that became established in them. Root pathogens that are specialized for certain plant species may promote plant diversity in situations where they induce the decline of dominant species (Van der Putten 2001), and this is analogous to the Janzen-Connell hypothesis (Janzen 1970; Connell 1978), which predicts greater tree diversity in tropical forests in the presence of specialist herbivores. This effect is especially apparent in the study of Packer and Clay (2000) (see chapter 5) in which pathogens prevented seedling establishment of *Prunus serotina* under adult *P. serotina* trees, encouraging the juveniles of other species to establish themselves instead. Mycorrhizal fungi can influence (and reduce) the intensity of competition between plant species (discussed in chapter 5), resulting in a greater coexistence of plant species being possible and therefore a greater diversity; positive effects of arbuscular mycorrhizae on grassland plant diversity have been demonstrated by Grime et al. (1987) and by Van der Heijden et al. (1998b). However, the reverse effect is also possible, and the studies of Hartnett and Wilson (1999) and Smith et al. (1999) in a tallgrass prairie in Kansas, U.S.A., showed that mycorrhizae promoted obligately mycorrhizal C4 grasses, resulting in competitive exclusion of facultatively mycorrhizal C3 species, reducing overall diversity. A similar mechanism appears to operate in some tropical rainforests, in which ectomycorrhizal trees competitively exclude arbuscular mycorrhizal tree species, leading to lower overall tree diversity (Connell and Lowman 1989).

Diversity Linkages across Trophic Levels

If habitat diversity is a determinant of the diversity of consumer organisms, then plant diversity may be able to influence that of higher trophic levels. Several studies have demonstrated correlations between plant diversity and diversity of both aboveground consumers (e.g., Murdoch et al. 1972; Southwood et al. 1979; Siemann 1998) and soil consumers (Wardle et al. 1999b; Hooper et al. 2000; but see Broughton and Gross 2000), although it is unclear in each case whether direct causation between plant diversity and consumer diversity is involved, or whether both plant and consumer diversity are merely responding in the same direction to an underlying third factor. Few studies have investigated the effect of directly manipulating plant diversity on consumer diversity. Siemann et al. (1998) and Knops et al. (1999) found for two field experiments in a prairie at Cedar Creek, Minnesota, U.S.A. (described by Tilman et al. 1996, 1997a), that plots sown with increasing numbers of plant species supported a higher species richness of aboveground consumers. However, as will be described later in this chapter, these Cedar Creek experiments have experimental design problems that may confound their conclusions. The issue of how plant diversity affects decomposer diversity has rarely been experimentally tested, although the studies of Hansen and Coleman (1998) and Kaneko and Salamanca (1999) provide evidence that increasing tree litter species diversity from one to three species can increase oribatid mite diversity (fig. 6.4b,c). For plant diversity to promote decomposer diversity requires that plant diversity promotes shifts in total resource quality, quality, or resource (and therefore habitat) diversity of those substrates entering the belowground subsystem. Whether diversity does alter these factors is unclear;

as discussed later in this chapter there is little undisputed evidence of increasing diversity significantly increasing the input of resources to the ecosystem. Further, little is known about whether plant diversity significantly increases the heterogeneity of resources entering the belowground subsystem. However, it does appear likely that plant species differ considerably in their effects on habitat diversity; one species of tree would probably be able to provide a wider range of habitats for decomposers than would several species of grasses.

The potential effects of diversity of soil organisms on plant diversity also remain largely uninvestigated. Van der Heijden et al. (1998b) conducted an experiment in which including a greater species diversity of arbuscular mycorrhizal fungal species resulted in a greater plant species diversity. However, this experiment employed the same type of design as that of Tilman et al. (1996) and as outlined later in this chapter unfortunately suffers from the same problems of interpretation.

THE ENIGMA OF SOIL DIVERSITY

As described earlier, the belowground substratum consists of an extremely diverse biota when compared with most other systems. This has led to the question of the "enigma of soil diversity" as posed by Anderson (1975), or the issue of how it is possible for such large numbers of species to apparently coexist without biotic mechanisms (e.g., competitive exclusion) reducing diversity. This is analogous to the "paradox of the plankton" presented by Hutchinson (1961) for aquatic systems, in relation to how it is possible for large numbers of species to coexist in the same environment and utilize the same resources without competitive exclusion operating. For soils at least, one possible explanation is the nature of habitat and substrate heterogeneity created by the physical environment, coupled with considerable resource partitioning among the organisms that occupy this environment (see Schoener 1974). The three-dimensional structure of soil may promote greater species coexistence by encouraging resource partitioning through providing a wide range of surface types, pore sizes, and microclimates at a spatial scale relevant to soil biota, as well as by reducing competitive exclusion through enabling the physical separation of organisms in the soil matrix. With regard to resource acquisition, while some soil biota are highly generalist consumers (e.g., bacterial-feeding microfauna, some saprophagous fauna) others can be quite selective (e.g., fungi, fungal-feeding micro- and mesofauna) (see chapter 2). Indeed, Scheu and Falca (2000) demonstrated that for certain groups of both soil mesofauna and macrofauna, the component species showed a continuum of $\delta^{15}N$ values, indicative of considerable resource partitioning with regard to food. Other axes of resource partitioning that have been found to be important for groups of soil organisms include phenological differentiation (e.g., Vegter 1987), depth distribution (e.g., Faber and Joose 1993), microclimatic properties (e.g., Widden 1984), and soil chemical properties (e.g., Van Straalen and Verhoef 1997).

It is therefore apparent that the belowground environment provides a range of niche axes along each of which species can specialize so as to minimize competition, and when all the possible combinations of positions on these axes is considered, an enormous diversity of soil organisms is possible. Further, conditions in the soil are frequently unfavorable for the activity of soil organisms, and as a result the majority of the soil microflora is usually dormant (Jenkinson and Ladd 1981) and many groups of soil organisms can undergo lengthy phases of inactivity [e.g., nematodes through anhydrobiosis (see Freckman and Womersley 1983) and arthropods (see Joose and Verhoef 1987)]. Coexistence of large numbers of species is therefore frequently possible at least for groups consisting of smaller soil organisms simply because many component organisms are often relatively inactive at any one time. When conditions suddenly become favorable (e.g., through rewetting dry soil; addition of a substrate pulse) these organisms rapidly activate and grow but are probably still at sufficiently low levels to avoid competitive exclusion. A further adaptation that promotes diversity, at least amongst microbes, is their ability to evolve rapidly to take advantage of new opportunities and changing conditions, and there is recent evidence that bacterial species can undergo adaptive radiation in a matter of days as substrate complexity is increased (Rainey and Travisano 1998; Rainey et al. 2000). In conclusion, the high levels of biodiversity characteristic of most soils can be explained by the heterogeneity of the habitat, the immense capacity for resource partitioning along several axes in multidimensional space, the spatial isolation of potentially competing organisms from each other in the soil matrix, and the ability of soil organisms to remain inactive during harsh conditions and rapidly adapt to changing conditions. These all ultimately work against competitive exclusion for most components of the soil biota, and therefore promote greater coexistence of species.

DIVERSITY OF SOIL ORGANISMS
OVER LARGER SPATIAL SCALES

It is well known that the species richness of many taxa increases toward the equator, and in the words of Taylor and Gaines (1999) this is "among the most prominent features of the natural world." Despite this, there is little agreement among scientists as to the mechanisms underlying this pattern. A range of explanations have been put forward, including species geographical ranges declining toward the equator (Rapoport 1982, modified by Stevens 1989), latitudinal gradients in habitat characteristics (e.g., habitat heterogeneity, productivity, soil fertility, nature and intensity of biotic interactions) (reviewed by Rohde 1992), increased speciation rates in the tropics (Rohde et al. 1993), and lower latitudes occupying a larger area than higher latitudes (Rosenzweig 1995; but see Chown and Gaston 2000). The issue remains unresolved, and further "the number of hypotheses about species diversity far outstrips the data available to test these hypotheses" (Huston 1999).

The above hypotheses are based on observations of mainly aboveground and marine biota. In contrast, the soil biota has been largely ignored in developing these theories, despite soils arguably containing the majority of terrestrial species. Data on how diversity of soil organisms varies with latitude remain scant and difficult to assess for many groups given the paucity of detailed taxonomic data available coupled with the fact that most species of soil organisms remain undescribed. This issue is emphasized by Rusek (1998), who shows that the number of species of Collembola described from Europe is currently increasing exponentially (fig. 6.5a). Based on the available evidence, it is clear that termite diversity declines along large latitudinal gradients (Eggleton and Bignall 1995; fig. 6.5e) but this pattern does not appear to hold for most other groups of soil-dwelling biota, such as ciliate protozoa (Foissner 1997; fig. 6.5b), or nematodes and earthworms (Lavelle et al. 1995; Boag and Yeates 1998; fig. 6.5d,f). Further, the Shannon-Weiner diversity index for fungal species diversity was found by Kjøller and Struwe (1982) to be no less for Arctic tundra soils than for temperate grassland and forest soils. For many groups the only apparent decline in diversity with increasing latitude occurs at the very high end of the latitudinal gradient (i.e., above 60°; see fig. 6.5d for nematodes and fig. 6.5f for earthworms). Further, diversity in sub-Antarctic and Antarctic areas declines sharply with decreasing temperature or increasing latitude for both protozoa (Smith 1996; fig. 6.5c) and nematodes (Wall and Virginia 1999).

Over the 0°–60° latitudinal range, why does the diversity

*more spp
described
over
time* →

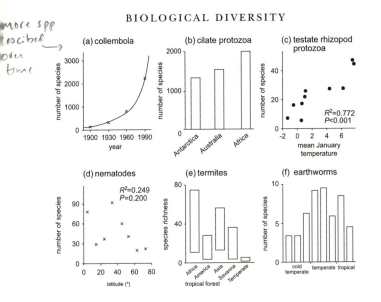

FIGURE 6.5. Diversity of soil faunal groups in relation to biogeography and latitude. (a) Total number of collembolan species described in Europe over the past century. From Rusek (1998). (b) Estimated numbers of species of ciliate protozoa in three continents. From Foissner (1997). (c) Numbers of species of testate rhizopod protozoa on sub-Antarctic and Antarctic islands (each island represented by a point), in relation to temperature (indicative of latitude). From Smith (1996). (d) Estimated species richness per standardized soil samples in relation to latitude, derived from a literature synthesis. From Boag and Yeates (1998). (e) Typical range of termite species richness in various ecosystem types. From Eggleton and Bignall (1995) and Lavelle et al. (1997). (f) Earthworm species richness measured in sites along a thermo-latitudinal gradient. From Lavelle et al. (1995).

of all soil organism groups (except termites) not decline with latitude in the manner shown for so many above-ground groups? I identify two reasons. First, with increasing latitude there is a general trend of organic matter accumulation, and higher amounts of carbon and nutrients are stored in the soil relative to the amount in the phytomass (Swift et al. 1979). Greater humus depth may promote both greater habitat heterogeneity and greater net amounts of nutrients present in the soil; both these factors should promote soil diversity (see the earlier discussion). Huston (1994, 1999) proposes that the decline of diversity with increasing latitude is explicable in terms of the hump-backed relationship between diversity and fertility, with fertility, on balance, being least at low latitudes, leading to greater species coexistence and therefore greater diversity. Diversity of belowground biota generally shows a monotonic increase, not a hump-backed relationship, with increasing fertility and substrate availability, and if the hypothesis of Huston is correct, then the diversity of these groups should therefore be promoted at higher latitudes. Secondly, diversity of soil organisms may be governed by local factors rather than by regional pool size (see Ghilarov 1977). Existing data are scarce, but analysis of community data of Collembola in Austrian grasslands points to communities being saturated, with local species richness being independent of regional pool richness (Winkler and Kampichler 2000). Further, for smaller soil organisms, notably microflora and microfauna, many (perhaps most) species have a pandemic distribution and can be transported globally. Finlay et al. (1999) determined that all but one of 85 species of ciliate protozoa found in the sediments of a crater lake in Victoria, Australia, are also known from Europe. It has been suggested that a high migration rate of soil bacteria exists across continents (Roberts and Cohen 1995). Despite the large numbers of undescribed nematode species, many are probably also cosmopolitan; recent work on criconematid nematodes in New Zealand indicates several species that are also known from specific locations in other parts of the world [one is known only from the Chatham Islands of New Zealand and the Falkland Islands (Wouts et al. 1999; Yeates 2001)]. These studies again point to many groups of soil biota [i.e., those less than 1 mm in body size according to Finlay et al. (1999)] being potentially ubiquitous globally, and again suggest that diversity of species on a local scale is more likely to be governed by local factors (e.g., substrate quality, habitat heterogeneity) than limited by the size of the regional pool. Some mechanisms that have been proposed as to why aboveground diversity increases toward the tropics could conceivably also apply to the soil biota, but the two reasons I give should work in the reverse direction, and it is the balance of these factors which may explain why diversity of most groups of soil organisms need not increase with decreasing latitude.

BIODIVERSITY AND ECOSYSTEM FUNCTION

Having discussed the factors that regulate biological diversity in an aboveground-belowground context, I will now discuss the functional significance of this diversity. There are four aspects of the diversity-function issue which are relevant to aboveground-belowground linkages, i.e., how plant diversity influences NPP (which then determines resource input to the soil), how plant diversity influences soil processes, how diversity of soil organisms affects processes both aboveground and belowground, and whether diversity (both aboveground and belowground) influences the stability of ecosystem properties and processes. Each of these issues is now considered in turn.

Plant Diversity and Ecosystem Productivity

A key issue in understanding whether and how species richness affects ecosystem productivity involves evaluation of whether communities have a high level of functional redundancy (Walker 1992) with species richness effects saturating at diversity levels below that occurring in most biological communities, or whether species richness may have effects that saturate at much higher species richness levels [cf. the "rivet popper hypothesis" (Ehrlich and Ehrlich 1981)] and therefore operates as an important ecosystem driver (fig. 6.6b) (discussed by Vitousek and Hooper 1993; Lawton 1994; Schläpfer and Schmid 1999; Schwartz et al. 2000). A

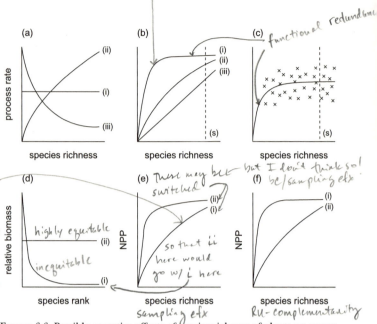

[handwritten, top] don't necessarily need a lot of spp for ↑ in process rate to saturate

[handwritten, near (c)] functional redundance

[handwritten, near (d), left margin] Because there is a high ° of inequitability, you need to take more samples (x→) before you randomly get the one(s) that are highly productive. Whereas w/ii species are highly equitable so smaller # of sample will be needed to include highly productive spp. See p.220. As equitability↑ sampling efx↓

[handwritten, plot a labels] (ii), (i), (iii)

[handwritten, plot b labels] (i), (ii), (iii), (s)

[handwritten, plot c labels] (s)

[handwritten, plot d labels] highly equitable (ii), inequitable (i)

[handwritten, near (e)] There may be but I don't think so! be/sampling efx

[handwritten, near (e)] switched

[handwritten, (e) plot] (ii), (i), so that ii here would go w/ i here

[handwritten, below (e)] sampling efx

[handwritten, below (f)] RU-complementarity

[handwritten, (f) plot labels] (i), (ii)

FIGURE 6.6. Possible causative effects of species richness of plants on ecosystem process rates. (a) Null effects models. (i) represents no effect of species richness on process rates; (ii) and (iii) are possible null models arising from randomly assembled communities which are explicable entirely in terms of "sampling effect." (ii) represents a situation in which the most productive species are also the most competitive (cf. Gaudet and Keddy 1988); (iii) occurs when the most competitive species are less productive than the other species, resulting in sampling effect causing reduced productivity through increasing domination by unproductive competitors as species richness increases (see Hector 2000; Troumbis et al. 2000). (b) Patterns of relationships between species richness and processes that may occur over and above "sampling effect," for example, through resource use complementarity or facilitation. (i) represents functional redundancy, in which species richness effects saturate at levels well below that which normally occurs in real communities (i.e., "S"). (ii) and (iii) represent situations in which species richness may operate as an important ecosystem driver. (c) A special case of the "redundancy model," i.e., the idiosyncratic hypothesis of Lawton (1994) in which addition of species (represented as crosses) can have either positive or negative effects depending upon their identity. Species richness effects are unimportant, but species identity becomes increasingly important with increasing variance of the crosses around the solid line. (d) Dominance-diversity relationships (*sensu* Whittaker 1975), ranging from highly unequitable with few species occupying most of the community biomass (i) to highly equitable (ii).

[handwritten, bottom] We don't necessarily know the relationship btwn spp equitability & functional redundancy (d & e lines ii)

further possible relationship involves the idiosyncratic hypothesis of Lawton (1994), which represents situations in which there is high functional redundancy, but where effects of adding new species can either promote or reduce process rates depending upon their identity (fig. 6.6c).

In ecology, one of the primary goals is to better understand pattern and process in nature. Therefore it seems intuitive that in evaluating whether NPP is influenced by plant species richness we should first look toward patterns in natural systems to determine whether there is apparent evidence for such a relationship. If species richness is a key driver of NPP then there should be several examples in nature in which such effects are apparent. However, the hump-backed model between species richness and productivity or biomass (Grime 1973a,b; see earlier discussion) would predict that plant communities that are the most diverse are frequently relatively unproductive, while low-diversity communities are often dominated by productive species that competitively exclude other species. This pattern appears to be widespread; indeed, a literature synthesis of observational data for real ecosystems by Waide et al. (1999) found that at "local" spatial scales (i.e., those below 20 km) over 40% of studies showed a unimodal relationship between plant diversity and NPP, with only 7% finding a positive relationship. Positive relationships between NPP and diversity were common only for studies considering continental to global spatial scales.

The hump-backed model would predict that there should be productivity ranges over which species richness and pro-

(e) Species richness–productivity relationships as manifested through "sampling effect," for situations in which the most productive species are also the most competitive. (i) and (ii) correspond to curves (i) and (ii) in panel (d). (f) Species richness effects on productivity as manifested through resource use complementarity with resource partitioning between species. (i) and (ii) correspond to curves (i) and (ii) in panel (d).

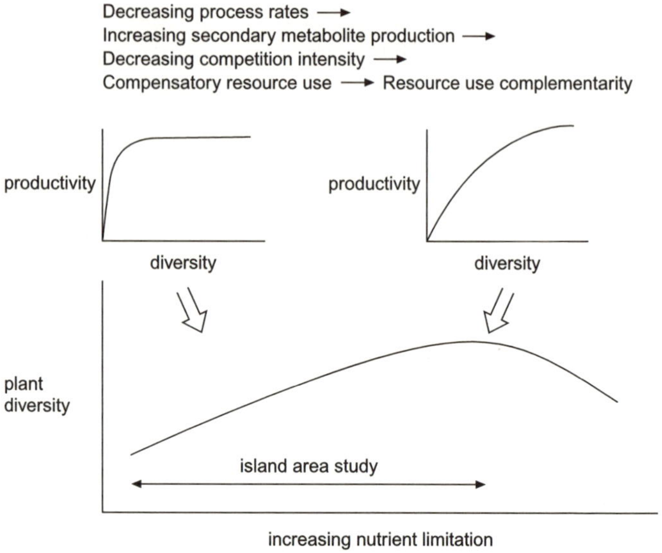

Decreasing process rates ⟶
Increasing secondary metabolite production ⟶
Decreasing competition intensity ⟶
Compensatory resource use ⟶ Resource use complementarity

increasing nutrient limitation

FIGURE 6.7. Relationships between diversity and ecosystem function in relation to the hump-backed model of Grime (1973a,b) and applied to the lake island study of Wardle et al. (1997b). Under low nutrient limitation, domination by faster-growing competitive species (and those that promote rapid ecosystem-level process rates) results in competitive exclusion of subordinate species and reduced diversity; increasing nutrient limitation reduces these competitive species, enabling greater species coexistence and therefore diversity, but domination by those species that retard ecosystem-level process rates. Note that island area decreases with increasing nutrient limitation; see text for explanation of study.

ductivity (and related processes) are negatively related. Evidence for this is apparent from the results of Wardle et al. (1997b) for a series of 50 lake islands in northern Sweden (fig. 6.7; see also chapter 3). Here, large islands that contain earlier successional vegetation support a greater standing plant biomass, greater process rates, and lower plant diversity, probably through competitive exclusion exerted by the dominant species. The smaller islands support later successional plant species which grow slowly and invest in production of phenolics, resulting in retardation of ecosystem process rates and a greater diversity of plant species, presumably as a result of reduced intensity of competition. This would support the view that plant diversity is unimportant as a mechanism in driving variation in ecosystem properties across these islands. Although interpretation of this particular study has been criticized on the basis of its correlational nature and the possibility of confounding factors therefore influencing the observed pattern (e.g., by Tilman et al. 1997b; Lawton et al. 1998b), the reality remains that the patterns of ecosystem functioning across these islands can be explained easily in terms of differences in ecophysiological traits of the dominant plant species present, and cannot be explained by any mechanism that involves the effects of species richness (Wardle et al. 1997c). In contrast, one recent study (Troumbis and Memtas 2000) claimed to provide evidence from a species-poor Mediterranean shrubland for diversity increasing productivity because the two were spatially correlated. Although no consideration was given as to what factors may have regulated diversity and productivity, the sites that they used were extremely unproductive (aboveground NPP for all sites was less than 20 g/m^2/yr) and well within the productivity range for which productivity and diversity should be correlated according to the hump-backed model (*sensu* Grime 1973a,b) simply because both variables should be limited by the same factors without a causative relationship existing between the two.

Further observational evidence for the lack of a causative relationship between plant species richness and biomass production is apparent in a recent study by Enquist and Niklaus (2001), based on an extensive global data set of forested ecosystems ranging from monospecific stands to some of the Earth's most diverse forests (Gentry 1988). This study provides compelling evidence that forest standing biomass and size-frequency distributions are entirely independent of tree diversity and are instead consistent with predictions of basic allometric principles operating at the level of the individual. In total, there is little evidence in natural communities that plant species richness promotes ecosystem process rates at local spatial scales, at least beyond very small numbers of species. Indeed, some of the world's most productive natural communities are dominated by a single species (Holdgate 1996).

Experimental investigation of the effects of increasing diversity at the lowest end of the diversity spectrum (i.e., from one to two species) has had a long history, especially in the agronomic literature (see Vandermeer 1990). Two-species mixtures appear to be generally more productive on average than monocultures but the effects are not strong; in an analysis of 344 published comparisons Trenbath (1974) found productivity of mixtures to be greater than that of the mean of the component species in monoculture in 60.2% of cases, and this was statistically significantly greater than the 50% of cases that we would expect by chance. Further, Joliffe (1997) found upon synthesizing the results of 54 published experiments that two-species mixtures performed better than monocultures in 38 experiments, and that mixtures were on average 12% more productive than monocultures. On balance, it therefore appears that increasing plant diversity from one to two species has positive though relatively weak effects on NPP. This would suggest support for curve (i) in figure 6.6b, although many of these experiments have limitations through being able to detect only short-term interactions and being restricted to herbaceous (and often agricultural) plant species. The positive effects that arise in such experiments occur either through resource use complementarity (with coexisting species having partially nonoverlapping niches) or through facilitation and positive interactions (especially when one of the two plant species is a nitrogen fixing legume; see Vandermeer 1990).

If increasing diversity from one to two species has generally weak effects, then we would expect species richness effects on productivity to saturate at very low species richness levels. However, the first experimental study to investigate the effects of species richness on ecosystem properties across a wide range of diversities (Naeem et al. 1994) instead claimed to provide evidence of large positive effects of increasing diversity across four trophic groupings (including increasing plant species richness from two to 16 species) on a range of properties related to NPP. The interpretation of these results has been intensely debated (e.g., André et al. 1994; Garnier et al. 1997; Huston 1997; Hodgson et al. 1998; Lawton et al. 1998b) but the main problem with this study is *Sampling efx* that highly productive plant species were present in all replicates of the most diverse treatment and no replicates of the lower-diversity treatment, making the outcome of the study inevitable (Huston 1997). Two further studies involved setting up field plots in a grassland at Cedar Creek, U.S.A. One of these (Tilman et al. 1996) involved random draws of between one and 24 species from a pool of 24 species, and claimed to present evidence of a causative effect of species richness on NPP and related variables. Another larger experiment at the same location (Tilman et al. 1997a), which was also based on random draws of species, also claimed to provide evidence of species richness affecting key ecosystem properties, although diversity and composition of plant functional types emerged as being of greater importance. Further, a glasshouse experiment based on conceptually the same approach claimed to find productivity of grassland plants to increase with increasing diversity from one to 16 species (Naeem et al. 1996).

Those experiments that have varied species richness through random draws from a pool of species have been criticized on the basis that their results can be explained entirely by "sampling effect," or the fact that the most productive species appear in an increasing proportion of plots as diversity is increased (Aarssen 1997; Huston 1997). Indeed, Huston (1997) showed through an analysis based on known plant sizes of all the plant species used in the studies of Naeem et al. (1996) and Tilman et al. (1996) that the results of both studies did not differ from what would be expected due to sampling effect alone. While Tilman et al. (1997c) present the sampling effect as a mechanism by which diversity effects are expressed in nature, this requires the biologically unrealistic assumption that plant communities are random assemblages of species (Wardle 1999b). As an extreme case, consider an experiment involving 20 plant species, one of which grows well and the others of which show only negligible growth. An experiment based on this species pool in which species richness is varied, and with treatments consisting of between one and 20 species drawn at random, would result in all the 20-species plots, but only 5% of the one-species plots, becoming heavily vegetated. This would generate a relationship between diversity and NPP entirely through sampling effect, which has little meaning in nature. It has recently been suggested that a "sampling effect" model based on randomly synthesized communities may actually serve as the appropriate "null model" for evaluating species richness effects (see Wardle 1999; Emmerson and Raffaelli 2000; Hector 2000). This creates a parallel with food web theory (e.g., Williams and Martinez 2000), in which randomly assembled communities may serve as null models which do not explain food web characteristics in the way that niche-based models do. While the null relationship between productivity and diversity has frequently been assumed to be flat (fig. 6.6a, curve i), the sampling effect model for testing diversity effects could be either a positive relationship [i.e., when the most productive species are also the most competitive, as shown for wetland plants by Gaudet and Keddy (1988)] (fig. 6.6a, curve ii) or a negative relationship [if the most competitive species are also less productive than the other species in the species pool (Hector 2000; Troumbis et al. 2000] (fig. 6.6a, curve iii).

Tilman et al. (2001) acknowledged that the results of Tilman et al. (1996) may have been due to sampling effect, but suggested that in subsequent years both in that experiment, and in the experiment of Tilman et al. (1997a), the relationships between species richness and NPP strengthened. Further, several chapters in the book edited by Kinzig et al. (2001), notably Tilman et al. (2001), now take the position that over time the positive effects of species richness on NPP have changed from being driven primarily by sampling effect initially in the experiment to being driven by resource partitioning in later years. However, these experiments may not be designed in such a way as to effectively separate these two explanations, especially at higher levels of species richness (fig. 6.6b, curve i vs. curves ii and iii). Tabulation of details of these experiments presented on http://www.lter.umn.edu/research/core1.html reveals that of the 24 plant species used by Tilman et al. (1996), 11 (46%) do not appear in any monoculture plots. Species entirely absent in monoculture (but present in all the highest-diversity plots) include two of the three largest-growing species out of the 24 [based on size data presented by Huston (1997)] (the third species appears only in one monoculture plot), and the yellow flowered *Rudbeckia hirta* which dominates all the heavily vegetated plots in a photograph of the site (see Kareiva 1996; Huston 1997). Because the productivity in the highest-diversity treatments could well be driven by species that are absent in monoculture, there are problems in using this experiment to test for effects of plant species richness on ecosystem or community response variables [including diversity effects on consumer trophic levels and susceptibility of communities to invasion as assessed in this experiment by Siemann (1998), Knops et al. (1999), and Naeem et al. (2000b)]. Further, for the larger experiment described by Tilman (1997a), 19 of the 37 species used (51%) appear in no monoculture plots, although a subset of these plots contains mixtures of up to 16 species in which all species appear in replicated monoculture plots. However, the pool of the 18 species from which random draws were made belong to five plant functional groups, and a strong relationship between productivity and diversity could arise simply through resource use complementarity among some functional groups (with no resource use complementarity between species within functional groups), with an increased probability of including plants from a greater number of functional groups in a given plot as the number of species selected is increased. Here, sampling effect would be operating by increasing the probability of including more functional groups in the mixture when more species are selected from the species pool.

An experimental study (BIODEPTH) which overcomes some of the problems that confound the interpretation of the above studies was established in each of eight locations throughout Europe (Hector et al. 1999). While this study was based on constrained random draws of species across an experimental diversity gradient (which ranged from one to between eight and 32 species depending on location) there was greater emphasis on including monoculture treatments, and indeed for three of the locations all species used were represented in replicated monoculture plots [see Huston et al. (2000); a fourth site includes 30 mixtures in which all species appear in monoculture]. Although Hector et al. (1999) concluded that productivity was related to diversity and that on average each halving of diversity reduced above-ground NPP by 80 g/m^2, alternative interpretations exist for these results. In particular, it is unclear to what extent overyielding (through resource use complementarity) vs. sampling effect contributes to this outcome, and an evaluation of these results by Huston et al. (2000) found that NPP in the highest-diversity treatments did not differ significantly from that of the most productive monocultures. While there is agreement that overyielding does occur in at least some of the plots (Hector et al. 2000a, 2001; Huston et al. 2000) (particularly those that include the nitrogen fixing legume *Trifolium pratense*), and Hector et al. (2001) provide four lines of evidence that "reject sampling effect as the *sole* explanation" of their results, the relative importance of sampling effect in influencing their results remains uncertain. It is still conceivable that true overyielding is restricted to the extreme low end of the diversity gradient (e.g., one vs. two or three species) and to mixtures containing specific combinations, with sampling effect explaining the entire remainder of the apparent relationship between diversity and productivity.

The effect of increasing species richness on NPP is likely to be greatest when coexisting species show significant resource use complementarity and low niche overlap. Therefore, multiple-species mixtures may be more productive when component species are more dissimilar (see Nijs and Roy 2000) and sourced from different functional groups. However, even increasing functional group diversity may not strongly affect productivity. In an experiment conducted in a serpentine grassland in southern California, U.S.A. (Hooper and Vitousek 1997; Hooper 1998), treatments were established that consisted of plots sown with between one and four plant functional groups [namely early season annual forbs (E), late seasonal annual forbs (L), perennial bunchgrasses (P), and nitrogen fixers (N)]; all four functional groups were represented in replicated monoculture plots. Resource use complementarity occurred in the EL mixture (through temporal resource partitioning) and in the LP mixture (probably through spatial resource partitioning). Despite this pattern, there was no overall relationship between functional group richness and productivity (the P monoculture treatment was more productive than all the mixtures), because of the dominant role of interspecific competition (and compensatory effects) in the experiment. A negative sampling effect (see fig. 6.6a, curve iii) has been suggested as a possible explanation of these results, in that productivity declined with increasing dominance by the least productive group (Hector 2000; Hector et al. 2001).

The study of Hooper (1998) points to the role of the balance of resource partitioning vs. competition in determining the nature of relationships that occur between biodiversity and productivity. This suggests that diversity-productivity relationships are likely to be context dependent and strengthened by environmental factors that promote low competition intensity and high resource partitioning [including habitat heterogeneity; see Cardinale et al. (2000)] and therefore greater resource use complementarity. If this is the case, then effects of species richness on ecosystem process rates (especially those related to productivity) should be greatest in unproductive environments (see earlier discussion) and along the most unproductive and species-rich portions of stress gradients (see fig. 6.7). Consistent with this, Austin and Austin (1980) found that ten-species mixtures of grasses were less productive than some of the component species grown in monoculture under high-nutrient conditions, but were more productive than all of the monocultures under low nutrient conditions.

Although the findings of Hooper (1998), Trenbath (1974), and Joliffe (1997) indicate that even when effects of diversity on NPP occur they should still saturate at low species richness levels, Loreau (1998) and Tilman et al. (1997c) utilized mathematical models to show that resource use complementarity should lead to diversity-function relationships that saturate at much higher species richnesses [i.e., comparable to those claimed in the experiments of Naeem et al. (1994) and Tilman et al. (1996)]. However, the species richness at which saturation occurs depends on model conditions, and these include the assumed nature of the diversity-dominance curve for the plant community. Increasing evenness (or equitability) of the dominance-diversity relationship of plant communities should maximize the effects of species richness on ecosystem productivity (see Nijs and Roy 2000; Wilsey and Potvin 2000). Indeed, as equitability of the plant community increases, the importance of sampling effect diminishes and the importance of resource

220

use complementarity increases (fig. 6.6d–f). Schwartz et al. (2000) used one of the two models presented by Tilman et al. (1997c) to show that Tilman's analysis assumed an unrealistically equitable dominance-diversity relationship, and that when the model is rerun using a biologically realistic relationship (i.e., with diminished evenness), saturation of resource utilization (used as an indicator of NPP) occurs at a much lower species richness (i.e., a situation represented by curve i rather than curve ii in fig. 6.6b). This is more consistent with the findings of observational studies, and with the results of those experiments that incorporate replicated monoculture treatments. There is therefore no undisputed evidence that plant species richness effects on ecosystem process rates are important beyond very low levels of species richness, or that species richness operates as a driver of ecosystem-level properties in nature.

Plant Diversity and Belowground Processes

Increasing species richness of live plants also has potential influences on belowground processes. If, as described above, species richness has detectable effects on NPP at the low end of the species diversity spectrum (one vs. two or three species), then this could conceivably promote those components of the soil food web that are regulated by productivity (see chapter 2) as well as the processes that they regulate. However, the available evidence is mixed. In the Hooper and Vitousek (1997, 1998) experiment described earlier, increasing plant functional group richness from one to two groups sometimes had positive effects on soil microbial biomass, immobilization of added ^{15}N (fig. 6.8a), and soil resource use by plants (indicative of nutrient supply from the soil), but no effects on soil nitrification. There was no effect of increasing richness beyond two functional groups, and it was concluded that functional group composition, and the combinations of functional groups present, were the drivers of belowground processes. Further, Hector

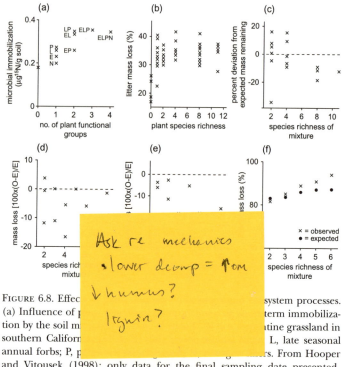

FIGURE 6.8. Effec[...] system processes. (a) Influence of [...] term immobilization by the soil m[...]tine grassland in southern Califor[...] L, late seasonal annual forbs; P, [...]rs. From Hooper and Vitousek (1998); only data for the final sampling date presented. (b) Mass loss of *Holcus mollis* litter added to field plots differing in live plant species richness and (c) decomposition rates of litter observed in multiple-species litter mixes relative to that which would be expected based on the decomposition of all component species in monoculture, in a grassland at Silwood Park, U.K. From Hector et al. (2000b). (d) decomposition of multiple species mixtures of tree leaf litter and (e) tree leaf and grass leaf litter, expressed as decomposition of observed (O) values relative to expected (E) values based on decomposition of all component species in monoculture, in an abandoned field in Hamilton, New Zealand. From Wardle et al. (1997a). (f) Decomposition rates of multiple-species mixtures of litter and expected decomposition rates based on all component species in monoculture, in a microcosm experiment. From Bardgett and Shine (1999).

et al. (2000b) found for the Silwood Park site of the BIODE-PTH experiment that litter mass loss rates of *Holcus mollis* litter placed in plots sown with between one and 11 plant species were statistically significantly but very weakly related to plot species richness (fig. 6.8b). However, sampling effect may explain part of the response to plot diversity, and these data also point to saturation of species richness effects at very low levels of diversity. For the Swedish BIODEPTH site, Mulder et al. (1999) found that the relationship between mass loss of cotton strips placed in the soil and plant species richness depended upon whether or not insecticides were sprayed on the plots to reduce invertebrate herbivores.

Other studies have compared the effects of plant mix-tures vs. monocultures on the levels of soil organisms that carry out soil processes. Christie et al. (1974, 1978) found that when pairs of plant species were grown in mixture, fun-gal and bacterial densities on their root surfaces were fre-quently greater than when they were grown in monoculture, although many exceptions existed. Aberdeen (1956) did not find fungal densities on the root surfaces of plant spe-cies grown in mixture to differ from those in monoculture. Further, Wardle and Nicholson (1996) conducted a glass-house experiment in which *Lolium perenne* was grown in mix-ture with each of nine grassland forb species, with all spe-cies used also grown in monoculture. Soil microbial respiration and biomass, and decomposition rates of added litter, were all sometimes significantly greater and some-times significantly less in the mixtures, based on what was observed in the monoculture treatments, and these effects were independent of belowground NPP of the mixtures rel-ative to those of the monocultures. In the Swiss site of the BIODEPTH experiment, Spehn et al. (2000) found soil mi-crobial biomass and decomposition of substrates to be un-affected by plant species richness, although inclusion of le-gumes stimulated both of these properties. These results collectively suggest that the relationships between plant species richness and soil organisms or processes are driven by species identity (and are often idiosyncratic) but that species richness per se has few effects even at very low levels of species richness.

Diversity of dead plant parts may also affect decomposer organisms and processes through "afterlife effects" (*sensu* Findlay et al. 1996), and this can be investigated through mixture experiments in which organism density and decomposer processes in multiple-species litter combinations are compared with those of all the component species in monoculture. Kaneko and Salamanca (1999) and Hansen (2000) both detected, using litter mix experiments, that population densities of saprophagous mites were greater in three-species litter mixtures than in litter monocultures of the three species. However, Chapman et al. (1988) found that litter in mixed (two-species) tree stands could support either greater or lesser populations of earthworms and enchytraeids than expected based on the populations of these organisms in monoculture stands of these species, with the direction of the effect being determined by the specific combination of species. Further, Blair et al. (1990) found levels of some groups of decomposer biota to be greater than expected, and others to be less than expected, in two- or three-species mixtures of tree leaf litters relative to corresponding monoculture litter types.

The apparently idiosyncratic effect of litter mixing on decomposer biota also explains the varied responses of decomposer processes (i.e., rates of litter decomposition and nutrient mineralization) to litter mixing (see Wardle et al. 1999a), and different studies have detected positive (e.g., Briones and Ineson 1996; McTiernan et al. 1997), neutral (e.g., Blair et al. 1990; Finzi and Canham 1998), and idiosyncratic (e.g., Fyles et al. 1993; Robinson et al. 1999) effects of mixing different litter types relative to what would be expected based on monocultures. Further, the three studies that have involved litter mixing of several (>3) species together also show varied responses. Hector et al. (2000b) found that mixing of litters of up to 11 grassland plant species in the proportions found in field plots resulted in faster decomposition rates of litter than expected based on monoculture values (fig. 6.8c). Wardle et al. (1997a), in contrast, found an idiosyncratic effect of mixing litters of up to eight species both within and across plant functional groups on both decomposition and nitrogen release from the litter, and there was no evidence of a unidirectional response to increasing species richness (fig. 6.8d,e). Meanwhile, Bardgett and Shine (1999) found that increasing litter diversity from one to six species caused a positive effect on decomposition rates relative to expected values derived from monoculture treatments (fig. 6.8f). It is clear that a range of responses to litter mixing are possible, and although a recent theoretical modeling study (Loreau 2001) predicts a negative effect of plant-derived substrate diversity on decomposers and decomposer processes, this is not supported by the available data at least if species diversity of plant litter is assumed to promote substrate diversity. Indeed, both positive and negative responses of decomposition to litter mixing are possible because high-quality litter may accelerate the decomposition rates of associated lower-quality litter types while litter of poor quality (e.g., litter with high concentrations of phenolics) may have the reverse effects (Seastedt 1984). This again points to the importance of composition of plant species, rather than species richness, in generally determining the functioning of the decomposer subsystem.

Functional Significance of Soil Biodiversity

Despite the functional role of soil organisms in driving belowground processes and regulating long-term ecosystem productivity, the issue of how the composition and diversity of the soil biota influence these processes remains little studied. Investigation of how diversity of belowground primary consumers (bacteria and fungi) affects process rates is especially difficult given that most microbial species have not been described and that the vast majority of soil bacteria cannot be cultured. To date, synthetic community experiments investigating microbial species effects on processes have been few and limited to fungi. Mixture experiments investigating the effects of increasing saprophytic fungi from one to two species have found varying results. Janzen et al. (1995) found no effect of combining two fungal species on CO_2 release from *Hordeum vulgare* litter relative to the performance of both species in monoculture (fig. 6.9a). Meanwhile, Wardle et al. (1993) determined for two mixtures of fungal species that one mixture showed overyielding in terms of biomass production relative to corresponding monocultures, while the other did not (fig. 6.9b,c). Further, Robinson et al. (1993) showed that all the possible pairwise combinations of four fungal species caused significantly greater CO_2 release from *Triticum aestivum* straw than would be expected based on fungal monoculture treatments (fig. 6.9e). This last result arises because the four fungal species used differed significantly in functional attributes, including their abilities to break down different substrates, indicative of resource use complementarity between the species.

This pattern of complementarity is not, however, apparent in the study of Cox et al. (2001). Here, sterilized *Pinus sylvestris* litter was inoculated with either *Trichoderma viride* or *Marasmius androsaceus*, or left uninoculated (control treatment), and placed in the field. The control and *M. androsaceus*–inoculated litters were colonized by a diverse range of fungal species (26 in total) while the *T. viride*–inoculated litter remained uncolonized by other species (as a result of the antagonistic ability of this species) and therefore had an extremely low diversity. After 12 months the litter inoculated by *T. viride* or *M. androsaceus* had decomposed more completely than the control litter (fig. 6.9d) and *T. viride* was clearly as capable as the diverse fungal assemblages present on the other litter types in utilizing structural polysaccharides. These results can be explained in

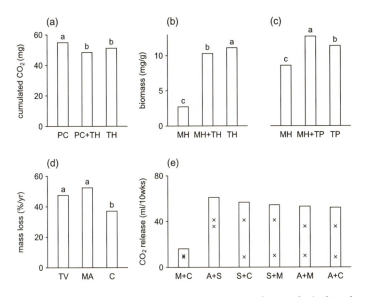

FIGURE 6.9. Potential ecosystem-level consequences of saprophytic fungal species diversity. (a) Predicted maximum cumulative CO_2 release based on experimental data for mixtures and monocultures of *Phanerochaete chrysosporium* (PC) and *Trichoderma harzianum* (TH) in microcosms containing straw of *Hordeum vulgare*. From Janzen et al. (1995). (b) Total fungal biomass in a sterilized agricultural soil inoculated with mixtures and monocultures of *T. harzianum* (TH) and *Mucor hiemalis* (MH), and (c) fungal biomass in a forest soil inoculated with mixtures and monocultures of *Trichoderma polysporum* (TP) and *M. hiemalis* (MH). From Wardle et al. (1993). (d) Decomposition rates of sterilized *Pinus sylvestris* litter inoculated with *Trichoderma viride* (TV) or *Marasmius androsaceous* (MA), or nonsterilized and colonized with a diverse fungal assemblage (C). From Cox et al. (2001). (e) CO_2-C release in microcosms containing *Triticum aestivum* straw in two-species mixtures of four fungal species, i.e., *Mucor hiemalis* (M), *Chaetomium globosum* (C), *Agrocybe gibberosa* (A), and *Sphaerobolus stellatus* (S). For each bar, the associated crosses represent the performance of the two component species of the mixture in monoculture. From Robinson et al. (1993). For (a) to (d), bars not topped with the same letters are significantly different at $P = 0.05$. For (e), the mixture values are significantly greater than expected based on the monoculture values for all six.

terms of fungal community composition, but not diversity, determining decomposition rates. Further, antagonistic interactions among fungi can conceivably even result in increased diversity reducing decomposition rates, and there is evidence that this may indeed be the case for wood-decomposer fungi (see Coates and Rayner 1985; Rayner 1994). Here, a fungal community composed of one genetic individual may effectively exploit the resource without impediment. However, when several genetically dissimilar individuals (of either the same or different species) occur together, the intensity of interactions between them may be sufficient to impair the ability of individual genets to effectively exploit the resource (Rayner 1994). This, combined with the likelihood that interference zones between adjacent genets would remain unused, should result in lower resource utilization. Such a mechanism suggests that greater diversity, either at the within-species genetic level, or across species, would slow decomposition rates.

Soil microbial communities also consist of bacteria, and it is not normally possible to adequately investigate the effects of bacterial species diversity or community composition on soil processes using the synthetic community approach. However, some studies provide indirect evidence that alteration of overall microbial composition (including that of bacteria) can affect process rates. For example, experiments by Degens (1998) and Griffiths et al. (2000) used fumigation of soil to alter microbial community composition, and also provided evidence that certain measures proposed as being indicative of microbial diversity were reduced by fumigation. In both these studies, differences between fumigated and nonfumigated soils arose with regard to rates of soil processes governed by the decomposer community; these differences were explicable in terms of fumigation effects on microbial community composition but not in terms of effects on microbial diversity. Griffiths et al. (2000) also found that the microbial community of the fumigated soil was less resistant to a second stress (copper addition) but whether this was a result of reduced diversity or altered community composition is unclear. It may be that in the vast majority of soils, diversity of microorganisms is never likely to be reduced to such a level that diversity may emerge as a plausible mechanism for affecting functioning of the decomposer subsystem. It is more likely that only those processes carried out by few microbial species, such as those involving specific nutrient transformations (e.g., nitrification, atmospheric nitrogen fixation) have the potential to be responsive to shifts in diversity (Wardle et al. 1999a). The effect of composition and diversity of these organisms on processes remains largely unexplored, although Cavigelli and Robertson (2000) provided evidence that denitrifier communities from two sites apparently differed sufficiently in composition to influence N_2O flux, and the study of Gulledge and Schimel (1998) suggested that different communities of CH_4 oxidizers across five sites in subarctic Alaska, U.S.A., may differ in their rates of CH_4 consumption.

There have been few attempts to investigate the effects of soil faunal diversity on the functioning of the decomposer subsystem, although it has often been suggested that there is considerable functional redundancy in soil faunal communities given the large number of trophically equivalent species usually present (Andrén et al. 1995; Lawton et al. 1996; Ettema 1998). Indeed, Andrén et al. (1995) showed that the mass loss of decomposing *Hordeum vulgare* straw over a two-year period followed a simple pattern that was predictable in terms of moisture and temperature changes over that period but was independent of temporal changes in the communities of decomposer fauna inhabiting the litter. Further evidence of a lack of consistent effect of soil faunal diversity on ecosystem processes arises from the study of Mikola and Setälä (1998c). Here, microcosms were set up that contained either six nematode species (three bacterial feeders and three fungal feeders) or two species (one bacterial feeder, one fungal feeder); three two-species treatments were set up so that all nematode species occurred in one two-species treatment. The relationships of rates of carbon and nitrogen mineralization to nematode diversity were idiosyncratic and were able to be predicted only by the characteristics of component species and their interactions.

There have been few attempts to evaluate how decomposer diversity affects aboveground processes such as NPP. However, in a model ecosystem study, Laakso and Setälä (1999b) set up replicated microcosms with each of 26 different soil food web structures, and planted a seedling of *Betula pendula* in each microcosm. Addition of saprophagous or fungal-feeding soil mesofauna, either singly or in combination, affected plant production, although addition of five species of each group had no detectable effect relative to adding only one species per group. There were few effects of species composition within feeding groups on plant production, other than that inclusion of one of the saprophages (the enchytraeid *Cognettia sphagnetorum*) had strong positive effects that none of the other species were capable of, indicative of low functional replaceability of that species. Naeem et al. (2000a) provided results from a synthetic community microcosm study using an aquatic system which is also relevant to this issue, and which claims to show that species richness of both bacteria (decomposers) and algae (producers) contributes jointly to NPP in a manner that the diversity of neither group can on its own. This study did not include proper monoculture treatments, and although the issue of sampling effect was not acknowledged, there is no evidence that the results differ from what could be expected in terms of a null model based solely on sampling effect operating over both trophic groups.

Two recent studies have also considered aboveground consequences of mycorrhizal fungal species diversity. In the first study, Van der Heijden et al. (1998b) set up mesocosms containing synthetic plant communities, and which were inoculated with between one and 14 arbuscular mycorrhizal fungal species drawn at random from a pool of 23 fungal species. They found that plant productivity, phosphorus uptake, and species diversity were all greater when more fungal species were present. However, these results can be explained by sampling effect, with those fungal species most likely to influence aboveground properties having a higher probability of occurring in the most diverse treatments (Wardle 1999b). Although Van der Heijden et al. (1999) maintain that sampling effect cannot explain these results on the grounds that mycorrhizal fungal species which have a disproportionately positive effect on plant growth do not exist in their study, this issue remains unresolved because the experiment did not include replicated monoculture treatments of the fungal species used in their study. In the second study, Jonsson et al. (2001) provide some evidence of positive effects of mycorrhizal fungal diversity on NPP. Here, *Betula pendula* and *Pinus sylvestris* seedlings were each inoculated with between one and eight ectomycorrhizal fungal species; all fungal species also occurred in replicated monoculture treatments. For *B. pendula*, fungal species richness had significant positive effects on NPP in a low-fertility but not in a high-fertility substrate, while for *P. sylvestris*, species richness negatively affected NPP in a high-fertility but not in a low-fertility substrate. This points to mycorrhizal fungal species richness effects being context dependent and influenced by both soil fertility and host plant species.

Biodiversity and Stability of Ecosystem Properties

Even if species diversity may be unimportant (at least beyond very low levels of species richness) in influencing ecosystem properties at a given point of time, the possibility remains that diversity could influence the temporal dynamics of ecosystem properties, through affecting their stability (i.e., resistance and resilience to perturbations; maintenance of low temporal variability). It is well known that different communities or ecosystems differ in their responses to variations in external factors (MacGillivray et al. 1995;

Sankaran and McNaughton 1999) and that these responses are likely to be influenced by community composition (McGillivray et al. 1995; Collins 2000; see chapters 2–4). It has been proposed that species richness in its own right may stabilize ecosystem properties, notably NPP, because of compensatory relationships among species in response to environmental fluctuations (see May 1973; McNaughton 1977). Observational studies and studies along gradients have shown that a greater stability of both plant community structure (Frank and McNaughton 1991) and biomass or productivity (McNaughton 1977; Tilman 1996) occurs when a greater richness of species is present. However, it is impossible to determine from such studies whether diversity itself has stabilizing effects, because different plant functional types may dominate in high- and low-diversity portions of the gradient. For example, the study by Tilman (1996) conducted at Cedar Creek, U.S.A., found along an experimental gradient of nitrogen fertilization that more diverse plots showed lower temporal variability of plant biomass and greater biomass resistance to drought. However, in that study, nitrogen fertilization simultaneously reduced plant diversity and selected for plant communities which were more likely to respond to temporal variation in moisture characteristics [given that when nitrogen fertilizer is added, the next most limiting factor, i.e., moisture (which is highly temporally variable) drives biomass production] (Huston 1997). Therefore diversity and stability appear correlated simply because both are driven by a third factor.

There have been few attempts through manipulation experiments to investigate whether species diversity of organisms influences the variability of ecosystem properties, including those relevant to aboveground-belowground linkages. Wardle et al. (1999b) found using a removal experiment in a perennial grassland near Hamilton, New Zealand, that while vegetation composition (i.e. the nature of functional groups removed) was important in determining both temporal and spatial varability of plant production, soil microbial biomass, and soil respiration, there was little evidence of plant species richness being important. A complementary glasshouse experiment (Wardle et al. 2000) involved growing synthetic grassland plant communities with four plant species (all from different functional groups) in monoculture and in two- and four-species mixtures. Although increasing plant diversity increased NPP (presumably through resource use complementarity), stability of a range of ecosystem properties (NPP, plant biomass, soil respiration, microbial biomass, decomposition of added substrates, and soil chemical properties), as assessed in terms of their resistance to an experimentally imposed drought, was not enhanced by increased diversity. Vegetation composition instead emerged as the primary driver of stability. This result is consistent with that of McGillivray et al. (1995), which points to the importance of functional composition of vegetation in determining the effects of experimentally imposed disturbances on both community-level and ecosystem-level attributes.

Even less is known about how diversity of consumers affects the variability of ecosystem properties, although trophic dynamic theory has the potential, largely unexplored, to assist our understanding about how species diversity and redundancy in food webs may contribute to ecosystem stability (see Johnson 2000). No data currently exist with regard to belowground consumers, but two relevant studies have been performed in synthetic aquatic communities. In the first, Naeem and Li (1997) used a microcosm experiment to investigate the variability of ecosystem properties [i.e., biomass of producers (algae) and decomposers (bacteria)] in response to varying species richness across four trophic groups. Here, variation between replicates at the end of the experiment with regard to autotroph and decomposer biomass diminished with increasing species richness, enhancing "ecosystem reliability" (*sensu* Naeem and Li 1997). However, species composition across replicates was also more similar for those treatments with more species present at the start of the experiment, making the observed outcome inevitable (Wardle 1998b) through the "variance reduction effect" (Huston 1997). In the second, McGrady-Steed et al. (1997) investigated the effects of varying species richness across four trophic groupings in microcosms and found "ecosystem predictability" (based on a combined measure of temporal and inter-replicate variability) of CO_2 flux to be greatest when more species were present. However, this study is confounded both by the same problem as that of the Naeem and Li (1997) study and by the fact that several species occurred in all the high-diversity microcosms that were absent in all the lower-diversity ones. The possibility therefore remains that greater "predictability" could have occurred in the most diverse treatments simply because of effects of domination by species that were absent in the lower-diversity treatments. The studies of Naeem and Li (1997) and of McGrady-Steed et al. (1997) therefore cannot be used as undisputed evidence that increasing species richness reduces variability of ecosystem properties, as was done in several chapters in the book edited by Kinzig et al. (2001).

Despite the dearth of unambiguous data about how diversity affects ecosystem stability, several recent studies have attempted to provide a theoretical framework through which such relationships might occur. It has been proposed that functioning of ecosystems with larger numbers of subordinate species may be better buffered against environmental changes because some of the subordinate species present may be able to dominate in future conditions in which current dominants would not thrive (e.g., Grime 1998; Petchey et al. 1999; Walker et al. 1999). Further, species that are functionally redundant prior to disturbance might serve as "spare wheels," maintaining ecosystem function following perturbation (Andrén et al. 1995). Note that this mechanism need not involve beneficial effects of greater species richness; the only requirement is that the appropriate combinations of species (including those with traits that are ap-

Wardle doesn't like Naeem.

propriate for the changed conditions) need to be present prior to the perturbation. It has been proposed that species diversity may enhance ecosystem stability through an "insurance effect" (Yachi and Loreau 1999; Loreau 2000); such an effect has been proposed as operating through two mechanisms. One mechanism is through the greater probability of including species that are adapted for changing conditions through the selection of extreme trait values as species richness is increased (Loreau 2000), but this has close conceptual similarities to the sampling effect model (the only major difference being that the model here is applied in the temporal rather than the spatial dimension) and therefore suffers from the same problems. Although simulation models based on randomly assembled communities can demonstrate this mechanism (Yachi and Loreau 1999), this model may be more appropriately considered as a null model (cf. fig. 6.6a curve ii) than as a mechanism by which diversity stabilizes ecosystem processes in nature. The second mechanism operates through resource use complementarity across species along the temporal niche axis (see Hooper 1998; Yachi and Loreau 1999; Loreau 2000), although it is unclear at what level of species richness species saturation would occur (cf. fig. 6.6b). This mechanism is based on species interactions, although it is still uncertain as to how important are species interactions such as competition in driving the relationship between species diversity and stability (see Ives et al. 1999; Hughes and Roughgarden 2000; Lehman and Tilman 2000; Loreau 2000; Cottingham et al. 2001). Further, increasing species richness can be expected to increase ecosystem stability through "variance averaging," in which a more diversified species pool should have a lower combined variance for biomass or productivity purely on statistical, not biological, grounds (Doak et al. 1998).

Some theoretical modeling studies considering the effects of diversity on ecosystem stability predict a positive relationship between species richness and stability of biomass (Loreau and Behera 1999; Lehman and Tilman 2000;

Loreau 2000) but these predictions are based on simulations of randomly assembled communities, and importantly do not enable distinction of whether species richness effects saturate at low or high levels of species richness. Indeed, as noted in a recent review by Cottingham et al. (2001) "most theoretical studies" investigating the biodiversity–ecosystem stability issue "are based on highly unrealisytic assumptions." For example, predictions such as those of Lehman and Tilman (2000) that biomass stability increases linerarly as species richness increases as a result of three mechanisms (overyielding, competitive interactions, and variance reduction effect) are dependent upon model conditions, and are likely to be influenced by the allocation of the same properties to all species other than their responses to a factor that varies temporally, as well as an unrealistic underlying distribution of species abundances. In this light, Schwartz et al. (2000) found that the variance reduction effect saturates at very low levels of species richness when realistic distributions of species abundance are used. In total, there is no undisputed evidence, theoretical, experimental, or observational, which suggests that species richness stabilizes ecosystem properties at least beyond very low levels of species richness (and below that which occurs in most biological communities; fig. 6.6 curve i vs. curves ii and iii) or that species richness is a key driver of ecosystem stability in nature.

SYNTHESIS

This chapter has considered two main issues, i.e., what regulates biological diversity in an aboveground-belowground context, and what the consequences of this biodiversity are for ecosystem functioning. Because of the immense diversity of soil organisms, and the fact that much of the soil biota remains undescribed, there are obvious logistic difficulties in addressing diversity-related questions in the soil using approaches that have been developed for aboveground biota. However, there is evidence that diversity of many soil biotic groups is promoted by habitat heterogeneity and by high levels of available resources. Further, it appears that diversity of most groups of soil biota does not show a hump-backed response to increasing stress or disturbance but rather a monotonically declining response, probably because of the relatively low importance of competition in relation to other factors in regulating much of the soil biota. Biotic controls of soil diversity operate through influencing soil resource levels and the physical habitat, and both plants and soil organisms can exert these effects. Feedbacks between the aboveground and belowground subsystems are also likely to be important with regard to diversity, given that soil organisms can also alter plant community structure and plant diversity. However, whether plant diversity per se can increase soil biodiversity (e.g., through promoting habitat heterogeneity), or vice versa, remains little studied.

competition prob not as imp. BG as AG.

The "enigma of soil diversity," or why such a high diversity of biota occurs in the soil, can be explained in terms of temporal and spatial resource partioning along a number of niche axes, driven in part by the tremendous habitat heterogeneity belowground. The spatial isolation of potential competitors from each other in the soil matrix may also be a contributing factor. At larger spatial scales, groups of soil organisms do not generally show the trend demonstrated by most other groups of biota of decreasing diversity with increasing latitude, except at latitudes above about 60°. This is probably because at higher latitudes humus depth (and therefore soil carbon content) increases, promoting both resource quantity and habitat heterogeneity, both of which enhance soil diversity. Another reason is that for smaller organisms at least, local diversity is probably not regulated by regional species pool size but rather by habitat factors.

Despite the recent emphasis placed on the diversity-function issue, much remains unresolved about the role of species richness of organisms, either aboveground or belowground, in influencing ecosystem properties and processes. There is little support from observational studies for a predictable relationship between species richness of organisms and ecosystem processes. Experimental studies indicate that some processes, notably NPP, can be influenced by plant species richness at the very low end of the diversity spectrum (one vs. two or three species) due to resource use complementarity and facilitation; however these effects are usually weak and saturate at levels of species richness below that occurring in most ecological communities. Further, little consistent evidence emerges for diversity of belowground organisms being important in determining most soil processes or plant growth, except at levels of diversity that are well below what would ever occur in most ecosystems. While some studies have claimed to present experimental evidence that increasing species richness of organisms to levels well beyond two or three species has positive effects on ecosystem processes (notably NPP), ecosystem stability, and diversity of other trophic levels, all experimental studies to date making such claims incorporate experimental designs and "hidden treatments" (*sensu* Huston 1997) that make the observed outcome statistically inevitable. Theoretical modeling studies provide some evidence that increasing species diversity may increase both NPP and ecosystem stability, but most of these studies are founded on biologically unrealistic assumptions, and it is unclear at what level of species richness saturation of species should occur. There is therefore no undisputed evidence, observational, experimental, or theoretical, to support claims such as that of Tilman (1999) that biodiversity "should be added to species composition, disturbance, nutrient supply, and climate as a major controller of population and ecosystem dynamics and structure." The main biotic controls of ecosystem function are instead most likely to be the key traits of the dominant species, community composition of the component organisms, and the nature of biotic interactions among organisms (see numerous examples in chapters 2–5).

CHAPTER SEVEN

Global Change Phenomena in an Aboveground-Belowground Context

The rapid growth of the human population over the past few centuries has had a major effect on the Earth's ecosystems. Over a third of the Earth's land surface, including most of that which is fertile and capable of supporting high NPP, has been transformed by human activity (Vitousek et al. 1997a). Colonization of new land surfaces by humans appears to have coincided with rapid extinctions of most of the larger vertebrate species in Australia, New Zealand, and North America. Further, current extinction rates of species are probably around 100 to 1000 times their prehuman levels (Pimm et al. 1995). There has also been a globalization of a significant subset of the world's biota, with alien organisms making up an increasing proportion of the resident biota on many of the Earth's land masses (Vitousek et al. 1997b). Modification of the atmosphere by human activity has been equally substantial, particularly since the beginning of the Industrial Revolution, with atmospheric CO_2 levels rising by 30% since the 1850s (Kates et al. 1990). Concomitant with this have been apparent shifts in the Earth's climate, with air temperatures expected to rise by 3°C on average over the next century (Schneider 1992). Further, nitrogen fixed through anthropogenic sources has become the major source of nitrogen input in many ecosystems, and on a global basis this nitrogen probably exceeds the amount of nitrogen that is fixed biologically (Galloway et al. 1995).

While there has been an increasing awareness of the significance of global change phenomena, much still remains unknown about the mechanisms by which they may influence the performance of the Earth's ecosystems, or the likely consequences of these influences in the long term. Explicit consideration of the linkages between the aboveground and belowground biota may therefore provide a tool through which our understanding of the consequences of global change for the Earth's ecosystems could be improved (see Wardle et al. 1998b; Wolters et al. 2000). In this chapter, I will draw upon the concepts developed in the preceding chapters to discuss the ecological consequences of four aspects of global change phenomena, i.e., species losses and gains in ecosystems; land use change; atmospheric CO_2 and nitrogen enrichment; and global climate change.

SPECIES LOSSES AND GAINS

The ecological consequences of losses or gains of species in a community are likely to be most extreme when those species that are removed or added differ significantly in their functional attributes from those of the remainder of organisms in the community. There is a vast literature considering the ecology of biological extinctions and invasions, but only a minority of studies have considered these issues specifically in an aboveground-belowground context. The issue of species loss and gain will now be discussed within such a context.

Species Losses from Ecosystems

When species are lost from a community as a result of anthropogenic factors, the process is not random, and species with certain suites of traits are lost before others. For example, human colonization of new land masses leads to rapid extinction of the larger animal species that provide the most food per unit hunting effort; logging of rainforest results in the first instance in the loss of timber-producing tree species and those organisms that are obligately dependent upon these species; and nitrogen deposition selectively filters out noncompetitive, stress-tolerant plant species. Further, Duncan and Young (2000) found, in an analysis of plant extinctions in the Auckland area of New Zealand over the 145 years since European settlement, that the proportion of total species which had become extinct differed considerably across both plant height classes and major plant taxonomic groups. Therefore, synthetic community experiments based on random assemblages of species, such as those described in the second half of chapter 6 are not, as claimed by some (e.g. Naeem et al. 1994; Kareiva 1996; Tilman 1999), particularly relevant for understanding how species loss in nature affects community and ecosystem properties. These questions are more appropriately addressed by removal experiments, in which species (or groups) of species are selectively removed from a community in order to simulate extinction of that species, and the effects of loss of that species are then monitored [see Lamont (1995) and Petchey (2000) for a theoretical elucidation of this]. Losses of species are most likely to have important effects when extinction results in loss of a dominant that is not functionally replaceable (Grime 1998), or species that have keystone attributes (i.e., those species that have a disproportionately large functional role relative to their biomass contribution within their trophic level).

Although removal experiments have long been used in plant ecology to investigate the significance of interspecific interactions in plant communities (see Aarssen and Epp 1990; Goldberg and Barton 1992), few studies have explicitly used this approach to investigate species loss effects on ecosystem properties. However, it is apparent from the plant interaction literature that species loss effects on NPP are usually inversely related to the degree of resource use over-

241

lap with the other species (Goldberg and Barton 1992). In a model experimental system, Symstad et al. (1998) found that effects on NPP of deletion of species from assemblages depended primarily upon the identity of the species that were deleted and the composition of the community from which they were excluded. However, ecosystem nitrogen retention was unresponsive to these factors. Further, Wardle et al. (1999b) conducted a removal experiment in a perennial grassland near Hamilton, New Zealand, in which different functional groups of the flora were continually removed over a three-year period. While there were few effects of removals on NPP (other than negative effects of removing functionally irreplaceable C4 grasses over the summer months), the community composition of most groups of soil biota (microbes, arthropods, three trophic groupings of nematodes) was often affected by which plant groups were removed. There was, however, little evidence of removals affecting key processes such as decomposition or CO_2 release rates from the soil, unless all plants were removed. On balance, it is expected that plant species loss effects on ecosystem properties are likely to be context specific, and based on the material presented in chapter 3, the magnitude of such effects (both above- and belowground) will be driven by the degree of dissimilarity in suites of functional traits between those species that are lost and those that remain. Loss of species that grow slowly, invest carbon into secondary metabolites, and produce litter of poor quality will therefore clearly have opposite consequences to loss of those that grow rapidly and produce high-quality litter.

Some of the most significant ecosystem-level consequences of anthropogenically induced extinctions probably involve loss of the largest herbivore species present on land masses colonized relatively recently by humans. For example, the extinction of the New Zealand moa species 500–800 years ago (see Anderson 2000; Holdaway and Jacomb 2000), the megaherbivore fauna of North America and Beringia

10000–12000 years ago (see Mosimann and Martin 1975; Zimov et al. 1995), and the Australian megamarsupials over 40000 years ago (see Flannery 1994) all coincided with the period of first human settlement of these areas. These extinctions have almost certainly profoundly affected the functional composition of the vegetation of these areas, which should in turn influence belowground organisms and processes. For example, Flannery (1994) has proposed that human-induced megamarsupial extinction in Australian forests caused an increased incidence of wildfire resulting from greater fuel buildup (due to less plant material being consumed), leading to replacement of non-fire-adapted rainforest tree species by fire-adapted *Eucalyptus* species and sclerophyllous vegetation. These fire-adapted species are well known to produce very recalcitrant litter; this should in turn lead to diminished soil microbial activity and lower rates of nutrient mineralization, leading to reduced rates of nutrient supply for plants (fig. 7.1a). Further, Zimov et al. (1995) provide evidence that megaherbivore extinctions in northern Russia and western Alaska have led to replacement of steppe grassland with wet moss tundra in which the dominant plant species produce litter of vastly diminished quality. This, together with associated increased waterlogging of the soil, should again lead to reduced nutrient mineralization, providing a feedback that maintains dominance of the tundra vegetation (fig. 7.1b). Based on the material presented in chapter 4, it is inevitable that loss of aboveground consumers should have significant effects on both aboveground and belowground components of ecosystems. This is supported by the results of field exclosure experiments aimed at removing the effects of larger mammals. Such studies have demonstrated that these animals can exert both positive and negative effects on populations of decomposer organisms (fig. 4.5) and the soil mineralization processes that they carry out (fig. 4.6) because of the variety of mechanisms by which herbivores can affect soil biota (see chapter 4).

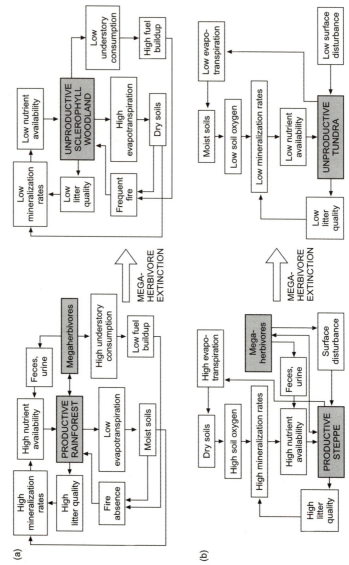

(a)

(b)

Little is known regarding the community- or ecosystem-level effects of species loss of belowground biota, although for smaller decomposer organisms at least (i.e. those less than 1 mm in width), many species are probably both globally ubiquitous and functionally redundant (see chapter 6), making extinctions of these organisms unlikely, and functionally nonsignificant even if they do occur. As outlined in chapter 6, loss of species that perform specialist functions carried out by few other species (e.g., nitrifying and nitrogen fixing bacteria) is likely to be of greater functional significance, as is shown, for example, by literature demonstrating adverse effects of heavy metals on processes carried out by these organisms (see Giller et al. 1998). The functional significance of species loss of larger organisms in the decomposer food web (e.g., earthworms, insects) has seldom been specifically addressed, although numerous examples presented in chapters 2 and 5 would suggest that loss of at least some of these organisms should have major consequences on both belowground mineralization processes and vegetation characteristics. This includes not just the larger saprophagous invertebrates (see fig. 2.9) but also those species that serve as top predators and can induce trophic cascades in soil food webs (fig. 2.6).

Aboveground Invasive Organisms

Invasions of exotic plants into natural ecosystems are most likely to affect aboveground-belowground feedbacks when the invading species has vastly different ecophysiological attributes from those of the native flora. Attempts to

---

FIGURE 7.1. (a) Hypothesized consequences of megafauna extinction in Australian rainforest more than 40000 years ago for feedbacks between wildfire, vegetation, and nutrient cycling. (b) Consequences of megaherbivore extinction in northern Russia and western Alaska in the late Pleistocene for feedbacks between vegetation and nutrient cycling, as hypothesized by Zimov et al. (1995).

demonstrate differences in functional traits between alien and native floras have produced mixed results (see Lodge 1993; Rejmánek 1996; Thompson et al. 1995; Williamson and Fitter 1996), although Baruch and Goldstein (1999) did find that in the Hawaiian archipelago invasive species as a group had leaf traits linked to greater resource capture (i.e., higher specific leaf areas and CO_2 assimilation rates, and greater concentrations of foliar nutrients) relative to those of native species. If invasive plants generally differ from native plants in such traits, then this should lead to predictable consequences of dominance by invaders for quality of litter return to the soil, soil food web characteristics, soil microbial activity, and plant nutrient supply rates (see chapter 3 and fig. 3.8).

Few studies have presented relevant data about how alien plant species affect aboveground-belowground linkages in natural communities. Vitousek et al. (1987) and Vitousek and Walker (1989) did, however, show that invasion by the actinorhizal shrub *Myrica faya* into early successional stands of *Metrosideros polymorpha* in Hawaii resulted in ecosystem nitrogen input increasing by over fourfold simply because of its ability to fix nitrogen symbiotically (fig. 7.2a). This would be expected to have important effects on ecosystem productivity, given that fertilization experiments at that site have shown nitrogen to be the limiting nutrient for productivity of *M. polymorpha*. The effects of invasive plants on decomposer processes or the organisms that conduct these processes have been little investigated. Saggar et al. (1999) provided data suggesting that invasion of *Hieracium pilosella* into tussock grasslands in New Zealand promoted soil carbon and nitrogen buildup, and enhancement of soil microbial biomass and carbon (but not nitrogen) mineralization rates (fig. 7.2b). This is consistent with the high quality of litter produced by *H. pilosella* relative to that of the resident vegetation. A further study (Gremmen et al. 1998) provided evidence that the invasion of the European grass *Agrostis*

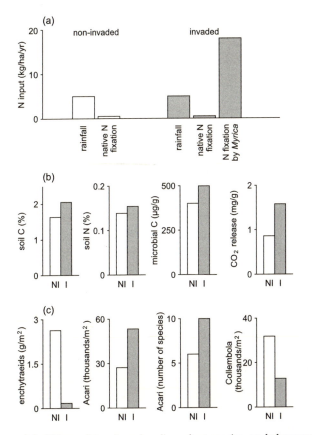

FIGURE 7.2. Effects of invasions by alien plant species on belowground properties. (a) Sources of nitrogen in stands of *Metrosideros polymorpha* in Hawaii not invaded and invaded by the actinorhizal shrub *Myrica faya* originating from the Canary Islands and Azores. From Vitousek et al. (1987). (b) Belowground properties in tussock grassland in New Zealand, under patches of the invasive European herb *Hieracium pilosella* (I) and adjacent herbfield (NI). From Saggar et al. (1999); the CO_2 release data are based upon 38 weeks of incubation. (c) Soil invertebrate densities in the subantarctic Marion Island, in natural vegetation dominated by the dwarf shrub *Acaena magellanica* (NI), and in vegetation dominated by the invasive grass *Agrostis stolonifera* (I). From Gremmen et al. (1998).

stolonifera into natural dwarf shrub communities in subantarctic Marion Island had important effects on many components of the soil decomposer fauna. However, these effects
were not unidirectional, with some groups being enhanced
by *A. stolonifera* but others being suppressed (fig. 7.2c).

There is also evidence that invasive plant species can increase the importance of disturbance regimes in ecosystems,
and D'Antonio and Vitousek (1992) and Mack and D'Antonio (1998) describe several examples in which invasion of
natural ecosystems by exotic grasses increases the incidence
of fire. For example, invasion of perennial C4 grasses in Hawaiian submontane forest results in accumulation of highly
flammable dead material which facilitates fires that kill most
of the native dominant vegetation. The grasses recover
quickly from fire, leading to a positive feedback between the
occurrence of fire and dominance by exotic grasses. The
consequences of dramatic ecosystem-level shifts such as this
are likely to significantly affect soil organisms and processes;
Ley and D'Antonio (1998) showed in Hawaii that rates of
net nitrogen fixation were much lower in a grassland dominated by alien grasses than in an adjacent natural woodland
because the main substrate upon which nitrogen fixation
occurs, i.e., leaf litter from native woody species, is absent
from the grassland.

While invasive plants may affect belowground properties,
supply rates of nutrients in the soil may in turn determine
the susceptibility of plant communities to invasion. It has
been proposed that plant communities become more susceptible to invasion when the amounts of unused resources
present are increased (Huston 1994; Davies et al. 2000) and
that invasion is most likely to be enhanced when intermittent resource enrichment coincides with the availability of
invading propagules (Davies et al. 2000). Factors that influence the degree of utilization of resources by the plant community will therefore determine the susceptibility of the
community to invasion; these include vegetation composi-

tion (e.g., Van der Putten et al. 2000), disturbance regime (e.g., Burke and Grime 1996) and grazing of the plant community (Olff and Ritchie 1998), and presence of appropriate root microbial symbionts [e.g., mycorrhizal fungi, nitrogen fixing bacteria (reviewed by Richardson et al. 2000)]. It has also been proposed that plant diversity per se reduces invasibility through reducing resource availability. However, the least diverse plant communities in nature are often those that are the most resistant to invasion (Wiser et al. 1998; Stohlgren et al. 1999), especially when low-diversity communities are dominated by plant species that are capable of competitive exclusion. Results of recent experimental studies based on synthetic communities of random assemblages of species, which claim to find that species richness reduces invasibility (e.g., Knops et al. 1999; Naeem et al. 2000b; Prieur-Richard et al. 2000), can be explained entirely in terms of "sampling effect" (see chapter 6), in which the most diverse experimental units are likely to be dominated by the most competitive species in the entire species pool, and therefore those species most likely to competitively exclude the invader (Wardle 2001). Further, the discrepancy between the results of observational and experimental studies appears to arise because whatever effects plant diversity might have in reducing invasability are not of sufficient strength to be manifested in natural ecosystems (see Levine 2000).

Introduction of herbivorous mammals to new areas has frequently resulted in their invasion of natural ecosystems, and while there are several instances in which these animals have undergone rapid population growth and greatly modified the native vegetation, there have been few attempts to understand their effects in an ecosystem context. Introduction of pigs (*Sus scrofa*) to new land masses has resulted in a vastly altered disturbance regime through their foraging for plant roots and this has been shown to result in significant plant death, soil organic matter loss, increased mineraliza-

249

tion rates, and reduced retention of nutrients in the ecosystem (see Mack and D'Antonio 1998). Further, removal of feral pigs from a Hawaiian forest was shown to promote populations of major soil mesofaunal groups over the following seven years (Vtorov 1993); over this time there was a 2.5-fold increase in total microarthropod biomass and a six-fold increase in the biomass of springtails. Introduction of rabbits (*Oryctolagus cuniculus*) to Australia and New Zealand has caused major modifications of both natural and managed grasslands; exclusion of rabbits by exclosure fences in the McKenzie Basin of New Zealand was found by McIntosh et al (1998) to cause a 62% increase in total root biomass and a 14% increase in soil nitrogen concentration. Introduction of deer and goats to areas lacking these animals (e.g., Hawaii, New Zealand) has induced dramatic effects on the composition of forest vegetation, and fenced exclosure studies in New Zealand have demonstrated that these animals change forest understory composition from dominance by faster-growing palatable broadleaved shrubs to dominance by unpalatable ferns, shrubs, and monocotyledonous species (e.g., Wallis and James 1972; Jane and Pracy 1974). These modifications could be expected to influence the functioning of the decomposer subsystem, and a recent study (Wardle et al. 2001) assessed a range of aboveground and belowground properties inside and outside each of 30 exclosure plots located in natural forests throughout New Zealand in order to investigate this. They found that the effects of browsing mammals on soil nutrients, microflora, microfauna, and carbon mineralization were often strong, although idiosyncratic, with both positive and negative effects on these variables being common. This suggests that different mechanisms through which browsing mammals affect soil organisms and processes operate in different forest communities (see chapter 4). Browsing mammal effects on larger soil invertebrates (e.g., mites, springtails, opilionids, millipedes, gastropods) were, in contrast, usually highly negative, and probably due to increased disturbance in the litter layer associated with the activity of these mammals.

Belowground Invasive Organisms

Most investigations of the ecological consequences of invasion by soil organisms of natural ecosystems have focused on soil macrofauna, although it is unclear whether larger soil organisms are inherently more invasive or if larger organisms are simply more likely to be noticed when they invade. The most significant demonstrated effects of invasive soil organisms involve earthworms. As outlined in chapter 5, invasion of the earthworm *Pontescolex corethurus* in Brazilian Amazonia is favored by clearing tropical rainforest, and the casting behavior of this species leads to reduced soil macroporosity and severe soil compression, which adversely affect native soil macrofauna and the potential for woody plants to recolonize (see Chauvel et al. 1999). Ecological impacts of earthworm invasion, both aboveground and belowground, are also apparent in several North American studies. Much of the North American earthworm fauna was eliminated by the most recent glaciations, and following European settlement many European earthworm species have become widespread. The European litter-dwelling earthworm *Dendrobaena octaedra* has invaded the natural forests of the Kananaskis Range of Alberta within the past 15 years. McLean and Parkinson (1998a,b, 2000a,b) have shown, through both mesocosm and field manipulation experiments, that this species significantly alters the community composition of both the soil mesofauna and soil fungi. Addition of earthworms to field plots was found to reduce the abundance of many groups of soil mites and favor domination by faster-growing fungal taxa. Further, Scheu and Parkinson (1994) set up a field experiment in which experimental chambers containing reconstructed soil profiles were placed in the field with and without the addition of *D. octa-*

edra. The earthworms exerted small but positive effects on soil mineral nitrogen concentrations, and stimulated microbial biomass in the LF layer but reduced it in the H layer. Shoot biomass and the shoot-to-root ratio of a phytometer (the grass *Agropyron trachycaulum*) were both enhanced by earthworm activity, suggesting that earthworms had the potential to increase nutrient supply rates to plants. Another example involves invasion by the Asian earthworm *Amynthas hawayanus* of a deciduous forest at Milbrook, New York, U.S.A. (Burtelow et al. 1998). Here, patches containing the earthworm had greatly elevated levels of microbial biomass and denitrifier enzyme activity when compared to adjacent noninvaded patches, apparently through earthworm effects on soil structure. This was associated with a 36% loss of organic matter from the soil O horizon. Native tallgrass prairie in North America contains an indigenous earthworm fauna whose species may be functionally distinct from the European earthworm species that invade it (James 1992). James and Seastedt (1986) found, through a pot experiment, that the native earthworm *Diplocardia* spp. and the European-sourced earthworm *Aporrectodea turgida* had comparable effects on nitrogen mobilization, although the native earthworm had a greater positive effect on root growth of *Andropogon gerardii* than did the exotic one.

Invasions by predatory soil-associated macrofauna may also be ecologically significant, although few studies have investigated this. Accidental introduction of the predatory New Zealand flatworm *Arthurdendyus triangulata* to the Faeroe Islands and the British Isles appears to have significantly reduced populations of the lumbricid earthworms upon which it preys, and Boag et al. (1999) found in a field site in Scotland that spatial abundance of earthworms was inversely related to spatial abundance of flatworms. The loss of earthworms due to flatworm predation in invaded Scottish grasslands appears to have adversely affected soil porosity and drainage, contributing to greater waterlogging

(Haria et al. 1998), and leading to greater vegetative domination by rushes (*Juncus* spp.) and reduced populations of moles (*Talpa europaea*) (Boag 2000). Further, the loss of earthworms may be expected to have important effects on soil microflora, invertebrates, and the processes that they carry out, ultimately affecting plant-available nutrient supply. Another example involves the introduction to subantarctic Marion Island of the house mouse (*Mus musculus*). Here, the endemic flightless moth *Pringleophaga marioni* functions as the primary macrofaunal detritivore and may process 50% of the annual litter production; predation of this moth by introduced mice has been estimated to reduce litter processing by this moth by up to 40% (Crafford 1990).

Other invasive predators may also compete with, and even displace, native predators. For example, the invasion of the South American red fire ant (*Solenopsis invicta*) in the southern U.S.A. has reduced the densities of native ant species at local spatial scales as well as the co-occurrence patterns of native species at wider scales (Gotelli and Arnett 2000). Niemelä et al. (1997) performed field manipulation experiments to study the effects of the invasive European carabid beetle *Pterostichus melanarius* on native carabid beetle assemblages in a deciduous forest in Alberta, Canada, and found that while the invasive species did not affect populations of the native species, there were apparent behavioral responses of the most abundant native species to *P. melanarius*.

LAND USE CHANGES

Land transformation by humans is probably the largest single terrestrial driver of global change phenomena. Over a third of the Earth's surface has been directly altered by human activity; land transformation emerges as the most important of all global change factors in determining biodiversity change (Sala et al. 2000), and contributes about 20% of current anthropogenic CO_2 emissions (Vitousek et al.

253

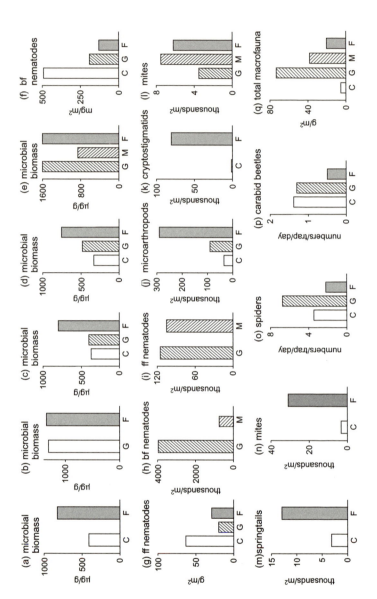

1997a). Land use change and the type of use adopted both have important effects on aboveground-belowground feedbacks through two primary mechanisms, i.e., by determining the nature of vegetative cover and therefore the quantity and quality of resources entering the soil, and by modifying disturbance regimes.

Comparisons of Forest, Grasslands, and Cropping Systems

Natural and human-transformed ecosystems often differ tremendously in terms of belowground properties. The majority of areas currently under agriculture were once forested, and conversion of natural forests to production systems has important effects on belowground organisms and processes. Several studies have demonstrated large differences in the composition of soil biota between adjacent forest, grassland, and cropping systems (fig. 7.3). The soil microbial biomass, representing the mass of the primary consumer in the soil food web, is usually much greater in forests than in cropping systems, and sometimes greater than in grassland systems (e.g. Srivastava and Singh 1991;

FIGURE 7.3. Densities or biomasses of soil biota in response to different land uses (C, cropping; G, grassland; M, managed or plantation forest; F, undisturbed forest). (a) Marodera, Zimbabwe. From Kirchmann and Eklund (1994); only 0–5 cm soil depth data presented. (b) Amazonia, Brazil. From Luizao et al. (1992); only 0–5 cm soil depth data presented. (c) Orissa, India. From Basu and Behera (1994). (d) Uttar Pradesh, India. From Srivastava and Singh (1991). (e) Near Rororua, New Zealand. From Ross et al. (1999); only 0–10 cm soil depth data used. (f) and (g) Throughout Poland. From Wasilewska (1979); bf indicates bacterial feeding, ff, fungal feeding. (h) and (i) Near Rotorua, New Zealand. From Yeates et al (2000), only data from highest planting density for plantation forest presented. (j) Near Guelph, Canada. From Winter et al. (1990). (k) Near York, England. From Sgardelis and Usher (1994). (l) Ibadan, Nigeria. From Badejo and Tian (1999). (m) and (n) Ibadan, Nigeria. From Critchley et al. (1979). (o) and (p) Northwest Poland. From Kajak and Lukasiewicz et al. (1994); data are for number of catches per pitfall trap per day. (q) Estimates for the humid tropics. From Lavelle et al. (1994).

Luizao et al. 1992; Basu and Behera 1994; Kirchmann and Eklund 1994; Ross et al. 1999; see fig. 7.3a–e). Microbe-feeding nematodes are frequently more abundant in arable and grassland systems than in forest (e.g.,Wasilewska 1979; Yeates et al. 2000; see fig. 7.3f,g), while forests support much greater numbers of microarthropods (e.g., Critchley et al. 1979; Winter et al. 1990; Badejo and Tian 1999; see fig. 7.3j–n). Soil macrofauna show a variety of responses depending upon the group of organisms present, but are frequently most numerous in grasslands (e.g., Kajak and Lukasiewicz 1994; Lavelle et al. 1994; fig. 7.3o–q). These trends are largely consistent with the predictions in chapter 3. Thus undisturbed systems are dominated by longer-lived plants, which have traits associated with lower litter quality and higher secondary metabolite production (i.e., trees), and lead to domination by fungal-based food webs in which microarthropods are important. In contrast, regularly disturbed systems dominated by plant species with relatively high litter quality (i.e., many agricultural crop species) tend to promote the bacterial-based energy channel and bacterial-feeding microfauna.

The low levels of some components of the soil biota in cropping relative to grassland and forest systems stem from disturbances due to tillage (discussed later), and loss of total soil carbon and nitrogen, resulting in fewer resources being available for those components of the soil food web that are bottom-up regulated. Loss of organic matter when natural systems are converted to tillage agriculture arises because decomposition rates exceed the rates of carbon input to the soil; cultivation of soils that have never previously been tilled often results in an initial reduction of total soil organic matter on the order of 20–40 % (Davidson and Ackerman 1993). The net result is likely to be a shift from bottom-up-regulated food webs dominated by the fungal energy channel to food webs in which the bacterial energy channel and the role of predation become increasingly important.

Comparisons between adjacent forested and managed grassland systems provide fewer consistent trends with regard to the relative levels of the various groups of soil biota that they support. This may be because there are some situations in which soil carbon and nitrogen levels are greater in forests than in grasslands and other situations in which the reverse is true; indeed, studies comparing carbon or nitrogen stocks in grassland vs. forest soils have found a variety of results (see Reiners et al. 1994; Veldkamp 1994; Neill et al. 1996, 1997; Corre et al. 1999; Rhoades et al. 2000). When a forest is converted to pasture (or vice versa), whether or not net soil organic matter loss occurs depends on the balance between carbon loss through mineralization and carbon gain through photosynthesis in the pasture. This is ultimately driven by the traits of the dominant plant species present, and whether the dominant pasture species have traits that are associated with encouraging the buildup of soil organic matter such as the production of recalcitrant litter (see chapter 3). In managed grasslands, another important consideration is the role of livestock; the proportion of NPP processed by herbivores is likely to be much greater in pasture than in forest. Given that grazing animals can affect both the soil biota and soil processes by a variety of mechanisms, many of which have opposite effects to each other (see chapter 4), a range of responses of the decomposer subsystem to conversion to grassland through the effects of livestock may be expected.

Intensification of Land Use

Although a substantial proportion of the Earth's land surface is devoted to production of resources for human use, the level of intensification of management within these production systems is itself a major contributor to global change. As management becomes more intensified, those processes governed by soil organisms become increasingly replaced by mechanical inputs and chemicals (Altieri 1991; Giller et al. 1997). Several experiments that have compared agricultural systems with differing levels of intensification provide clear evidence that intensity of management influences both the levels of soil biota supported and the processes that they regulate. A five-year field study at Kjittslinge, Sweden (Andrén et al. 1990; Paustian et al. 1990) involved comparisons of four agricultural systems of varying management intensity, i.e., barley (*Hordeum vulgare*) with and without nitrogen fertilizer, a meadow ley with nitrogen fertilizer, and an unfertilized lucerne (*Medicago sativa*) ley. Although the soil microflora did not show any consistent response to increasing intensification, biomasses of most groups of soil fauna were much greater in the lucerne treatment than in the barley-planted treatments; this was especially apparent for earthworms and microarthropods, suggesting that larger soil animals are disproportionately affected by intensification. Further, the contribution of soil fauna to carbon and nitrogen mineralization in the lucerne ley treatment was over double that in the barley treatments, indicating the greater importance of soil animals in contributing to plant nutrient supply when management intensity is reduced. Comparison of an integrated and a conventional farming experimental field (Lovinkhoeve) in the Netherlands (De Ruiter et al. 1993a,b) showed that intensification caused a reduction in the densities of microflora and microfauna, although the mesofauna were more variable in their responses. Earthworms had the highest biomass of all faunal groups and made the greatest contribution of all groups to nitrogen mineralization in the integrated treatment, but not in the conventional (i.e., more intensively managed) treatment (De Ruiter et al. 1993b, 1994). A seven-year field experiment in New Zealand involved independently manipulating three aspects of agricultural intensification, i.e., residue addition, cultivation, and herbicidal weed control, in both an annual and a perennial cropping system (Wardle et al. 1999c,d; Yeates et al. 1999b). Generally, the microflora and soil arthropod populations were promoted by those treatments that encouraged high levels of surface residues, and after the third year by treatments that allowed the presence of weeds at levels below the economic threshhold for crop production. Top predatory nematodes, but not microbe-feeding nematodes, were promoted by additions of residue (sawdust) to the soil surface, indicative of a tritrophic effect in which resource availability promoted microbes and top predators, but with top predators exerting top-down effects on microbe-feeding nematodes. Those treatments that promoted greater microbial biomass, arthropods, and top predatory nematodes (and which involved reduced intensification) also resulted in faster rates of decomposition of added litter and soil carbon mineralization.

One component of intensification of land use for production which is relevant to the soil biota is the selection of plant species; plant species differ in their effects on the amounts and quality of resources added to the soil (see chapter 3) as well as the degree of disturbance associated with their management. Several experimental studies have shown that planting different crop species, or using crop rotations that differ in the plant species included, can have strong effects on soil microbial biomass (e.g., Biederbeck et al. 1984; Granatstein et al. 1987; Drury et al. 1991), elements of the soil fauna (e.g., Fyles et al. 1988; Hance et al. 1990), and both the storage and mineralization rates of soil carbon and nitrogen (e.g., Odell et al. 1984; Wood and Edwards 1992). The use of different tree species in silvicultural and agroforestry systems also demonstrates the importance of species-specific effects on the soil microflora (e.g., Mao et al. 1992) and fauna (e.g., Badejo and Tian 1999; Vohland and Scroth 1999).

Multiple-species cropping, such as is done through intercropping (increasing diversity over space) or crop rotation (increasing diversity over time), may have beneficial effects on soil biota relative to monoculture cropping. Monoculture production usually involves the exploitation of organic matter for production (as occurs, for example, through the continuous cultivation of cereals). The effects of this can be at least partially mitigated by inclusion of species which have characteristics that contribute to organic matter buildup and the functioning of the decomposer subsystem (see Biederbeck et al. 1984; Odell et al. 1984; Anderson and Domsch 1990). A compelling example of the benefits of crop rotation over monocultures for the functioning of the decomposer subsystem is presented by Anderson and Domsch (1990). In that study, soils from 20 monoculture cropping plots and 21 plots under crop rotation were collected from Germany, Denmark, and Poland. While organic carbon and microbial biomass levels did not vary significantly between the two sets of soils, the microbial metabolic quotient (respiration loss per unit microbial biomass) was on average much greater for the soils from the monoculture plots, indicative of greater inefficiency of the soil microbial biomass (fig. 7.4). The ratio of soil carbon loss per unit microbial biomass was also greater for the monoculture soils, pointing to the potential of monoculture cropping to accelerate soil organic matter loss through selecting for microbial communities that are less effective at conserving carbon. It is important to note that these effects of multiple-species cropping need not necessarily be due to beneficial consequences of increasing plant species richness (cf. chapter 6); plant traits of those species grown in continual monoculture (e.g., cereals) differ markedly from the traits of those species added when intercropping or crop rotation is employed (e.g., green manuring species).

Enhancement of the intensity and frequency of disturbance is a major component of land use by humans. In agricultural systems, conventional tillage represents a major per-

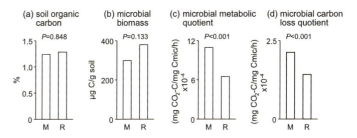

FIGURE 7.4. Soil carbon and microbial properties from an analysis of 41 soils from cropping monocultures (M) and rotations (R) in Germany, Denmark, and Poland. (a) and (b) are calculated from data presented by Anderson and Domsch (1989); (c) and (d) are from Anderson and Domsch (1990).

turbation which often results in losses of topsoil through erosion, reduced moisture retention, organic matter loss, and loss of soil structure (House et al. 1984; Phillips and Phillips 1984; Unger and Cassell 1991). Conservation tillage and no-tillage agriculture, which are becoming increasingly widely adopted, are intended to reduce these effects. Comparison of tilled and nontilled systems points to the role of disturbance in influencing soil food web structure and belowground processes. In a comparison of conventionally tilled and nontilled field plots at Horseshoe Bend, Georgia, U.S.A., Hendrix et al. (1986) demonstrated that larger soil animals such as earthworms were disproportionately affected by cultivation, and that tillage appeared to favor enchytraeids and bacterial-feeding nematodes relative to the fungal-feeding fauna. Beare et al. (1992) performed a litter bag experiment at the same site, in which litter was both buried and placed on the surface of tilled and nontilled plots. All groups of soil microflora, microfauna, and mesofauna were promoted by litter burial. Further, the importance of the bacterial-based energy channel relative to the fungal-based channel was much greater for the buried litter in the cultivated plots than for the surface-placed litter in the nontilled plots, and litter mass loss rates were much greater for the buried litter. Given that most plant residues are buried under conventional tillage and remain on the soil surface under nontillage, these results suggest that tillage promotes bacterial-based food webs and losses of organic matter. A large number of studies have compared the effects of tillage and nontillage treatments on specific groups of soil organisms, and an analysis of 106 published studies by Wardle (1995) revealed that across these studies larger soil animals are consistently more adversely affected by tillage than are smaller ones (fig. 2.10). This is consistent with tillage-induced disturbance promoting the bacterial-based energy channel in which nematodes and protozoa play an important role, and with lack of disturbance favoring the fungal-based energy channel and a greater involvement of microarthropods and earthworms.

Disturbances associated with production forestry also have important effects on belowground organisms and processes. In particular, clear-cutting and tree harvesting reduce the longer-term input of resources to the soil through both root production and litter return (even though there is usually a short-term pulse of residues), and these management practices are therefore most likely to adversely affect those components of the soil biota that are regulated by bottom-up control. Tree harvesting often has negative effects on the soil microbial biomass, at least during the period of the forest management cycle when resource addition is reduced (e.g., Bååth 1980; Pietikäinen and Fritze 1995; Chang and Trofymow 1996; but see Smolander et al. 1998); in forests most of the microbial biomass is fungal and therefore likely to be directly regulated by resource amount (see chapter 2). In contrast, soil nematodes, most of which are not strongly bottom-up regulated, show varied responses to tree harvesting, although significant changes in nematode community structure can occur (e.g., Armendáriz and Arpin 1996; Sohlenius 1996, 1997). Further, Sohlenius (1996) found clear-cutting to promote bacterial-feeding nematodes at the expense of fungal-feeding nematodes, pointing to disturbance favoring the bacterial-based energy channel. Larger soil animals, such as micro- and mesoarthropods, can be strongly adversely affected by clear-cutting (Blair and Crossley 1988; but see Setälä et al. 2000) and residue removal (Bengtsson et al. 1997); many of these arthropods derive their energy from the fungal-based energy channel, suggesting that multitrophic consequences of reduced resource availability are propagated through the fungal-based branch of the food web. Disturbances associated with clear-cutting, and consequent effects on the soil food web, can greatly increase microbial respiration per unit biomass and mineral nitrogen formation (Martikainen 1996; Smolander et al. 1998) although the reverse also occurs when reductions in decomposer biota following clear-cutting result in the retardation of process rates (Blair and Crossley 1988).

The examples of land use described above all point to intensification causing shifts in decomposer food web structure and soil processes, and an increasing role of the bacterial-based energy channel and microfauna at the expense of the fungal-based energy channel, arthropods, and larger soil invertebrates. Often the net result is the development of more open, leaky decomposer systems in which organic matter and nutrients are lost from the system at an increased rate. Vitousek and Matson (1984) found, through a ^{15}N addition experiment in a recently harvested forest in North Carolina, U.S.A., that microbial uptake of nitrogen during decomposition was the most important process for nitrogen retention in the ecosystem. Further, Tiessen et al. (1994) demonstrated that retention of soil organic matter was fundamental for sustaining long-term fertility and agricultural production, and was a key determinant of the length of time that agriculture could be maintained without supplementary fertilization. The effect of land use management intensification on the constituent organisms of the soil food web, and the consequences of this for soil processes, play a key role in determining the retention of nutrients and organic matter in the soil and therefore the productivity of management systems in the long term.

Land Use Abandonment

When land use intensification is relaxed, it is intuitive that the changes that occur will work in the opposite direction to those that took place when the land use was first introduced. Cessation of land use, such as occurs when agricultural land is abandoned, results in succession that often leads to forest development; this is especially apparent over much of eastern North America (Williams 1990). Successional development of vegetation that occurs on abandoned land, and on land that is being reclaimed or restored, is accompanied by the buildup of organic matter (e.g., Zak et al. 1990a; Wali 1999), soil microbial biomass (e.g., Insam and Domsch 1988; Sparling et al. 1994), and soil invertebrate populations (see Curry and Cotton 1983; Behan-Pelletier 1999). It is expected that there should also be a shift toward dominance by fungal-based (vs. bacterial-based) energy channels and invertebrates with larger body sizes (see chapter 3). However, when cultivated land is being restored the recovery of organic matter and associated soil biota may proceed quite slowly, given that there can be a considerable lag between plant production and storage of soil carbon (Burke et al. 1995). For example, Compton and Boone (2000) sampled coniferous and hardwood forests in the Harvard Forest, Massachusetts, U.S.A., which had been previously used for cropland, grassland, or woodlots, but where land use had been discontinued 90–120 years previously. They found that there were strong belowground legacy effects of the earlier land use, in that sites previously under cultivation had less carbon stored belowground, had greater amounts of phosphorus and nitrogen in the mineral soil layer, had lower soil carbon-to-nutrient ratios, and supported greater rates of nitrogen mineralization and especially nitrification, when compared with those sites that had always supported woodland (fig. 7.5). Belowground properties generally respond much more slowly to changes in environmental conditions than do aboveground properties, and the slower recovery of belowground properties relative to aboveground properties following abandonment of land use is therefore to be expected.

CARBON DIOXIDE ENRICHMENT
AND NITROGEN DEPOSITION

Anthropogenic activity is currently significantly influencing both the global carbon and nitrogen cycles. First, the concentration of atmospheric CO_2 is continually increasing largely through the burning of fossil fuels, and is expected to reach double preindustrial levels by the latter part of this century (Rotty and Marland 1986). Secondly, human activity has approximately doubled the rate at which nitrogen is entering the terrestrial nitrogen cycle, with the rate expected to continue increasing (Vitousek et al. 1997c). Fertilization of terrestrial ecosystems by both carbon and nitrogen as components of global change has the potential to significantly alter belowground communities and plant-soil feedbacks at the levels of both the individual plant and the plant community, and these effects will now be discussed.

Atmospheric CO_2 Enrichment and
Aboveground-Belowground Linkages

Increasing atmospheric CO_2 concentrations can affect the soil biota via three main mechanisms which will be considered in turn, i.e., alteration of the quantity of resources entering the soil through affecting NPP, modification of plant litter quality, and changes in the functional composition of plant communities. Direct effects of CO_2 enrichment on soil organisms are probably of little consequence because of the

265

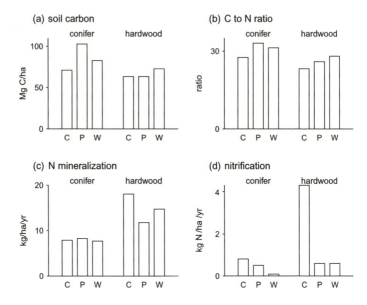

FIGURE 7.5. The effects of prior land use (C, cultivated; P, pasture; W, woodlot) in areas currently under coniferous or hardwood forest in the Harvard Forest, Massachusetts, U.S.A., on belowground properties and processes. Land use was discontinued and the land abandoned 90 to 120 years before present. (a) Soil carbon storage in the forest floor and top 15 cm of mineral soil; (b) carbon-to-nitrogen ratio of the forest floor; (c) rate of soil nitrogen mineralization; (d) net rate of soil nitrification. For all four panels, overall effects of prior land use are statistically significant at $P = 0.05$. Derived from Compton and Boone (2000).

high levels of CO_2 already present in soils (Van Veen et al. 1991) and will not be considered here.

Carbon dioxide enrichment can significantly promote NPP, including the total amount of litter return, the production of fine root biomass (Bazzaz 1990; Körner and Arnone 1992; Norby 1994), and the degree of rhizodeposition (Paterson et al. 1996; Rouhier et al. 1996; but see Schortemeyer et al. 2000). However, enhancement of belowground carbon input by increased CO_2 is not a universal pattern, and whether it occurs at all can depend on nitrogen availability.

For example, Pregitzer et al. (2000) found in a field experiment that elevated CO_2 increased the production and turnover of fine roots of *Populus tremuloides* seedlings under high soil nitrogen but had no effect under low soil nitrogen. Further, there is little consistency across studies as to how CO_2 elevation affects root vs. shoot allocation; in a review of literature for crop plants Rogers et al. (1996) found that 59.5% of studies found that increased CO_2 concentrations enhanced plant shoot-to-root ratios while 37.5% of studies found the reverse trend. Increased resources in the root zone caused by CO_2 elevation could be expected to enhance those rhizosphere organisms that are regulated by bottom-up control. For example, Vose et al. (1995) found CO_2 elevation to cause over a fivefold increase in fungal hyphal density in the rooting zone of *Pinus sylvestris* seedlings, and Rouhier et al. (1996) determined that microbial rhizosphere respiration rates per unit root mass of *Castanea sativa* seedlings were increased by elevated CO_2. However, Gorissen and Cotrufo (1999) detected little evidence of greater levels of rhizosphere microbial biomass or microbial immobilization of ^{15}N when three C3 grasses were subjected to greater CO_2 levels, and Schortemeyer et al. (2000) found that CO_2 enrichment did not significantly promote microbial activity or bacterial populations in the rhizosphere of *Quercus geminata* seedlings. Enhancement of carbon allocation to roots due to increased CO_2 also has the potential to enhance mycorrhizal colonization of plant roots, and although several studies have investigated this, the evidence is again mixed, with some studies finding positive effects [e.g., Klironomos et al. (1996) and Rillig and Allen (1998) for arbuscular mycorrhizae; Rygiewicz et al. (2000) for ectomycorrhizae] and others finding neutral effects [e.g., Monz et al. (1994) for arbuscular mycorrhizae; Gorissen and Kupyer (2000) for ectomycorrhizae].

Carbon dioxide enrichment may also affect soil organisms through altering the quality of litter returned to the soil.

The majority of studies have found CO_2 elevation to increase leaf or leaf litter carbon-to-nitrogen ratios (see syntheses by Coûteaux et al. 1999; Gifford et al. 2000; Körner 2000), but there are several exceptions, and effects can be positive or negative depending on plant species (Franck et al. 1997). Further, increasing nitrogen availability can reduce the negative effects of CO_2 on tissue nitrogen concentrations (Coûteaux et al. 1999), and in a field experiment, Hartwig et al. (2000) found that elevated CO_2 significantly reduced leaf nitrogen concentrations of *Lolium perenne* when it was grown alone but not when it was grown with the nitrogen fixing legume *Trifolium repens*. Carbon-to-nitrogen ratios for roots are often less affected by elevated CO_2 than are those for shoots; Billes et al. (1993) and Schenck et al. (1995) both found CO_2 enrichment to enhance shoot but not root nitrogen concentrations. Further, tissue lignin concentrations and lignin-to-nitrogen ratios can both be increased by elevated CO_2 (Cotrufo et al. 1994; see also Moore et al. 1999), but several studies have also found neutral or even negative effects of CO_2 enrichment (Coûteaux et al. 1996), and the literature synthesis of Coûteaux et al. (1999) found no evidence of a consistent trend. Increased carbon availability resulting from CO_2 enrichment may also increase plant allocation to soluble secondary compounds (e.g., phenolics) and tannins (Peñuelas and Estiarte 1998) and soluble sugars (e.g., Randlett et al. 1996).

It is apparent that CO_2 enrichment can therefore influence litter quality both by altering nitrogen concentrations of tissues, and by determining the relative amounts of different carbon compounds synthesized by the plant; these effects are well known to serve as important controls of decomposer organisms and processes. Several studies have compared the decomposition of litter produced under ambient and elevated CO_2, and again the effects are inconsistent, with some finding decomposition to be retarded by increased CO_2 (e.g., Coûteaux et al. 1991; Boerner and Rebbek 1995), but others finding little effect. For example, Hirschel et al. (1997) investigated the decomposition rates of litter of several species produced under ambient and elevated CO_2 conditions in each of three ecosystems, i.e., an alpine grassland and a lowland calcareous grassland in Switzerland, and a tropical rainforest in Panama. For nearly all species considered, there was no effect on decomposition of CO_2 enrichment. Further, Franck et al. (1997) compared the decomposition of litter produced under ambient and elevated CO_2 for each of four grassland species in a serpentine soil in California, U.S.A. Both positive and negative effects of CO_2 enrichment on both litter mass loss and nitrogen release were detected, and the direction of effects was highly species specific.

Although CO_2 enrichment effects on the quality and decomposition of plant litter produced by plants presumably operate through litter quality affecting the activity of the decomposer biota (see Körner 2000), few studies have explicitly investigated this link. However, Coûteaux et al. (1991, 1996) conducted an experiment in which *Castanea sativa* litter was produced under ambient and elevated CO_2 conditions, and both litter types were then incubated with each of four different decomposer food webs of increasing faunal complexity. They found that addition of fauna had much greater effects on CO_2 and nitrogen release from litter produced under elevated CO_2 conditions than that from litter produced under ambient CO_2 (fig. 7.6a). In particular, at the end of the experiment, decomposition of CO_2-enriched litter with only microbes and protozoa was much less than that for the nonenriched litter, while for the most complex food web assemblages which also included saprophages the reverse was true. However, Cotrufo et al. (1998) found that addition of saprophagous isopods accelerated the decomposition of *Fraxinus excelsior* litter produced under ambient CO_2 conditions but not that produced under elevated CO_2 (fig. 7.6b). Meanwhile, Hättenschwiler et al.

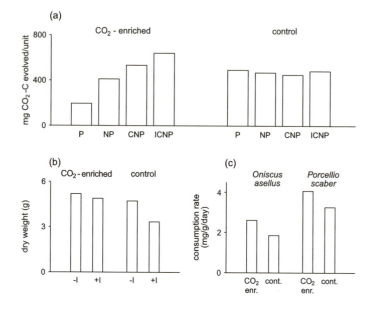

FIGURE 7.6. Effects of soil faunal additions on decomposition of plant litter prepared under ambient (control) and elevated CO_2 concentrations. (a) Net CO_2 evolution over 24 weeks from experimental units containing *Castanea sativa* litter and amended with different soil faunal assemblages. P, protozoa; N, nematodes; C, Collembola; I, isopods. From Coûteaux et al. (1991). (b) Mass remaining of leaf litter disks of *Fraxanus excelsior* after three days incubation without (−I) and with (+I) the isopod *Oniscus asellus*. From Cotrufo et al. (1998). (c) Consumption rate of *Fagus sylvatica* leaf litter by two different isopod species. From Hättenschwiler et al. (1999).

(1999) found that two isopod species each consumed greater amounts of *Fagus sylvatica* litter produced under elevated than under ambient CO_2 conditions (fig. 7.6c). In this case, the litter produced under elevated CO_2 had a lower concentration of nitrogen than that produced under ambient CO_2, and the animals presumably therefore needed to consume more litter in order to acquire sufficient nitrogen. This suggests that while CO_2-enriched litter may decompose more slowly in the absence of saprophagous fauna ("litter transformers"; see chapter 2), greater communition of CO_2-enriched litter by these animals may actually accelerate decomposition of this litter relative to that of nonenriched litter. This would also explain why addition of saprophagous animals increased carbon and nitrogen release from CO_2-enriched litter relative to that of nonenriched litter in the study of Coûteaux et al. (1991).

While the mechanisms described above explain how physical and chemical changes at the individual plant level in response to CO_2 enrichment may affect the soil biota, the effects of increased CO_2 on plant community structure could also conceivably have belowground consequences. Few studies have explicitly investigated how plant community structure respondes to CO_2 enrichment (see Bazzaz 1990), or what plant traits may be relevant in determining the relative responses of different species in a community to elevated CO_2 (Díaz 1995). However, experimental studies on grassland plant communities have shown CO_2 enrichment to favor legumes at the expense of nonlegumes (Newton et al. 1995; Lüscher 1996), early successional and forb species at the expense of grasses (Potvin and Vasseur 1997), and C3 forbs at the expense of C3 grasses (with C4 grasses showing no response) (Owensby et al. 1993). In an arid ecosystem in the Mojave Desert, Nevada, U.S.A., Smith et al. (2000) found that free air CO_2 enrichment favored an adventive annual grass (*Bromus tectorum*) at the expense of native annuals, which could lead to an accelerated fire cycle that further maintains the adventive community. Modeling studies have predicted CO_2 enrichment to favor dominance by plant species with high nutrient use efficiency (Herbert et al. 1999a), and C3 grasses over C4 grasses as a result of the greater photosynthetic efficiency of the former under enriched CO_2 (Collatz et al. 1998). Several of the above studies point to CO_2 enrichment favoring faster-growing

271

Effect of CO$_2$ enrichment (% of control)

PRIMARY CONSUMERS

Microbial biomass
- (a) herb community — *
- (b) *Populus gradidentata* — *
- (c) pasture turves — NS
- (d) *Lolium perenne Trifolium repens* — * / NS
- (e) *Artemisia tridentata:* low N — NS; high N — *
- (f) subalpine grassland: low N — NS; high N — NS
- (g) *Danthonia richardsonii:* low N — NS; high N — *
- (h) *Quercus germinata* — NS
- (i) *Populus tremuloides:* low N — NS; high N — NS

SECONDARY CONSUMERS

Protozoa
- (j) *Brassica nigra* — NS
- (k) sandstone grassland — NS; serpentine grassland — NS

Total nematodes
- (e) *Artemisia tridentata* low N — *; high N — *

m.f. nematodes
- (k) sandstone grassland — NS; serpentine grassland — NS
- (l) cotton plants — P=0.09

b.f. nematodes
- (m) pasture turves — *

f.f. nematodes
- pasture turves — *

Total microarthr.
- (n) *Populus x euramericana* — NS

m.f. microarthr.
- (e) *Artemisia tridentata* low N — NS; high N — *

Collembola
- (o) *Populus tremuloides* low N — *; high N — NS

m.f. mites
- *Populus tremuloides* low N — NS; high N — NS

TERTIARY CONSUMERS

Predatory nematodes
- (m) pasture turves — *

Predatory microarthropods
- (e) *Artemisia tridentata* low N — NS; high N — NS

Predatory mites
- (o) *Populus tremuloides* low N — NS; high N — NS

plant species, which are known to produce superior litter quality, which could in turn work to stimulate certain components of the soil biota.

Given the variety of mechanisms by which CO_2 enrichment can affect the quality and quantity of resources that plants add to the soil, it may be expected that soil organisms also show a variety of responses to enriched CO_2 as mediated by the plant. The available evidence suggests that this is indeed the case (fig. 7.7). Several studies have found neutral responses of the soil microbial biomass to CO_2 enrichment (Ross et al. 1995; Niklaus and Körner 1996; Schortemeyer et al. 2000; Zak et al. 2000a), while others have found strong positive effects (e.g., Díaz et al. 1993; Zak et al. 1993), or large positive effects under high-nitrogen but not low-nitrogen conditions (Klironomos et al. 1996; Lutze et al. 2000). These effects can be propagated through to higher trophic levels; secondary and tertiary consumers have been found to be either stimulated (Klironomos et al. 1996; Yeates et al. 1997) or largely unaffected (Klironomos et al. 1997; Treonis and Lussenhop 1997; Lussenhop et al. 1998; Hungate et al. 2000) by CO_2 enrichment, and two studies have even found fungal-feeding fauna to be significantly reduced by elevated CO_2 (Klironomas et al. 1997; Yeates et al. 1997). The variety of responses of the soil biota to elevated

FIGURE 7.7. Effect of CO_2 enrichment on biomasses or populations of consumer trophic levels in the decomposer food web. (a) Díaz et al. (1993); (b) Zak et al. (1993); (c) Ross et al. 1995; (d) Schenck et al. (1995), highest-planting-density data presented only; (e) Klironomos et al. (1996); (f) Niklaus and Körner (1996); (g) Lutze et al. (2000); (h) Schortemeyer et al. (2000); (i) Zak et al. (2000a), 1996 data presented only; (j) Treonis and Lussenhop (1997); (k) Hungate et al. (2000), final sampling date data presented only; (l) Runion et al. (1994), August data presented only; (m) Yeates et al. (1997); (n) Lussenhop et al. 1998; (o) Klironomos et al. (1997). Notes: b.f. indicates bacterial feeding; f.f., fungal feeding; m.f. microbe-feeding; * and NS, CO_2 enrichment effect significant and nonsignificant, respectively, at $P = 0.05$.

CO_2 in turn results in considerable variation across studies in terms of mineralization processes regulated by the soil biota. In a synthesis of published literature, Zak et al. (2000b) found that while most studies showed CO_2 enrichment to promote microbial and total soil respiration, there was considerable variation across studies in terms of effects on microbial immobilization and mineralization of nitrogen.

There are four likely reasons for the variety of responses shown by the soil biota to CO_2 enrichment. First, as described earlier, enhancement of CO_2 exerts different effects on different plant species, which will in turn result in the soil biota associated with different plant species varying in their responses. Evidence for this is presented by the study of Hungate et al. (1996a), in which the response of total soil microbial nitrogen to CO_2 enrichment was influenced by which of six grassland plant species was grown. Secondly, the effect of increased CO_2 on the plant-soil system is likely to be affected by the availability of nutrients, particularly nitrogen. Enrichment by CO_2 is more likely to promote plant productivity, and therefore potentially the soil biota, in situations where nitrogen is not limiting to plant growth. The studies of Klironomos et al. (1996) and Lutze et al. (2000) suggest that this is indeed the case; soil organisms were promoted by CO_2 enrichment only when nitrogen was also added (fig. 7.7). However, other studies have found the effects of CO_2 enrichment to be independent of nitrogen addition (Niklaus and Körner 1996; Zak et al. 2000a) (fig. 7.7). Thirdly, different subsets of the soil biota are regulated to varying degrees by top-down and bottom-up control (see chapter 2); only those organisms that are regulated by resource availability are likely to respond strongly to CO_2 enrichment. For example, the studies of Yeates and Orchard (1993) and of Yeates et al. (1997) found that CO_2 enrichment of NPP of pasture turves resulted in little increase in the soil microbial biomass, modest increases in microbe-

feeding nematodes, and very large increases in top predatory nematodes. This suggests that there were very strong
top-down controls of the primary and secondary consumers,
with only the tertiary consumer significantly increasing its
total biomass as a result of increased basal resource availability. The fourth reason is that CO$_2$ enrichment often pro- (4)
motes increases of some species within a given trophic
grouping and reductions in others, resulting in no net
change necessarily occurring at the coarser levels of taxonomic resolution used in most studies. Enhancement of
CO$_2$ has been shown to change community structure for soil
microbes (e.g., Zak et al. 1996; Elhottová et al. 1997), nematodes (e.g., Yeates et al. 1997), protozoa (Treonis and Lussenhop 1997), and microarthropods (e.g., T. H. Jones et al.
1998). In this last study, elevated CO$_2$ increased soil concentrations of dissolved organic carbon by over 50%, and while
this did not affect total soil microbial biomass, there were
significant effects on fungal community structure (and on
the functional composition of the fungal community) which
resulted in corresponding changes in the species composition of fungal-feeding springtails.

Carbon dioxide enrichment effects on the plant-soil system may also induce longer-term ecosystem-level changes
through affecting soil carbon sequestration. Carbon storage
is enhanced when the producer subsystem adds carbon to
the soil at a rate faster than the rate at which it is mineralized by the decomposer subsystem. Positive effects of CO$_2$
elevation on both NPP and the carbon-to-nitrogen ratios of
organic matter input to the soil could therefore be expected
to enhance carbon sequestration in the soil, and this is supported by the predictions of several modeling studies (see
Post et al. 1992; McKane et al. 1997; Lloyd 1999; Schimel et
al. 2000). Some experimental studies have also found elevated CO$_2$ to promote soil carbon storage (e.g., Owensby et
al. 1993; Rice et al. 1994; Wood et al. 1994), and the magnitude of these increases is greater than can be explained

solely in terms of increased NPP (Hungate et al. 1996b). In contrast, in a three-year field experiment in both a grassland and a serpentine system in coastal California, U.S.A., Hungate et al. (1997) found that CO_2 enrichment did not cause a significant increase in soil carbon storage but instead increased allocation of carbon to rapidly cycling belowground carbon pools, leading to a greater flux of CO_2 from the soil. This may arise through microorganisms shifting from consuming older soil organic carbon under ambient CO_2 to consuming easily degradable rhizodeposits resulting from enhanced root production under elevated CO_2 (Cardon et al. 2001). Further, in a model ecosystem study of tropical plants (Körner and Arnone 1992), CO_2 enrichment was observed to actually decrease soil carbon storage as a result of stimulation of rhizosphere activity. This is consistent with a literature synthesis of 41 previously published comparisons, which found that CO_2 enrichment increased soil respiration on average by 45% (Zak et al. 2000b).

There are two reasons why some experimental and modeling studies may overestimate the degree of enhancement of carbon storage due to elevated CO_2. First, most experiments involve a step increase in CO_2 concentration from ambient to elevated, while in real ecosystems the increase in CO_2 is very gradual (Wardle et al. 1998b). In a modeling study, Luo and Reynolds (1999) predicted that introducing a step increase in CO_2 concentration could greatly overestimate the initial rate of ecosystem sequestration of carbon. Secondly, buildup of soil carbon under elevated CO_2 ultimately depends on promotion of NPP being sustained in a high-CO_2 environment. However, biotic feedbacks may work against this in the long term because elevated CO_2 can enhance carbon-to-nitrogen ratios of organic matter, leading to reduced nutrient availability. There is evidence for this in some studies; for example, Díaz et al. (1993) found in a growth chamber experiment that elevated CO_2 initially stimulated NPP and therefore soil microbial biomass, which appeared to lead to the microbial biomass in turn effectively competing with the plant community for nutrients and thus damping further stimulation of NPP. However, in contrast, Hu et al. (2001) recently provided evidence from a field experiment in a grassland in coastal California, U.S.A., that elevated CO_2 may favor utilization of nitrogen by plants over that of microbes, and proposed that this would lead to increasing ecosystem carbon accumulation through slowing decomposition. Comparison of the results of Díaz et al. (1993) with those of Hu et al. (2001) therefore suggests that the effect of CO_2 enrichment on the competitive balance between plant biomass and soil microbial biomass for available soil nitrogen may be an important determinant of whether elevated CO_2 promotes carbon sequestration in the soil.

Nitrogen Deposition and Plant-Soil Feedbacks

The ecological consequences of nitrogen deposition make an interesting comparison with those of CO_2 enrichment because fertilization by nitrogen and by carbon have opposite implications for the carbon-to-nitrogen balance of the ecosystem. Although both nitrogen and carbon fertilization usually enhance NPP, increasing nitrogen availability favors the production of shoot relative to root tissue (see Chapin 1980), promotes reduction of carbon-to-nitrogen ratios of plant tissues (e.g., Wedin and Tilman 1996), and favors plant species that grow faster and are more competitive in the presence of high concentrations of mineral nitrogen. Increased nitrogen availability causes major functional shifts in vegetation composition, for example through favoring grasses and forbs over nitrogen fixers in grassland (e.g., Huenneke et al. 1990), grasses over other species in alpine tundra (Theodose and Bowman 1997), grasses and forbs over dwarf shrubs in boreal forest understory (Turkington et al. 1998), and herbs and deciduous shrubs over evergreen shrubs in Arctic tundra (Jonasson et al. 1999). High nitrogen deposition has also been shown to promote successional change; in the Netherlands, which experiences some of the world's highest levels of nitrogen deposition, increased nitrogen availability causes shifts from infertile and unproductive species-rich heathlands to productive species-poor grasslands and forests characteristic of more fertile habitats (Aerts and Berendse 1988). The effects of nitrogen deposition on individual plants, plant community structure, and plant succession should all operate to enhance both the amounts and availability of resources entering the decomposer subsystem.

Despite the relative predictability of plant and plant community responses to nitrogen enrichment, the responses of belowground organisms and processes are far less predictable. Fertilization studies in a range of ecosystems have found nitrogen addition to have effects on soil microbial biomass (i.e., the primary consumer of the decomposer food web) that are positive (e.g., Hart and Stark 1997; Scheu and Schaefer 1998), variable (e.g., Bardgett et al. 1999a; Ettema et al. 1999a), and negative (e.g., Söderström et al. 1983; Smolander et al. 1994). Positive effects of fertilizer addition on microbial biomass are indicative of nitrogen limitation of the microflora, and neutral effects are explicable in terms of the microflora being more limited by carbon than by nitrogen; microcosm studies have shown that microbial production responds much more strongly to nitrogen and carbon in combination than to either element alone (e.g. Bååth et al. 1978; Elliott et al. 1983; Vance and Chapin 2001). Why nitrogen addition should have an adverse effect on the microbial biomass is less clear, but it may result from the influence of nitrogen addition on osmotic potential, pH, or the formation of recalcitrant substrates (see Fog 1988; Hart and Stark 1997). The variable effect of nitrogen addition on soil microflora is in turn reflected by the response of decomposition rates to added nitrogen, which can also be positive (e.g., Hunt et al. 1988), negative (e.g., Nohrstedt et al. 1989), and neutral (e.g., Koopmans et al. 1998) (see the review by Fog 1988). Further, it has been shown using litter bag studies that chronic nitrogen addition can both stimulate tree litter decay rates through promoting microbial cellulase activity, and reduce decay rates through suppressive effects on ligninolytic enzyme activity (Carreiro et al. 2000).

The variable effects of nitrogen addition on the microflora are also apparent in the higher trophic levels of the soil food web. Ettema et al. (1999b) showed, through a fertilization experiment in a riparian forest in Georgia, U.S.A., that nitrogen addition initially promoted bacterial-feeding (but not fungal-feeding) nematodes, but this effect was transitory, and followed by an increase in top predatory nematodes. This is indicative of the effects of nitrogen fertilization being propagated through three consumer trophic levels of the bacterial-based energy channel. However, there was little evidence of nitrogen fertilization of a barley crop influencing soil nematodes, mites, springtails, or earthworms during a five-year field study at Kjittslinge, Sweden (Andrén et al. 1990). Further, Scheu and Schaefer (1998) found that addition of nitrogen to plots in a *Fagus sylvatica* forest near Göttingen, Germany, had little consistent effect on most components of the saprophagous and predatory soil macrofauna, despite microbial biomass being promoted in the litter layer. Addition of nitrogen may be expected to affect those components of the soil food web that are regulated by bottom-up control, and variable responses of the soil fauna may therefore be expected. Based on the material in chapter 2, we would expect certain groups of soil biota to be more responsive to improved nitrogen availability (i.e., groups regulated by bottom-up control, such as fungi, top predatory nematodes, microarthropods) than other groups (i.e., those regulated by predation, such as bacteria and microbe-feeding microfauna). Evidence that fungal biomass is indeed more responsive than bacterial biomass to nitrogen addition has been presented in two experimental studies

(Entry and Backman 1995; Bardgett et al. 1999a), and strong top-down regulation of lower trophic levels by top predatory nematodes following nitrogen addition is apparent in the results obtained by Ettema et al. (1999b). In the longer term, however, increased soil fertility and corresponding vegetation changes resulting from sustained nitrogen addition may be expected to lead to domination by the bacterial energy channel at the expense of the fungal channel, and consequently more rapid cycling of nutrients (Bardgett et al. 1996; see also chapter 3).

Effects of nitrogen deposition on the plant-soil system have the potential to promote carbon sequestration. If, as has been suggested, nitrogen is often more important in influencing production than decomposition (Hunt et al. 1988; Vitousek et al. 1997c; Hobbie and Vitousek 2000), then nitrogen addition would be expected to promote net accumulation of carbon both in standing plant biomass and in the soil. Consistent with this, experimental studies point to nitrogen addition promoting net carbon accumulation (Billings et al. 1984; Wedin and Tilman 1996; Johnson et al. 2000) although the extent of this accumulation is likely to be determined by plant community composition (Wedin and Tilman 1996). Further, there are likely to be strong interactive effects of nitrogen deposition and CO_2 enrichment in promoting carbon sequestration; in a modeling study Lloyd (1999) predicted that total carbon sequestration in boreal and temperate forests would be enhanced considerably more by elevated CO_2 and nitrogen deposition operating in combination than by either factor operating on its own. It has also been proposed that ecosystems subjected to high nitrogen deposition rates could serve as large localized sinks for CO_2 (Schimel 1995; Lloyd 1999) and may represent a major component of the missing carbon sink in global carbon cycling models (Schimel 1995). The importance of this mechanism has been disputed at least for northern temperate forests; Nadelhoffer et al. (1999) suggested, based on a ^{15}N tracer study for each of nine such forests, that current levels of elevated nitrogen deposition would be insufficient to make a major contribution to the missing sink. However, it has also been suggested that the methodologies and calculations used in that study could both lead to significant underestimates of the level of carbon sequestration likely to result from nitrogen deposition (Jenkinson et al. 1999; Sievering 1999). Although nitrogen deposition is able to promote carbon sequestration in many, perhaps most, ecosystems in which it occurs, the extent to which this is likely to be important in mitigating atmospheric CO_2 increases due to anthropogenic activity remains unclear.

GLOBAL CLIMATE CHANGE

Global CO_2 enrichment is predicted by most models to result in a mean increase in global temperature of around 3°C over the next century, as well as a greater incidence of extreme climatic events and enhanced temporal variability in patterns of rainfall (see Mearns et al. 1984; Schneider 1992; Timmermann et al. 1999). This may be expected to influence the quantity and quality of resources entering the decomposer subsystem through affecting NPP, the chemical composition of organic materials produced by plants, and vegetation composition. Increased temperature can promote NPP both directly through influencing plant growth, and indirectly by increasing the mineralization rates of plant-available nutrients (see Nadelhoffer et al. 1992; Schimel et al. 1994). Further, increased mineralization rates of organic matter resulting from higher temperatures may lead to greater net uptake of nutrients by plants (see Jonasson et al. 1999), potentially influencing the quality of litter that the plants produce (Nakatsubo et al. 1997).

In the longer term, the most important effects of climate change on the soil biota will probably occur indirectly via shifts in the functional composition of vegetation (Anderson 1991; Wardle et al. 1998b), especially given the role of climate in governing plant distributions (Woodward 1987) and the functional composition of vegetation (Box 1981, 1996; Walter 1985), both spatially and temporally. Díaz and Cadibo (1997), Díaz et al. (1998), and Grime et al. (2000) have all provided evidence that characterization of the functional type composition of plant communities may have value in predicting vegetation responses to climate change, and Díaz et al. (1998) demonstrated for a range of vegetation types along an elevational gradient in Argentina that there were clear shifts in key functional traits of the dominant vegetation which corresponded to changes in both temperature and precipitation. In particular, cooler, moister areas selected for plant species with larger specific leaf areas and shorter leaf lifespans, which, based on the material presented in chapter 3, could have important implications for leaf litter quality and ultimately the performance of the decomposer subsystem. Modeling and experimental studies also provide evidence of large shifts in vegetation likely to result from global warming that are relevant for the soil biota. For example, the modeling study of Starfield and Chapin (1996) predicted that a temperature increase of 3°C would be sufficient to induce some tundra communities to convert to coniferous forest within a time frame of 150 years, with the proportion of deciduous trees subsequently increasing as a result of greater fire frequency. Further, Pastor and Post (1988) used simulation models to show that as a result of global warming, boreal mixed spruce-hardwood forests in northeastern North America would be replaced by deciduous forests within three to four centuries, resulting in increasing domination by plant species that produce litter of superior quality. However, global warming may also promote plant species with poorer litter quality. For example, in temperate grasslands higher temperatures are expected to favor C4 grasses over C3 grasses (Hattersley 1983; Campbell et al. 1999); C4 species generally produce lower-quality litter than C3 species. Elevated temperature has also been shown through experimental field studies to exert important effects on vegetation functional composition, for example by favoring sedges and shrubs over grasses and non-vascular plants in an Arctic tundra (Hobbie et al. 1999), shrubs at the expense of forbs in a subalpine meadow (Harte and Shaw 1995), and short-lived fast-growing species over perennial grasses in a fertile (but not in an infertile) temperate grassland (Grime et al. 2000).

Although there is evidence that the elevated temperatures which result from global warming may influence the amounts and quality of resources entering the decomposer subsystem, there is little consistent evidence from global data sets that the biomasses or densities of the main groups of soil organisms will respond directly to a temperature increase of the order of 3°C. Syntheses of data from broad climatic ranges indicate only relatively weak relationships between densities of soil organisms and mean annual temperature. With regard to the primary decomposer of the decomposer food web, Insam (1990) found microbial biomass to be negatively correlated with temperature for a range of cropping soils throughout North America (fig. 7.8a), and in a global literature synthesis Wardle (1992) found microbial biomass to be significantly but weakly negatively correlated to temperature in arable and grassland, but not forested, ecosystems. However, in these studies, data were compiled from a very broad range of temperatures, and they provide little evidence to suggest that the effects of a temperature increase on the order of 3°C would be particularly significant. Similarly, literature syntheses for major soil faunal groups [e.g., enchytraeids (Didden 1993) (fig. 7.8b); springtails and mites (Petersen and Luxton 1982) (fig. 7.8c,d)] do not suggest that the strength of the relationship between temperature and population density is sufficient for the sorts of temperatures associated with global warming to

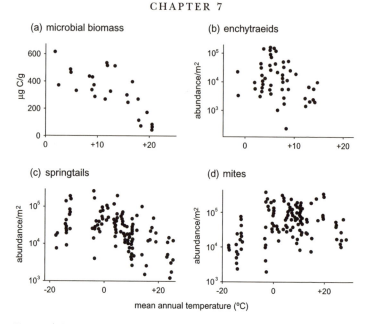

FIGURE 7.8. Changes in biomasses or densities of soil biota along large temperature gradients. (a) Agricultural sites from throughout North America. From Insam (1990). (b) Global data set from a range of natural and managed ecosystems. From Didden (1993). (c) and (d) Global data set from a range of natural ecosystems. From Petersen and Luxton (1982).

strongly influence their densities. Larger soil invertebrates, such as geophagous earthworms and termites, become much more numerous at lower latitudes (Lavelle et al. 1995) but again the rate of change of their densities in response to latitude is probably insufficient to result in a 3°C increase of temperature being of much significance, except perhaps at the periphery of the climatic ranges of particular species.

Few experimental studies have investigated the effects of temperature increases of the order of 3°C on soil organisms, and there is little consistency in the results obtained among those that have. In experimental studies using model ecosystems, elevation of temperature has been found to exert either variable or nonexistent effects on the soil microbial biomass (Sarathchandra et al. 1989; Kandeler et al. 1998; Bardgett et al. 1999c) and community composition (Kandeler et al. 1998; Bardgett et al. 1999c). In the study of Bardgett et al. (1999c) no microbial response was observed despite elevated temperature increasing plant productivity. However, Arnold et al. (1999) found that elevation of temperature by 5°C in both field plots and laboratory incubation experiments significantly reduced the microbial biomass, possibly through increased temperature either reducing soil moisture or accelerating the mineralization and loss of labile organic matter. With regard to soil animals, experimental studies on microarthropods in arctic tundra in West Spitzbergen, Svalbard, have shown that increased temperatures would be unlikely to cause large changes in their populations, and it is apparent that these arthropods are relatively insensitive to moderate changes in temperature (Hodkinson et al. 1996; Webb et al. 1998). However, experimental warming of plots in tundra at Abisko, Sweden, was found to significantly promote microbial biomass, and both bacterial- and fungal-feeding nematodes (Ruess et al. 1999). Further, Harte et al. (1996) showed that experimental heating of field plots in a subalpine meadow in Colorado, U.S.A., reduced biomass of soil meso- and macrofauna in dry zones of plots but enhanced biomass in moist zones. Transplantation of intact soil profiles to areas of differing climatic regime also provide some evidence that elements of global climate change, including global warming, may influence densities of soil fauna. Briones et al. (1997) transplanted soil cores collected from a site in Cumbria, England, to a lower elevation, resulting in the cores being subjected to an increase in mean annual temperature of 2.5°C and drier conditions. This change in climate resulted in reduced densities of soil dipteran larvae, and elevated densities of some species of enchytraeids at the expense of others. In another transplant experiment,

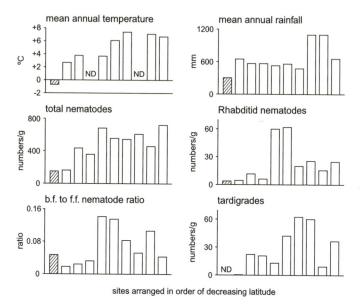

FIGURE 7.9. Results of a transplant study in which soil cores were removed from a site at Abisko in the Swedish Arctic, and transferred to nine other sites throughout Sweden of varying latitude and macroclimate. The Abisko site (68°22′N) is represented on each panel by a shaded bar. The other nine sites range in latitude from 63°48′N to 55°39′N. b.f. indicates bacterial feeding; f.f., fungal feeding; ND, not determined. Derived from Sohlenius and Boström (1999).

Sohlenius and Boström (1999) moved soil cores from Abisko, Sweden to each of nine sites of contrasting macroclimate located throughout Sweden. Warmer and moister sites tended to promote total nematode populations (and in particular rhabditid nematodes) and tardigrades, as well as the ratio of fungal-feeding to bacterial-feeding nematodes (fig. 7.9). The increased role played by fungal-feeding nematodes in some of the higher-temperature sites is indicative of a shift from the bacterial-based energy channel to the fungal-based energy channel being caused by warmer conditions.

Those experimental studies that have found the strongest

soil biotic responses to increased temperature tend to involve situations in which soil moisture content probably covaries with temperature. Global climate change is likely to involve shifts in rainfall patterns, and elevated temperatures should lead to greater evapotranspiration; both are likely to impact upon soil moisture levels and the frequency of soil wetting-drying cycles. Manipulation of soil moisture status has been shown in several studies to influence the soil microbial biomass, and regular wetting-drying cycles therefore promote greater turnover of the soil microbial biomass (e.g., McGill et al. 1986; Ross 1987). This may result either from desiccation following drying (Bottner 1985) or from osmotic stress created by rewetting dry soil (Kieft et al. 1987). However, microbial biomass in sites that are regularly dry appears to be more resistant to wetting-drying cycles than that in sites that are consistently moist, indicative of some degree of preadaption of the microbial community (West et al. 1988). Higher trophic levels of the soil food web are also responsive to alterations in moisture regimes. In a field study in Kjittslinge, Sweden (Schnürer et al. 1986), moistening of a droughted soil by either irrigation or natural rainfall caused large increases in both fungal hyphal length and bacterial densities, which was followed by increases in naked amoebae and nematodes. Steinberger and Sarig (1993) found in the Negev Desert, Israel, that diurnal fluctuations of soil moisture served as the primary determinant of both the soil microbial biomass and estimated populations of soil nematodes. In contrast, Bakonyi and Nagy (2000) determined through manipulation of temperature and moisture in field plots in a grassland in Hungary that the strongest treatment effects resulted from increasing soil moisture, which reduced soil nematode densities. Further, soil moisture has differential effects on different components of the soil biota. Bacteria, nematodes, and protozoa are dependent upon moisture films for their mobility while fungi and microarthropods are not (Wardle et al. 1998b; see chapter 2), so the fungal-based energy channel is less likely to be deactivated through desiccation. Microfauna are in general more capable of tolerating drying events than are larger soil animals, because of their ability to exhibit effective survival strategies through cryptobiosis (Freckman 1978; Demure et al. 1979). Microarthropods can demonstrate a range of physiological and behavioral adaptations to enable them to tolerate or avoid desiccation (reviewed by Joose and Verhoef 1987). However, resistance to desiccation of larger soil animals is generally relatively poor and many species appear to lack effective strategies for the conservation of moisture under dry conditions (Lavelle et al. 1997).

Global warming phenomena may also have important effects in cooler areas through influencing the density of snowpacks. A deep snow layer has insulating effects and can elevate soil surface temperatures to close to 0°C (Geiger 1950), and therefore well within the range suitable for the activity of many soil organisms. The presence of a deep snowpack has been shown in alpine tundras in the Colorado Rocky Mountains, U.S.A., to promote activity of soil microarthropods (Addington and Seastedt 1999), soil microbial biomass (Brooks et al. 1998), and decomposer processes (Brooks et al. 1996; O'Lear and Seastedt 1994). A thinner or nonexistent snowpack, such as may result, for example, from increased thawing events due to global warming, causes the soil surface to become less protected, resulting in lower soil surface temperatures and consequently greater ice cover. This may adversely affect soil organisms. For example, Coulsen et al. (2000) found that experimental creation of ice on the soil surface of field plots in a tundra in Svalbard reduced populations of microarthropods by over 50%, with springtails being more affected than mites. Soil organisms are especially sensitive to freeze-thaw cycles, and these cycles may make a major contribution to the turnover of the soil microbial biomass (Tearle 1987; Schimel and Clein 1996). Consumer organisms in areas prone to freezing are adapted to tolerate these conditions through extended periods of inactivity, e.g., through anhydrobiotic coiling in the case of nematodes in the Antarctic dry valleys (Treonis et al. 2000), and through the production of high-molecular-weight antifreeze proteins in the case of microarthropods (Joose and Verhoef 1987).

Some of the more significant effects of global climate change on soil organisms may arise through a greater frequency of extreme events, such as has been predicted to occur over the next few decades (see Timmermann et al. 1999). Although the issue of how extreme events influence decomposers has seldom been addressed, and it would be reasonable to expect that many soil organisms are sufficiently resilient to disturbance to recover quickly from the direct effects of such events, extreme events can also have a profound effect on the productivity and functional composition of vegetation, and therefore presumably the associated soil biota and soil processes in the longer term. For example, El Niño Southern Oscillation episodes, which normally last for about a year, can cause major vegetation shifts in some regions through extremely high rainfall, and in other regions through severe droughting (reviewed by Holmgren et al. 2001); this can in turn have important effects on consumer organisms, and the sources and amounts of detritus entering the decomposer subsystem (Polis et al. 1997b). Even individual extreme events can significantly alter the functional composition of vegetation; experimental imposition of single extreme heating and rainfall events on three New Zealand perennial pastures was found to favor fast-growing C4 grasses at the expense of C3 grasses (White et al. 2000), probably through reducing the competitiveness of the C3 grass community (White et al. 2001). Greater dominance of C4 relative to C3 grasses may be expected to significantly alter decomposer communities through diminished quality of litter input (Wardle et al. 1999b).

Global climate change may be expected to have impor-

289

tant effects on decomposer processes. Differences in decomposition rates across locations are widely regarded as being influenced by three interlinked factors: macroclimate, resource quality, and composition of the soil biota (see Swift et al. 1979; Aerts 1997b; Seastedt 2000). Increases in temperature of the magnitude expected through global warming are generally expected to directly increase carbon and nitrogen mineralization rates, and there is evidence for this from controlled incubation experiments (Hobbie 1996), field experiments involving warming treatments (Jonasson et al. 1999; Johnson et al. 2000), and decomposition rates of standardized substrates along gradients of latitude or macroclimate (Berg et al. 1993; Moore et al. 1999; Gholz et al. 2000) and elevation (Vitousek et al. 1994; Scowcroft et al. 2000). Other elements of global climate change can also enhance decomposition rates, although these have been less well studied. However, soil moisture is an important determinant of soil activity (Orchard and Cook 1983), and standardized substrates decompose more rapidly under higher precipitation (e.g., Moore et al. 1999; Austin and Vitousek 2000). Further, increased frequencies of wetting-drying cycles and freezing-thawing cycles both act to increase rates of litter mass loss (Taylor and Parkinson 1988a,b), probably through increasing turnover of the microbial biomass and disrupting the physical structure of the litter.

Increased temperatures may affect decomposition rates not only directly but also indirectly through influencing the other two drivers of decomposition, i.e., litter quality and the decomposer community. With regard to litter quality, higher temperatures can increase nutrient mineralization rates, plant nutrient uptake, and therefore plant tissue nutrient concentrations. For example, Nakatsubo et al. (1997) found that along an elevational gradient in Japan, nutrient concentrations of the moss *Hylocomium splendens* increased with increasing temperature. Effects of increased temperatures on quality of litter input may also occur over a longer time scale through affecting the functional group composition of the vegetation, and this can work in both directions. For an Arctic tundra in Alaska, U.S.A., Hobbie (1996) found that plant species likely to increase as a result of global warming (i.e., deciduous shrubs) produce a greater amount of slower-decomposing litter such as woody stems than do those plant species not expected to increase. In contrast, the model of Pastor and Post (1988) predicts global warming to promote replacement of conifers that produce poor litter quality with deciduous tree species that produce higher-quality litter. With regard to the decomposer community, warmer conditions promote a relatively greater abundance of larger soil animals (Seastedt 2000), and the major groups of soil animals differ in their functional importance under different climatic regimes. For example, Heneghan et al. (1999) found that decomposition rates of a standardized litter were strongly influenced by microarthropods at two tropical forest sites but not at a temperate forest site. Further, earthworms in tropical areas have a greater role in breaking down recalcitrant components of soil organic matter than do those in cooler climates (Lavelle et al. 1995). However, as discussed earlier, the temperature changes required to cause significant shifts in the major groups of decomposer biota are probably likely to be much greater than those predicted to result from global warming.

Global warming is expected to stimulate decomposition rates of organic substrates more than NPP (Kirchbaum 1995; Seastedt 2000), and this is in turn likely to result in a greater net loss of CO_2 from the soil (Jenkinson et al. 1991), faster turnover of soil organic matter (Trumbore et al. 1996), and therefore less storage of carbon in the soil (Oechel et al. 1993; Cao and Woodward 1998). In contrast, a recent global synthesis of data by Giardina and Ryan (2000a) demonstrated that the decomposition rate of organic matter in mineral soils of forests was not correlated with temperature, meaning that global warming may not necessarily accelerate losses of carbon from the mineral soil layer. However, these analyses and interpretations have been queried on the basis of inappropriate methodology and sampling design, as well as a failure to consider the indirect effects of temperature on decomposition (see the exchange between Davidson et al. 2000 and Giardina and Ryan 2000b). While elevated temperature may reduce soil carbon stocks through directly stimulating soil respiration and decomposition rates, these effects could in some cases be partially offset by changes in the functional composition of the vegetation (see Shaver et al. 1992; Jobbagy and Jackson 2000). For example, Seastedt et al. (1994) used simulation models to show that while a 2°C temperature increase may cause substantial carbon loss from a temperate grassland dominated by C3 grasses, domination by C4 or mixed C3–C4 grass communities (likely to be favored under a warmer climate) would help to mitigate loss of soil carbon because of increased NPP. Further, experimental studies in a tundra community in Alaska, U.S.A. (Verville et al. 1998), in which temperature was manipulated and vegetation removal treatments were performed, were used to suggest that climate change was more likely to affect soil CO_2 and CH_4 release (and therefore soil carbon storage) through inducing large-scale changes in vegetation composition rather than through direct effects of temperature on soil respiration.

SYNTHESIS

Global change phenomena are very relevant to the issue of aboveground-belowground feedbacks because they involve direct alteration of specific components of the ecosystem, and enable determination of how other components of the ecosystem are affected by this. Gains of species through invasion, or species loss through extinction, can exert important effects on organisms and processes on the other side of the aboveground-belowground interface, especially when these species exert some dominating influence and also differ in fundamental traits from trophically equivalent species. Land use change and intensification have major impacts on the type of vegetative cover present as well as the disturbance regime; both these effects impact significantly upon the composition of soil food webs and ultimately patterns of nutrient supply from the soil. In the longer term, anthropogenically induced changes in atmospheric CO_2 concentrations, nitrogen deposition, and climate all influence belowground organisms and processes mainly through altering the quality and quantity of resources produced by plants; these effects arise through global change phenomena influencing both the ecophysiology of individual plants and the functional composition of the vegetation.

The global change phenomena described in this chapter can all affect the composition of the soil food web, and with the exception of disturbances resulting from some aspects of land use intensification, are likely to operate mainly indirectly through changes that occur in the aboveground biota. Consistent with the material presented in chapters 2 and 3, vegetation community changes resulting from global change can influence the relative importance of top-down and bottom-up forces in soil food webs, the balance between the bacterial-based and fungal-based energy channels, and the relative abundances of different subsets of the soil fauna. These shifts will in turn influence decomposer processes and therefore nutrient supply rates for plants, further reinforcing feedbacks between the aboveground and belowground communities. Further, global change phenomena are likely to affect different plant communities to differing degrees, and assessment of how different plant functional groups should respond to global change factors may be powerful predictors of not only the degree of change in plant community structure but also the likely shifts in the composition and performance of the decomposer subsystem and ultimately the ways in which the ecosystem functions.

A key question regarding global change phenomena is how the sequestration of carbon in terrestrial ecosystems is likely to be affected; whether there is a net carbon gain or loss will ultimately be determined by how global change affects the balance of NPP and decomposition. Generally, land use intensification and global warming will probably contribute to net losses of stored terrestrial carbon; nitrogen deposition and CO_2 enrichment should lead to net gains. However, there are likely to be many local exceptions to this, largely because global change phenomena may lead to shifts toward different vegetation types which may either enhance or reduce carbon storage, and these effects are likely to be context specific. Effects of global change on soil organic matter levels are likely to exert major effects on decomposer communities and processes in the long-term perspective, since they will determine the resource base available for soil organisms. Finally, it is highly apparent that anthropogenic activity is having an enormous impact on the Earth's terrestrial ecosystems at all temporal and spatial scales, and explicit consideration of both the producer and decomposer subsystems (as well as the nature of linkages between them) is essential in better understanding the long-term environmental impacts of the global change phenomena that result from this activity.

CHAPTER EIGHT

Underlying Themes

A dominating theme throughout this book has been the significance of biotic interactions. Traditionally, ecology has focused on interactions involving competition and predation, and indeed the ecological literature is replete with examples of these sorts of interactions. As outlined in this book, top-down regulation by predators and bottom-up regulation involving competition are clearly important regulatory forces for many groups of biota both above- and belowground. In the decomposer subsystem this is especially apparent in the microfood-web subsystem involving microorganisms, nematodes, and protozoa. However, there are a variety of other types of interaction between organisms that, while not featured as prominently in the general ecological literature, are at least as important (and at some spatial scales far more important) when proper consideration is given to the belowground subsystem. One key interaction is mutualism, and throughout this book I have presented numerous examples of mutualisms which operate at a range of scales of resolution. At the level of the whole ecosystem, the entire producer and entire decomposer subsystems are obligately dependent upon one another. Some studies have even presented evidence that plant species may select for decomposer communities that are adapted for mineralizing the litter that they produce. Further, at the individual plant level, plants undergo a variety of interactions with soil organisms in their immediate rooting zone and many of these are mutualistic; these include plant associations with mycorrhizal fungi, nitrogen fixing bacteria, bacteria that produce plant growth stimulatory compounds, and possibly the microbial loop in the rhizosphere. There are also instances in which mycorrhizal hyphal connections between plants promote coexistence of plant species and reduce the intensity of plant competition. Herbivores often maximize plant productivity at intermediate levels of feeding (probably through their effects on nutrient cycling), and fungal-feeding fauna can optimize fungal growth when present at intermediate densities. Within the soil food web, mutualisms between microorganisms and soil animals become increasingly important with increasing body size of the fauna; while the soil microfood-web is based on predator-prey relationships (with mutualism existing between primary and tertiary consumers mainly only through trophic cascades), the "litter transformer" fauna (*sensu* Lavelle et al. 1995) undergo mutualistic associations with microbes via the external rumen in fecal pellets, and the larger soil invertebrates such as termites and earthworms enter more intimate mutualisms with microbes in their gut cavities (see chapter 2). More specialized mutualisms involving soil fauna also exist, for example through ants and termites maintaining fungus gardens in their nests. There is also theoretical evidence to suggest that mutualistic interactions could conceivably increase diversity of organisms within trophic levels (Bever 1999). In total, a combined aboveground-belowground approach to understanding community and ecosystem processes requires the recognition of mutualism as a major ecological driver at all spatial and temporal scales of resolution.

Another set of interactions concerns those that operate indirectly between organisms. Recognition by ecologists working in aquatic and aboveground systems of the considerable importance of indirect interactions in communities is relatively recent (e.g., Wootton 1994; Menge 1995), but these sorts of interactions are ubiquitous in the belowground subsystem and in aboveground-belowground associations, and have long been acknowledged as such by soil biologists. Indeed, the majority of interactions discussed in this book operate indirectly. There is a variety of types of these interactions, which may have either positive or negative consequences for the organisms affected by the interaction under consideration. While plants provide the resource base for the soil food web, indirect feedbacks occur through soil organisms mineralizing nutrients and thus determining plant nutrient supply (see Bengtsson et al. 1995). Indirect effects are also propagated through multitrophic effects in soil food chains, both via energy transfer to higher trophic levels, and through trophic cascades regulating lower trophic levels. Trophic relationships in soil food webs can also affect linkages between aboveground and belowground biota, for example, through microarthropod grazing on mycorrhizal fungal hyphae affecting plant nutrient acquisition and growth. Mobile soil fauna can operate as dispersal agents of propagules of organisms that interact with plants (e.g., spores of pathogenic or mycorrhizal fungi; nitrogen fixing bacteria), as well as plant seeds, all of which can conceivably influence plant growth and community structure. Some of the most important indirect effects in terrestrial ecosystems involve habitat modification, including ecosystem engineering. For example, many soil meso- and macrofaunal organisms create physical structures that provide habitats for smaller soil organisms, and structures created by some larger soil animals (e.g., termites, ants, earthworms) contribute to the spatial heterogeniety of plant communities by supporting different species of plants from those that otherwise dominate. Soil organisms such as fungi and earthworms, as well as plant roots, can stabilize soil structure through the exudation of polysaccharides, and this can profoundly affect soil porosity, compaction, aeration, and hydrology, therefore affecting other biota of the ecosystem. Plants produce litter which provides a habitat for decomposer microflora and fauna, and the structural attributes of this litter, as well as its residence time on the ground, influence both the community structure and activity of this biota. The degree of litter breakdown of litter on the soil surface by the decomposer biota indirectly affects the extent to which litter influences germination and recruitment of new plants to the community. Over a longer time scale, the extent to which plants can propagate disturbances such as wildfire and factors that contribute to forest decline have major effects on the quality and quantity of resources that may enter the decomposer subsystem. Further, there are important indirect linkages between aboveground and belowground consumers as mediated by the plant. As discussed in chapter 4, there are a variety of mechanisms by which aboveground herbivores indirectly affect decomposer organisms and their activity through altering the amounts and quality of organic matter produced by the plant. In a similar light, the handful of studies that have sought to investigate the indirect influence of belowground consumers on invertebrate herbivore performance have generally detected significant effects (see chapter 5). These examples all serve to illustrate the point that indirect interactions have major consequences for both the aboveground and belowground communities (as well as the feedbacks between the two), and may emerge as being more important than direct trophic interactions as the key ecological regulators in terrestrial communities.

The belowground subsystem is driven by the quantity and quality of plant-derived organic materials, and this is ultimately determined by the ecological strategies of the dominant plant species present. Most plant species that dominate in fertile habitats allocate resources to rapid growth, while other species that are more adapted for infertile conditions grow more slowly and allocate resources to secondary metabolites (e.g., phenolics). These strategies are associated with suites of ecophysiological traits which include a range of photosynthetic, nutrient use, and leaf and root characteristics (see chapter 3). As a result, plant species that grow more rapidly generally produce higher-quality foliage (with higher nutrient concentrations, and lower concentrations of lignin, structural carbohydrates, and secondary metabolites); this results in a higher proportion of fixed carbon entering the herbivore food web relative to the decomposer food web, as well as a higher quality of those organic materials that do enter the decomposer subsystem. These plant species effects have important consequences for the structure and function of decomposer communities. Plant species that produce high-quality litter appear to favor soil food webs that are dominated by bacterial-based energy channels, enchytraeids, and earthworms, and this in turn facilitates greater turnover of nutrients. The rapid rate of mineralization of nutrients from litter produced by such plants results in high rates of supply of plant-available nutrients from the soil, and relatively shallow soil organic layers. In contrast, plant species that produce poorer-quality litter generally favor domination by fungal-based energy channels, micro-arthropods (notably mites), and larger soil arthropods. Associated with this are slower rates of mineralization of plant litter, accumulation of organic residues on the soil surface, and the formation of moroid humus profiles. These fundamental differences in plant species effects in turn lead to different strategies of nutrient uptake by plants. Faster-growing plants in more fertile habitats tend to preferentially take up mobile forms of nitrogen, have arbuscular mycorrhizal associations, and may compete intensely for nutrients. Slower-growing plants in less fertile situations are more adapted for taking up less mobile organic forms of nitrogen, are frequently reliant on ectomycorrhizal or ericoid mycorrhizal fungi that can degrade organic matter, and (as discussed in chapter 3) may be expected to compete less intensely for nutrients. As a result, highly fertile sites often support fewer plant species than infertile sites, probably because in fertile conditions plants are actively competing for mobile forms of nutrients and light, leading to a greater capacity for competitive exclusion. Meanwhile, under infertile conditions, low transport rates of nutrients in soil and the strategies that plants use to access immobile forms of nutrients reduce the intensity of competition, leading to greater coexistence of species being possible.

The differences among plant species in their effects on consumer organisms ultimately contribute to across-ecosystem differences both above- and belowground. Throughout this book are numerous examples of how ecosystems which are dominated by plants that invest in rapid growth differ from ecosystems dominated by slower-growing plant species that invest in secondary defense compounds; these ecosystems differ in terms of the effects of herbivory on plants and nutrient cycling, the composition and activity of decomposer communities, the nature of nutrient supply for plants, the degree of sequestration of carbon and nitrogen in the soil, and the rates of nitrification and nitrogen fixation. These differences are especially apparent across ecological gradients such as those involving shifts in successional status and changes in NPP. Ecosystem succession involves the buildup and (given time) the decline of soil fertility; this has important effects on the plant functional types that are supported, and therefore key ecosystem properties and processes including those driven by belowground communities such as nitrogen fixation, transformations of major nutrients, and supply rates of plant-available nutrients. Gradients of NPP have often been studied by aboveground ecologists especially with regard to foliage-based food webs and intensities of biotic interactions. Increasing NPP generally results in corresponding increases in herbivore mass (and hence herbivore impact on ecosystem properties) and may also promote some components of the decomposer food web, particularly those that are bottom-up regulated such as fungi and subsets of the soil fauna. However, there are difficulties in assessing the influence of NPP along natural gradients on aboveground and belowground properties because of the effects of factors that covary with NPP along such gradients, and because plant species with different functional attributes frequently dominate different portions of the gradient. It appears likely that across-ecosystem differences in belowground communities and processes, and in aboveground-belowground feedbacks, are governed by the combined effects of differences in NPP (resource quantity) and differences in the quality of resources produced by the plants; both these effects are likely to be strongly correlated with each other across ecosystems because both are probably regulated by similar suites of plant ecophysiological traits.

A key issue addressed in this book involves the biotic controls of ecosystem functioning. Those processes that involve the gains, losses, and transformations of carbon and nitrogen in terrestrial ecosystems are mainly biotically driven, but the rates of these processes are strongly influenced both by the types of organisms present and by the sorts of interactions that they undergo with one another. Rates of belowground processes such as decomposition of organic materials and mineralization of nutrients are influenced by the composition of the soil food web, including the body size distribution of the soil fauna (i.e., the relative abundance of microfauna, mesofauna, and macrofauna), the relative contribution of the bacterial-based and fungal-based energy channels, the composition of communities within trophic levels, and the relative abundance of predators. Biotic interactions involving soil microbial—faunal associations are especially important in governing decomposition and mineralization processes; while most decomposition is directly carried out by bacteria and fungi, those animals that interact with microbes either as predators or as mutualists frequently promote greater rates of microbial turnover and consequently more rapid rates of decomposition of organic materials and nutrient mineralization. Predator-prey interactions involving higher trophic levels have also been shown to influence decomposer processes through promoting trophic cascades that influence microbial activity. The effects

301

of soil food web interactions on nutrient mineralization in turn affect nutrient supply rates for plant growth and therefore plant nutrient acquisition and NPP. Further, both aboveground community composition and the nature of biotic interactions among aboveground species are major determinants of aboveground and belowground processes. As described above, different plant functional types differ tremendously in terms of the amounts and the quality of resources that they provide to the belowground subsystem, and therefore their indirect effects on soil process rates. The quantity and quality of this resource input can in turn be strongly influenced by aboveground trophic associations; as outlined in chapter 4 there are a variety of mechanisms by which aboveground herbivores can indirectly affect processes driven by the soil biota over a range of temporal and spatial scales. Above all, it is apparent that the producer and decomposer subsystems are obligately dependent upon one another, and a complete consideration of the ecosystem functions carried out by each of these subsystems requires explicit recognition of the controls exerted on it by the other subsystem. Finally, while community composition and food web structure are both clearly major drivers of ecosystem level processes, there is also a school of thought which maintains that species richness within functional groups is also a major driver of ecosystem processes, notably NPP. However, the position taken in this book is that in most cases the effects of species richness probably saturate at much lower levels of diversity than that normally occurring in real communities, and that most of the experimental evidence presented to date suggesting that species richness effects are important at higher levels of diversity is based on designs that are confounded by "hidden treatments" (*sensu* Huston 1997) that make such an outcome inevitable.

The influence of composition of species and biotic interactions on the relative performance of the decomposer and producer subsystems ultimately governs the amount of carbon sequestration in the biosphere. Storage of carbon in terrestrial ecosystems has been proposed as an agent of potential importance in mitigating increases in atmospheric CO_2 resulting from anthropogenic activity (Prentice and Fung 1990), although the role of the total terrestrial biosphere as a carbon sink under elevated CO_2 may be more likely to result from changes in storage in the vegetation than in the soil (Schlesinger 1990). Plant species differ in the extent to which they promote terrestrial sequestration of carbon; ecosystems dominated by plants that invest in rapid growth, produce litter of high quality, and favor decomposer activity tend to store much less carbon in the soil than those dominated by slower-growing plant species which produce recalcitrant litter. In addition, unproductive plant species dominating in nutrient-poor conditions are frequently less palatable to herbivores than are productive plant species, resulting in a greater proportion of NPP being added to the soil as plant litter, and further contributing to greater sequestration of carbon in the soil. The issue of carbon sequestration in ecosystems has become a prominent topic in the general area of ecological impacts of global change phenomena, and land use change, nitrogen deposition, increased CO_2 concentrations, and global climate change are all predicted to influence this (see chapter 7). In particular, the relative response of different plant functional types to global change factors is likely to result in large scale shifts in major vegetation types which will in turn affect whether net increases or reductions of carbon storage occurs, and the direction of these effects should be context dependent.

One issue that remains generally little understood is what controls patterns of spatial variability of organisms and processes in an aboveground—belowground context. Geostatistical approaches provide some evidence that spatial distributions of decomposer biota can be determined by the spatial distribution of both vegetation and soil properties. Different plant species coexisting in the same community

303

can promote very different communities of soil biota especially when individual plants are sufficiently spatially discrete for each plant to have primary control of the resources (e.g., litter, root materials) that enter the decomposer subsystem in the patch that it occupies. There is evidence for this both in forests, and in deserts where individual shrubs create "islands of fertility." In unproductive environments, in particular, each plant has primary control of the resources under it (see chapter 3; Huston and De Angelis 1994), indicating the presence of a feedback between plant heterogeneity and soil heterogeneity. This feedback works to reduce competition among plant species, maintaining greater spatial heterogeneity by maintaining a greater diversity of vegetation. Aboveground herbivores influence the spatial pattern of return of plant-derived resources, for example, by concentrating dung and urine in specific portions of the landscape, by transporting resources across habitats, and by influencing the structure of the plant community. On a finer spatial scale, plant roots create local zones of depletion and resource addition, influencing spatial variability at a level of resolution relevant to the soil microfoodweb. Further, soil organisms can also influence the spatial heterogeneity of plant communities (and therefore their diversity) through selective root herbivory and pathogenicity, by influencing competitive interactions among plant species (e.g., through decomposers affecting plant nutrient supply; shared mycorrhizal connections among plants), and by creating spatially discrete physical structures which promote different plant species and functional types from those that otherwise dominate.

The issue of what controls temporal variability of organisms and processes in relation to aboveground—belowground linkages in terrestrial ecosystems remains similarly little explored. However, there is some evidence that this is influenced to a large extent by the dominant species of organisms present. Different plant species differ in their temporal patterns of resource addition to the soil, both across seasons and across years. Further, plant species differ in the extent to which they propagate the effects of intermittent severe disturbances (e.g., wildfire, hurricanes), and this will have major consequences for the quantity and quality of resources entering the decomposer subsystem in the short term, as well as effects over the longer term resulting from the alteration of vegetation successional pathways. These will all influence decomposer biota and ultimately the temporal patterns of nutrient supply for plants. Aboveground herbivores which show temporal patterns of abundance (e.g., through cyclical dynamics or through intermittent outbreaks) can operate as major drivers of temporal variability of plant community composition and therefore soil biota and processes. With regard to the belowground environment, different subsets of the decomposer food web respond differentially to abiotic fluctuations in environmental conditions, and there is some evidence that these differences are related to organism size (with larger soil fauna being less resistant than soil microfauna and microflora to these changes). The temporal dynamics of soil organisms in response to abiotic fluctuations in turn induces pulses in the supply of nutrients for plants. The structure of food webs, either aboveground or belowground, and hence the trophic interactions that take place, may also influence the responses to changes in environmental conditions of the processes driven by these food webs, although there is a dearth of experimental or observational investigations of this issue.

It appears reasonable that the temporal variability of ecosystem properties, both aboveground and belowground, as well as related measures of stability such as resistance and resilience to environmental perturbations, should be governed in the first instance by the traits or attributes of the species of organisms present in the community. There has been a preoccupation in some branches of theoretical ecology with evaluating how community diversity or complexity may influence the stability of communities or ecosystem properties, but the models that underlie investigations of this issue frequently involve simulations of randomly constructed communities, while most biological communities are probably not random assemblages of species (see chapter 6). Unambiguous experimental or observational evidence in support of the view that diversity or complexity predictably affects the stability of ecosystem properties or processes is currently lacking. There has traditionally been far less interest among ecologists in investigating how community composition and attributes of organisms in a community may affect ecosystem stability, which is unfortunate because the fragmentary pieces of evidence available would suggest that these are more likely to be the key drivers of stability. This issue is also very relevant in relation to understanding the ability of communities and ecosystems to resist anthropogenically driven global change phenomena such as global warming, atmospheric CO_2 enrichment, nitrogen deposition, and ingress by invasive species. While some experimental studies which have used synthetic community approaches have been presented as evidence that community diversity determines ecosystem responses to some of these global change factors, the results of those studies to date making such claims are open to alternative interpretations. In contrast, approaches based on the relative responsiveness of different plant functional types to global change phenomena, and the effects of plant traits on this, combined with appropriate consideration of the feedbacks that exist between the producer and decomposer subsystems, are already emerging as having tremendous potential in predicting how global change may affect the performance of terrestrial communities and ecosystems.

Finally, it is apparent through the examples presented throughout this book that aboveground-belowground linkages are key drivers of both community- and ecosystem-level attributes. For this reason, consideration of those interactions and processes that occur on one side of the soil/ surface interface in isolation from those that occur on the other side leads to an incomplete understanding of how communities and ecosystems operate. Traditionally, aboveground ecology has tended to pay insufficient attention to those belowground organisms that through their interactions regulate the supply of nutrients for aboveground organisms, while soil biologists have frequently insufficiently acknowledged the role of aboveground community processes in determining the quantity and quality of those substrates upon which the soil biota depends. However, a combined aboveground-belowground approach is essential if we are to better understand the ecological consequences over time and space of individual species and their interactions with each other and their environment, or to develop a balanced view about how human-induced global change phenomena affect the Earth's communities and ecosystems.

References

Aarssen, L. 1997. High productivity in grassland ecosystems: effected by species diversity or productive species? *Oikos* 80: 183–184.

Aarssen, L., and Epp, G. A. 1990. Neighbor manipulations in natural vegetation: a review. *Journal of Vegetation Science* 1: 13–30.

Aberdeen, J. E. C. 1956. Factors influencing the distribution of fungi and plant roots. Part I. Different host species and fungal interactions. *Papers of the Department of Botany of the University of Queensland* 3: 113–124.

Abrams, P. A. 1995. Monotonic or unimodal diversity-productivity gradients: what does competition theory predict? *Ecology* 76: 2019–2027.

Addington, R. N., and Seastedt, T. R. 1999. Activity of soil microarthropods beneath snowpack in alpine tundra and subalpine forest. *Pedobiologia* 43: 47–53.

Aerts, R. 1997a. Nitrogen partitioning between resorption and decomposition pathways: a trade-off between nutrient use efficiency and litter decomposability? *Oikos* 80: 603–606.

———· 1997b. Climate, leaf litter chemistry and leaf litter decomposition in terrestrial ecosystems: a triangular relationship. *Oikos* 79: 439–449.

Aerts, R., and Berendse, F. 1988. The effect of increased nutrient availability on vegetation dynamics in wet heathland. *Vegetatio* 76: 63–69.

Aerts, R., and Chapin, F. S. III 2000. The mineral nutrition of wild plants revisited: a re-evaluation of processes and patterns. *Advances in Ecological Research* 30: 1–67.

Aerts, R., and Van der Peijl, M. J. 1993. A simple model to explain the dominance of low-productive perennials in nutrient poor habitats. *Oikos* 66: 144–147.

Aerts, R., et al. 1990. Competition in heathland along an experimental gradient of nutrient availability. *Oikos* 57: 310–318.

Agrawal, A. A. 1998. Induced responses to herbivory and increased plant performance. *Science* 279: 1201–1202.

———· 1999. Induced responses to herbivory in wild radish: effects on several herbivores and plant fitness. *Ecology* 80: 1713–1723.

Allen-Morley, C. R., and Coleman, D. C. 1989. Resilience of soil biota in various food webs to freezing perturbations. *Ecology* 70: 1127–1141.

Alphei, J., et al. 1996. Protozoa, Nematoda and Lumbricidae in the rhizosphere of *Hordelymus europaeus* (Poaceae): faunal interactions, response of microorganisms and effects on plant growth. *Oecologia* 106: 111–126.

Altieri, M. A. 1991. How can we best use biodiversity in agriculture? *Outlook on Agriculture* 20: 15–23.

Andersen, D. C. 1987. Below ground herbivory in natural comunities: a review emphasizing fossorial animals. *Quarterly Review of Biology* 62: 261–286.

Anderson, A. 2000. Less is moa. *Science* 289: 1472–1473.

Anderson, J. M. 1975. The enigma of soil animal species diversity. In J. Vanek, ed., *Progress in Soil Zoology*, 51–58. Prague: Academia.

———. 1978. Inter- and intra-habitat relationships between woodland Cryptostigmata species diversity and the diversity of soil and litter microhabitats. *Oecologia* 32: 341–348.

———. 1991. The effects of climate change on decomposition processes in grasslands and coniferous forests. *Ecological Applications* 1: 326–347.

Anderson, J. P. E., and Domsch, K. H. 1978. A physiological method for the quantification of microbial biomass in soil. *Soil Biology and Biochemistry* 10: 215–221.

Anderson, T.-H., and Domsch, K. H. 1989. Ratios of microbial biomass carbon to total organic carbon in arable soils. *Soil Biology and Biochemistry* 21: 471–479.

———. 1990. Application of eco-physiological quotients (qCO_2 and qD) on microbial biomasses from soils of different cropping histories. *Soil Biology and Biochemistry* 22: 251–255.

Anderson, W. B., and Polis, G. A. 1999. Nutrient fluxes from water to land: seabirds affect plant nutrient status on Gulf of California islands. *Oecologia* 118: 324–332.

André, H. M. 2001. Skilled eyes are needed to go on studying the richness of soil. *Nature* 409: 761.

André, M., et al. 1994. Biodiversity in model ecosystems. *Nature* 371: 565.

Andrén, O., et al. Editors. 1990. *Ecology of Arable Land. Organisms, Carbon and Nitrogen Cycling*. Copenhagen: Ecological Bulletins.

Andrén, O., et al. 1995. Biodiversity and species redundancy among litter decomposers. In H. P. Collins, G. P. Robertson, and M. J. Klug, eds., *The Significance and Regulation of Soil Biodiversity*, 141–151. Dordrecht: Kluwer.

Angers, D. A., and Caron, J. 1998. Plant-induced changes in soil structure: processes and feedbacks. *Biogeochemistry* 42: 55–72.

Armendáriz, I., and Arpin, P. 1996. Nematodes and their relationship to forest dynamics: 1. Species and trophic groups. *Biology and Fertility of Soils* 23: 405–413.

Arnold, S. S., et al. 1999. Microbial response of an acid forest to experimental soil warming. *Biology and Fertility of Soils* 30: 239–244.

Arshad, M., and Frankenberger, W. T. 1998. Plant growth regulating substances in the rhizosphere: microbial broduction and functions. *Advances in Agronomy* 62: 45–151.

Augustine, D. J., and McNaughton, S. J. 1998. Ungulate effects on the functional species composition of plant communities: herbivore selectivity and plant tolerance. *Journal of Wildlife Management* 62: 1165–1183.

Austin, A., and Vitousek, P. M. 2000. Precipitation, decomposition and litter decomposibility of Metrosideros polymorpha in native forests on Hawai'i. *Journal of Ecology* 88: 129–138.

Austin, M. P., and Austin, B. O. 1980. Behavior of experimental plant communities along a nutrient gradient. *Journal of Ecology* 68: 891–918.

Bååth, E. 1980. Soil fungal biomass after clear-cutting of a pine forest in central Sweden. *Soil Biology and Biochemistry* 12: 495–500.

Bååth, E., et al. 1978. The effect of carbon and nitrogen supply on the development of soil organism populations and pine seedlings: a microcosm experiment. *Oikos* 31: 153–163.

Badejo, M. A., and Tian, G. 1999. Abundance of soil mites under four agroforestry tree species with contrasting litter quality. *Biology and Fertility of Soils* 30: 107–112.

Baker, K. F., and Cook, R. J. 1974. *Biological Control of Plant Pathogens*. San Francisco: Freeman.

Bakonyi, G., and Nagy, P. 2000. Temperature- and moisture-induced changes in the structure of the nematode fauna of a semiarid grassland—patterns and mechanism. *Global Change Biology* 6: 697–707.

Bamforth, S. S. 1973. Population dynamics of soil and vegetation protozoa. *American Zoologist* 13: 171–176.

Bardgett, R. D., and Chan, K. F. 1999. Experimental evidence that soil fauna enhance nutrient mineralization and plant nutrient uptake in montane grassland ecosystems. *Soil Biology and Biochemistry* 31: 1007–1014.

Bardgett, R. D., and Shine, A. 1999. Linkages between plant litter diversity, soil microbial biomass and exosystem function in temperate grasslands. *Soil Biology and Biochemistry* 31: 317–321.

Bardgett, R. D., et al. 1996. Changes in fungal:bacterial ratios following reductions in the intensity of the management of an upland grassland. *Biology and Fertility of Soils* 22: 261–264.

———· 1997. Seasonality of the soil biota of grazed and ungrazed hill grasslands. *Soil Biology and Biochemistry* 29: 1285–1294.

———· 1998a. Dynamic interactions between soil animals and microorganisms in upland grassland soils amended with sheep dung: a microcosm experiment. *Soil Biology and Biochemistry* 30: 531–539.

———· 1998b. Linking above-ground and below-ground interactions: how plant responses to foliar herbivory influence soil organisms. *Soil Biology and Biochemistry* 30: 1867–1878.

———· 1999a. Plant species and nitrogen effects on soil biological properties of temperate upland grasslands. *Functional Ecology* 13: 650–660.

———· 1999b. Below-ground herbivory promotes soil nutrient transfer and root growth in grassland. *Ecology Letters* 2: 357–360.

———· 1999c. Below-ground microbial community development in a high temperature world. *Oikos* 85: 193–203.

Barker, G. M., and Mayhill, P. C. 1999. Patterns of diversity and habitat relationships in terrestrial mollusc communities of the Pukemaru Ecological District, northeastern New Zealand. *Journal of Biogeography* 26: 215–238.

Baruch, Z., and Goldstein, G. 1999. Leaf construction cost, nutrient concentration and net CO_2 assimilation of native and invasive species in Hawaii. *Oecologia* 121: 183–192.

Basu, S., and Behera, N. 1994. The effect of tropical forest conversion on soil microbial communities. *Biology and Fertility of Soils* 16: 302–304.

Bateman, G. L., and Kwasna, H. 1999. Effects of number of winter wheat crops grown successively on fungal communities on wheat roots. *Applied Soil Ecology* 13: 271–282.

Bauhus, J., and Barthel, R. 1995. Mechanisms for carbon and nutrient release and retention in beech forest gaps. II. The role of soil microbial biomass. *Plant and Soil* 168: 585–592.

Bauhus, J., et al. 1993. The effects of fire on carbon and nitrogen mineralization and nitrification in an Australian forest soil. *Australian Journal of Soil Research* 31: 621–639.

Bazzaz, F. A. 1979. The physiological ecology of plant succession. *Annual Review of Ecology and Systematics* 10: 351–371.

———· 1990. Response of natural ecosystems to the rising global CO_2 levels. *Annual Review of Ecology and Systematics* 21: 167–196.

Beard, J. 1991. Woodland soil yields a multitude of insects. *New Scientist* 131 (1784): 14.

Beare, M. H., et al. 1990. A substrate-induced respiration (SIR) method for measurement of fungal and bacterial biomass of plant litter. *Soil Biology and Biochemistry* 22: 585–594.

———· 1991. Characterization of a substrate induced respiration method for measuring fungal, bacterial and total microbial biomass on plant residues. *Agriculture, Ecosystems and Environment* 34: 65–73.

———· 1992. Microbial and faunal interactions and effects on litter nitrogen and decomposition in agroecosystems. *Ecological Monographs* 62: 569–591.

Beggs, J. 2001. The ecological consequences of social wasps (*Vespula* spp.) invading an ecosystem that has an abundant carbohydrate resource. *Biological Conservation* (in press).

Behan-Pelletier, V. M. 1999. Oribatid mite biodiversity in agroecosystems: role for bioindication. *Agriculture, Ecosystems and Environment* 74: 411–423.

Belovsky, G. E., and Slade, J. B. 2000. Insect herbivory accelerates nutrient cycling and increases plant production. *Proceedings of the National Academy of Sciences, U.S.A.* 97: 14412–14417.

Belsky, A. J. 1986. Does herbivory benefit plants? A review of the evidence. *American Naturalist* 127: 870–892.

———· 1987. The effects of grazing: confounding of ecosystem, community and organism scales. *American Naturalist* 129: 777–783.

Bengtsson, G., et al. 1994. Food- and density-dependent dispersal: evidence from a soil collembolan. *Journal of Animal Ecology* 63: 513–520.

Bengtsson, J. 1994. Temporal predictability in forest soil communities. *Journal of Animal Ecology* 63: 653–665.

Bengtsson, J., et al. 1995. Food webs in soil: an interface between population and ecosystem ecology. In C. G. Jones and J. H. Lawton, eds., *Linking Species and Ecosystems*, 159–165. New York: Chapman and Hall.

———· 1997. Long-term effects of logging residue addition and removal on macroarthropods and enchytraeids. *Journal of Applied Ecology* 34: 1014–1022.

Bentley, B. L. 1984. Nitrogen fixation in termites: fate of newly fixed nitrogen. *Journal of Insect Physiology* 30: 653–655.

Berendse, F. 1990. Organic matter accumulation and nitrogen mineralization during secondary succession in heathland ecosystems. *Journal of Ecology* 78: 413–427.

———. 1994a. Litter decomposability: a neglected component of plant fitness. *Journal of Ecology* 82: 187–190.

———. 1994b. Competition between plant populations at low and high nutrient supplies. *Oikos* 71: 253–260.

———. 1998. Effects of dominant plant species on soils during succession in nutrient poor ecosystems. *Biogeochemistry* 42: 73–88.

Berendse, F., and Aerts, R. 1984. Competition between *Erica tetralix* and *Molinia caerulea* as affected by availability of nutrients. *Acta Œcologica, Œcologica Plantarum* 5: 3–14.

———. 1987. Nitrogen use efficiency: a biologically meaningful definition? *Functional Ecology* 1: 293–296.

Berendse, F., et al. 1987. A comparative study on nutrient cycling in wet heathland ecosystems. I. Litter production and nutrient losses from the plant. *Oecologia* 74: 174–184.

———. 1998. Adaptation of plant populations to nutrient poor environments and their implications for soil nutrient mineralization. In H. Lambers, H. Poorter, and M. M. I. van Vuuren, eds., *Inherent Variation in Plant Growth: Physiological Mechanisms and Ecological Consequences*, 503–514. Leiden: Backhuys.

Berg, B., and Ekbohm, G. 1991. Litter mass loss and decomposition patterns in some needle and leaf litter types. Long term decomposition in a Scots pine forest. *Canadian Journal of Botany* 69: 1449–1456.

Berg, B., et al. 1993. Litter mass loss rates in pine forests of Europe and eastern United States: some relationships with climate and litter quality. *Biogeochemistry* 20: 127–159.

Berlow, E. L., et al. 1999. Quantifying variation in the strengths of species interactions. *Ecology* 80: 2206–2224.

Bernier, N., and Ponge, J. F. 1994. Humus form dynamics during the sylvogenetic cycle in a mountain spruce forest. *Soil Biology and Biochemistry* 26: 183–220.

Bethenfalvy, G. J., et al. 1991. Nutrient transfer between the root zones of soybean and maize plants connected by a common mycorrhizal mycelium. *Physiologia Plantarum* 82: 423–432.

Bever, J. D. 1994. Feedback between plants and their soil communities in an old field community. *Ecology* 75: 1965–1977.

———. 1999. Dynamics within mutualism and the maintenance of diversity: inference from a model of interguild frequency dependence. *Ecology Letters* 2: 52–61.

Bever, J. D., et al. 1997. Incorporating the soil community into plant population dynamics: the utility of the feedback approach. *Journal of Ecology* 85: 561–573.

Biederbeck, V. O., et al. 1984. Effect of crop rotation and fertilization on some biological properties of a loam in southwestern Saskatchewan. *Canadian Journal of Soil Science* 64: 355–367.

Bignall, D. E. 1984. The arthropod gut as an environment for microorganisms. In J. M. Anderson, A. D. M. Rayner, and D. W. H. Walton, eds., *Invertebrate-Microbial Interactions*, 205–227. Cambridge: Cambridge University Press.

Billes, E., et al. 1993. Modification of the carbon and nitrogen allocations in the plant (*Triticum aestivum* L.) soil system in response to increased atmospheric CO_2 concentration. *Plant and Soil* 157: 215–225.

Billings, W. D., et al. 1984. Interaction of increasing atmospheric carbon dioxide and soil nitrogen on the carbon balance of tundra microcosms. *Oecologia* 65: 26–29.

Binkley, D., and Giardina, C. 1998. Why do trees affect soils? The Warp and Woof of tree-soil interactions. *Biogeochemistry* 42: 89–106.

Binkley, D., and Valentine, D. 1991. Fifty-year biogeochemical effects of green ash, white pine and Norway spruce in a replicated experiment. *Forest Ecology and Management* 40: 13–25.

Blair, J. M., and Crossley, D. A. 1988. Litter decomposition, nitrogen dynamics and litter microarthropods in a southern Appalachian hardwood forest 8 years following clearcutting. *Journal of Applied Ecology* 25: 683–698.

Blair, J. M., et al. 1990. Decay rates, nitrogen fluxed and decomposer communities in single and mixed species foliar litter. *Ecology* 71: 1976–1985.

Bloemers, G. F., et al. 1997. The effects of forest disturbance on diversity of tropical soil communities. *Oecologia* 111: 575–582.

Blomqvist, M. M., et al. 2000. Interactions between above- and belowground biota: importance for small-scale vegetation mosaics in a grassland ecosystem. *Oikos* 90: 582–598.

Boag, B. 2000. The impact of the New Zealand flatworm on earthworms and moles in agricultural land in western Scotland. *Aspects of Applied Biology* 62: 79–84.

Boag, B., and Yeates, G. W. 1998. Soil nematode biodiversity in terrestrial ecosystems. *Biodiversity and Conservation* 7: 617–630.

Boag, B., et al. 1999. Spatial distribution and relationship between the New Zealand flatworm *Arthurdentyus triangulata* and earthworms in a grass field in Scotland. *Pedobiologia* 43: 340–344.

Boddy, L. 2000. Interspecific combative interactions between wood-decaying basidiomycetes. *FEMS Microbiology Ecology* 31: 185–194.

Boerner, R. E. J., and Rebbek, J. 1995. Decomposition and nitrogen release from leaves of three hardwood species grown under elevated O_3 and/or CO_2. *Plant and Soil* 170: 149–157.

Bokhari, U. G., and Singh, J. S. 1974. Effects of temperature and clipping on growth, carbohydrate reserves and root exudation of western wheatgrass in hydroponic culture. *Crop Science* 14: 790–794.

Bond, W. J. 1989. The tortoise and the hare: ecology of angiosperm dominance and gymnosperm persistence. *Biological Journal of the Linnean Society* 36: 227–247.

Bonkowski, M., and Schaefer, M. 1997. Interactions between earthworms and soil protozoa: a trophic component in the soil food web. *Soil Biology and Biochemistry* 29: 499–502.

Bonkowski, M., et al. 2000. Substrate heterogeniety and microfauna in soil organic "hotspots" as determinants of nitrogen capture and growth of ryegrass. *Applied Soil Ecology* 14: 37–53.

Borkott, H., and Insam, H. 1992. Symbiosis with bacteria enhances the use of chitin by the springtail, *Folsomia candida* (Collembola). *Biology and Fertility of Soils* 9: 126–129.

Bottner, P. 1985. Response of microbial biomass to alternate moist and dry conditions in a soil incubated with ^{14}C and ^{15}N-labeled plant material. *Soil Biology and Biochemistry* 17: 329–337.

Bowen, G. D., and Rovira, A. D. 1999. The rhizosphere and its management to improve root growth. *Advances in Agronomy* 66: 1–102.

Box, E. O. 1981. *Macroclimate and Plant Form.* The Hague: Junk.

———. 1996. Plant functional types and climate at the global scale. *Journal of Vegetation Science* 7: 309–320.

Bradley, R. L., and Fyles, J. W. 1995. Growth of paper birch (*Betula papyrifera*) seedlings increases soil available C and microbial acquisition of soil nutrients. *Soil Biology and Biochemistry* 27: 1565–1571.

Bremner, J. M., and McCarty, G. W. 1988. Effects of terpenoids on nitrification in soil. *Soil Science Society of America Journal* 52: 1630–1633.

———. 1990. Reply to "Comments on 'effects of terpenoids on nitrification in soil'". *Soil Science Society of America Journal* 54: 296–298.

Breznak, J. A., and Brune, A. 1994. Role of microorganisms in the digestion of lignocellulose by termites. *Annual Review of Entomology* 39: 453–487.

Briones, M. J. I., and Ineson, P. 1996. Decomposition of *Eucalyptus* leaves in litter mixtures. *Soil Biology and Biochemistry* 28: 1381–1388.

Briones, M. J. I., et al. 1997. Effects of climate change on soil fauna—responses of enchytraeids, Diptera larvae and tardigrades in a transplant experiment. *Applied Soil Ecology* 6: 117–134.

Brooks, P. D., et al. 1996. Microbial activity under snowpacks, Niwot Range, Colorado. *Biogeochemistry* 32: 93–113.

——— 1998. Inorganic nitrogen and microbial biomass before and during spring snowmelt. *Biogeochemistry* 43: 1–15.

Broughton, L. C., and Gross, K. L. 2000. Patterns of diversity in plant and soil communities along a productivity gradient in a Michigan old-field. *Oecologia* 125: 420–427.

Brown, G. G. 1995. How do earthworms affect microfloral and faunal community diversity? *Plant and Soil* 170: 209–231.

Brown, J. C. 1958. Soil fungi of some British sand dunes in relation to soil type and succession. *Journal of Ecology* 46: 641–664.

Brown, M. J. F., and Human, K. G. 1997. Effects of harvester ants on plant species distribution and abundance in a serpentine grassland. *Oecologia* 112: 237–243.

Brown, V. K., and Gange, A. C. 1989a. Differential effects of above-ground and below-ground insect herbivory during early plant succession. *Oikos* 54: 67–76.

——— 1989b. Root herbivory by insects depresses plant species richness. *Functional Ecology* 3: 667–671.

——— 1990. Insect herbivory below ground. *Advances in Ecological Research* 20: 1–58.

——— 1992. Secondary plant succession: how is it modified by insect herbivory? *Vegetatio* 101: 3–13.

Brown, V. K., et al. 1988. Insect herbivory: effects on early old field succession demonstrated by chemical exclusion methods. *Oikos* 52: 293–302.

Brundrett, M. 1991. Mycorrhizas in natural ecosystems. *Advances in Ecological Research* 21: 171–313.

Bryant, J. P. 1987. Feltleaf willow-snowshoe hare interaction: plant carbon/nutrient balance and floodplain succession. *Ecology* 68: 1319–1327.

Bryant, J. P., and Riechart, P. B. 1992. Controls over secondary metabolite production in arctic woody plants. In F. S. Chapin, R. L.

Jeffries, J. F. Reynolds, G. R. Shaver, and J. Svoboda, eds., *Arctic Ecosystems in a Changing Climate: An Ecophysiological Perspective*, 377–390. New York: Academic.

Bryant, J. P., et al. 1993. Effects of mineral nutrition on delayed inducible resistance in Alaska paper birch. *Ecology* 74: 2072–2084.

Brzeski, M. W., and Szczech, M. 1999. Effect of continuous soil amendment with coniferous sawdust on nematodes and microorganisms. *Nematologia Mediterranea* 27: 159–166.

Buckland, S. M., and Grime, J. P. 2000. The effects of trophic structure and soil fertility on the assembly of plant communities: a microcosm experiment. *Oikos* 91: 336–352.

Burity, H, A., et al. 1989. Estimation of nitrogen fixation and transfer from alfalfa to associated grasses in mixed swards under field conditions. *Plant and Soil* 114: 249–255.

Burke, I. C., et al. 1995. Soil organic matter recovery in semiarid grasslands: implications for the conservation reserve programme. *Ecological Applications* 5: 93–801.

———. 1998. Plant-soil interactions in temperate grasslands. *Biogeochemistry* 42: 121–143.

Burke, M. J. W., and Grime, J. P. 1996. An experimental study of plant community invasibility. *Ecology* 77: 776–790.

Burtelow, A. E., et al. 1998. Influence of exotic earthworm invasion on soil organic matter, microbial biomass and denitrification potential in forest soils of the northeastern United States. *Applied Soil Ecology* 9: 197–202.

Butler, G. W., et al. 1956. Effects of shading and defoliation on the turnover of root and nodule tissue of plants of *Trifolium repens*, *Trifolium pratense* and *Lotus uliginosus*. *New Zealand Journal of Agricultural Research* 2: 415–426.

Callaway, R. M. 1995. Positive interactions among plants. *Botanical Reviews* 61: 306–349.

Callaway, R. M., and Walker, L. R. 1997. Competition and facilitation: a synthetic approach to positive interactions in plant communities. *Ecology* 78: 1958–1965.

Campbell, B. D., and Grime, J. P. 1989. A comparative study of plant responsiveness to the duration of episodes of mineral nutrient enrichment. *New Phytologist* 112: 261–267.

Campbell, B. D., and Grime, J. P. 1992. An experimental test of plant strategy theory. *Ecology* 73: 15–29.

Campbell, B. D., et al. 1991. A trade-off between scale and precision in resource foraging. *Oecologia* 87: 532–538.

———— 1999. Climate profiles of temperate C_3 and subtropical C_4 species in New Zealand. *New Zealand Journal of Agricultural Research* 42: 223–233.

Cao, M., and Woodward, F. I. 1998. Net primary and ecosystem production and carbon stocks of terrestrial ecosystems and their responses to climate change. *Global Change Biology* 4: 185–198.

Cardinale, B. J., et al. 2000. Linking species diversity to the functioning of ecosystems: on the importance of environmental context. *Oikos* 91: 175–183.

Cardon, Z. G., et al. 2001. Contrasting effects of elevated CO_2 on old and new soil carbon pools. *Soil Biology and Biochemistry* 33: 365–373.

Carey, P. D., et al. 1992. A field study using the fungicide benomyl to investigate the effect of mycorrhizal fungi on plant fitness. *Oecologia* 90: 550–555.

Carreiro, M. M., and Koske, R. E. 1992. The effect of temperature and substratum on competition among three species of forest litter microfungi. *Mycological Research* 96: 19–24.

Carreiro, M. M., et al. 2000. Microbial enzyme shifts explain litter decay responses to simulated nitrogen deposition. *Ecology* 81: 2359–2365.

Carson, P. W., and Peterson, C. J. 1990. The role of litter in an oldfield community: impact of litter quantity in different seasons on plant species richness and abundance. *Oecologia* 85: 8–13.

Cates, R. G., and Orians, G. H. 1975. Successional status and the palatability of plants to generalized herbivores. *Ecology* 56: 410–418.

Cavigelli, M., and Robertson, G. P. 2000. The functional significance of denitrifier community composition in a terrestrial ecosystem. *Ecology* 81: 1402–1414.

Cazares, E., and Trappe, J. M. 1994. Spore dispersal of ectomycorrhizal fungi on a glacier forefront by mammal mycophagy. *Mycologia* 86: 507–510.

Cebrian, J. 1999. Patterns in the fate of production in plant communities. *American Naturalist* 154: 449–468.

Cebrian, J., and Duarte, C. M. 1995. Plant growth-rate dependence of detrital carbon storage in ecosystems. *Science* 268: 1606–1608.

Cebrian, J., et al. 1998. The dependence of heterotrophic consumption and C accumulation on autotrophic nutrient content in ecosystems. *Ecology Letters* 1: 165–170.

Chang, S. X., and Trofymow, J. A. 1996. Microbial respiration and biomass (substrate induced respiration) in soils of old-growth forest and regenerating forests on north Vancouver Island, British Colombia. *Biology and Fertility of Soils* 23: 145–152.

Chapin, F. S. III 1980. The mineral nutrition of wild plants. *Annual Review of Ecology and Systematics* 11: 233–260.

Chapin, F. S. III, et al. 1993a. Evolution of suites of traits in response to environmental stress. *American Naturalist* 142 (supplement): 78–92.

———· 1993b. Preferential use of organic nitrogen for growth by a non-mycorrhizal arctic sedge. *Nature* 361: 150–152.

———· 1994. Mechanisms of primary succession following deglaciation at Glacier Bay, Alaska. *Ecological Monographs* 64: 149–175.

Chapman, K., et al. 1988. Metabolic and faunal activity in litter mixtures compared with pure stands. *Agriculture, Ecosystems and Environment* 34: 65–73.

Chauvel, A., et al. 1999. Pasture damage by an Amazonian earthworm. *Nature* 398: 32–33.

Chen, B., and Wise, D. H. 1999. Bottom-up limitation of predaceous arthropods in a detritus-based terrestrial food web. *Ecology* 80: 761–772.

Chen, J., and Ferris, H. 2000. Growth and nutrient mineralization of selected fungi and fungal-feeding nematodes on sand amended with organic matter. *Plant and Soil* 218: 91–101.

Chown, S. L., and Gaston, K. J. 2000. Areas, cradles and museums: the latitudinal gradient in species richness. *Trends in Ecology and Evolution* 15: 311–315.

Christensen, M. 1981. Species diversity and dominance in fungal communities. In D. T. Wicklow and G. C. Carroll, eds., *The Fungal Community. Its Organization and Role in the Ecosystem*, 201–232. New York: Marcel Dekker.

Christensen, S., et al. 1992. Huge increase in bacterivores on freshly killed barley roots. *FEMS Microbiology Ecology* 86: 303–310.

Christie, P., et al. 1974. Grassland plant species can influence the abundance of microbes on each other's roots. *Nature* 250: 570–571.

———· 1978. The influence of neighboring grassland plants on each others' endomycorrhizas and root surface microorganisms. *Soil Biology and Biochemistry* 10: 521–527.

Clarholm, M. 1985. Possible roles for roots, bacteria, protozoa and fungi in supplying nitrogen to plants. In A. H. Fitter, ed., *Ecological Interactions in Soil*, 355–365. Oxford: Blackwell.

———· 1990. Effects of plant-bacterial-amoebal interactions on plant uptake of nitrogen under field conditions. *Biology and Fertility of Soils* 8: 373–378.

Clements, F. E. 1928. *Plant Succession and Indicators.* New York: Wilson.

Coates, D., and Rayner, A. D. M. 1985. Fungal population and community development in cut beech logs. I. Spatial dynamics, interactions and strategies. *New Phytologist* 101: 183–198.

Coderre, D., et al. 1995. Earthworm populations in healthy and declining sugar maple forests. *Pedobiologia* 39: 86–96.

Cohen, J. E., et al. 1990. *Community Food Webs: Data and Theory.* New York: Springer-Verlag.

Coleman, D. C. 1994. The microbial loop as used in terrestrial soil ecology. *Microbial Ecology* 28: 245–250.

Coleman, D. C., and Crossley, D. A. 1995. *Fundamentals of Soil Ecology.* San Diego: Academic.

Coleman, D. C., et al. 1983. Biological strategies of nutrient cycling in soil systems. *Advances in Ecological Research* 13: 1–55.

Coley, P. D., et al. 1985. Resource availability and plant antiherbivore defense. *Science* 230: 895–899.

Collatz, G. J., et al. 1998. Effects of climate and atmospheric CO_2 partial pressure on the global distribution of C_4 grasses: present, past and future. *Oecologia* 114: 441–454.

Collins, S. L. 2000. Disturbance frequency and community stability in native tallgrass prairie. *American Naturalist* 155: 311–325.

Compton, J. E., and Boone, R. D. 2000. Long term impacts of agriculture on soil carbon and nitrogen in New England forests. *Ecology* 81: 2314–2330.

Connell, J. H. 1978. Diversity in tropical rainforests and coral reefs. *Science* 199: 1302–1310.

———. 1983. On the prevalence and relative importance of interspecific competition: evidence from field experiments. *American Naturalist* 122: 661–696.

Connell, J. H., and Lowman, M. D. 1989. Low diversity tropical rain forests: some possible mechanisms for their existence. *American Naturalist* 134: 88–119.

Connell, J. H., and Slayter, R. O. 1977. Mechanisms of succession in natural communities and their role in community stability and organization. *American Naturalist* 111: 1119–1144.

Cookson, W. R., et al. 1998. Effects of prior crop residue management on microbial properties and crop residue management. *Applied Soil Ecology* 7: 179–188.

Coomes, D. A., and Grubb, P. J. 2000. Impacts of root competition in forests and woodlands: a theoretical framework and review of experiments. *Ecological Monographs* 70: 171–207.

Copley, J. 2000. Ecology goes underground. *Nature* 406: 452–454.

Cornelissen, J. H. C. 1996. An experimental comparison of leaf decomposition rates in a wide range of temperate plant species and types. *Journal of Ecology* 84: 573–582.

Cornelissen, J. H. C., and Thompson, K. 1997. Functional leaf attributes predict litter decomposition rate in herbaceous plants. *New Phytologist* 135: 109–114.

Cornelissen, J. H. C., et al. 1996. Seedling growth, allocation and leaf attributes in a wide range of woody plants and types. *Journal of Ecology* 84: 755–765.

———. 1999. Leaf structure and defense control litter decomposition across species and life forms in two continents. *New Phytologist* 143: 191–200.

Corre, M. D., et al. 1999. Evaluation of soil organic carbon under forests, cool-season and warm season grasses in the northeastern U.S. *Soil Biology and Biochemistry* 31: 1531–1539.

Cortez, J., and Bouché, M. B. 1998. Field decomposition of leaf litters: earthworm-microorganism interactions—the ploughing-in effect. *Soil Biology and Biochemistry* 30: 795–804.

Cotrufo, M. F., et al. 1994. Decomposition of tree leaf litters grown under elevated CO_2: effects of litter quality. *Plant and Soil* 163: 121–130.

———. 1998. Elevated CO_2 affects field decomposition rate and palatability of tree leaf litter: importance of changes in substrate quality. *Soil Biology and Biochemistry* 30: 1565–1571.

Cottingham, K. L., et al. 2001. Biodiversity may regulate the temporal variability of ecological systems. *Ecology Letters* 4: 72–85.

Coulsen, S. J., et al. 2000. Experimental manipulation of the winter surface ice layer: the effects on a high Arctic soil microarthropod community. *Ecography* 23: 299–306.

Coûteaux, M.-M., et al. 1991. Increased atmospheric CO_2 and litter quality: decomposition of sweet chestnut leaf litter with animal food webs of different complexities. *Oikos* 61: 53–64.

———. 1996. Increased atmospheric CO_2: chemical changes in decomposing sweet chestnut (*Castanea sativa*) leaf litter incubated in microcosms under increasing food web complexity. *Oikos* 76: 553–563.

———. 1999. Influence of increased atmospheric CO_2 concentration on quality of plant material and litter decomposition. *Tree Physiology* 19: 301–311.

Cox, P., et al. 2001. Effects of fungal inocula on the decomposition of lignin and structural polysaccharides in *Pinus sylvestris* litter. *Biology and Fertility of Soils* (in press).

Crafford, J. E. 1990. The role of feral house mice in ecosystem function on Marion Island. In K. R. Kerry and G. Hempel, eds., *Antarctic Ecosystems: Ecological Change and Conservation*, 359–364. New York: Springer.

Crews, T. E., et al. 1995. Changes in soil phosphorus fractions and ecosystem dynamics across a long chronosequence in Hawaii. *Ecology* 76: 1407–1424.

Critchley, B. R., et al. 1979. Effect of bush clearing and soil cultivation on the invertebrate fauna of a forest soil in the humid tropics. *Pedobiologia* 19: 425–438.

Curry, J. P., and Cotton, D. C. F. 1983. Earthworms and land reclamation. In J. E. Satchell, ed., *Earthworm Ecology*, 215–228. New York: Chapman and Hall.

Cyr, H., and Pace, M. L. 1993. Magnitude and patterns of herbivory in aquatic and terrestrial systems. *Nature* 361: 148–150.

Dackman, C., et al. 1987. Quantification of predatory and endoparasitic nematophagous fungi in soil. *Microbial Ecology* 13: 89–97.

Daniel, O., and Anderson, J. M. 1992. Microbial biomass and activity in contrasting soil materials after passing through the gut of the earthworm *Lumbricus terrestris*. *Soil Biology and Biochemistry* 24: 465–470.

D'Antonio, C. M., and Vitousek, P. M. 1992. Biological invasions by exotic grasses, the grass/fire cycle and global change. *Annual Review of Ecology and Systematics* 23: 63–87.

Darbyshire, J. F., et al. 1994. Excretion of nitrogen and phosphorous by the soil ciliate *Colpoda steinii* when fed upon the bacteria *Arthrobacter* sp. *Soil Biology and Biochemistry* 26: 1193–1199.

David, J.-F., et al. 1999. Belowground biodiversity in a Mediterranean landscape: relationships between saprophagous macro-arthropod communities and vegetation structure. *Biodiversity and Conservation* 8: 753–767.

Davidson, D. W. 1993. The effect of herbivory and granivory on terrestrial plant succession. *Oikos* 68: 23–35.

Davidson, E. A., and Ackerman, I. L. 1993. Changes in soil carbon inventories following cultivation of previously untilled soils. *Biogeochemistry* 20: 161–193.

Davidson, E. A., et al. 2000. Soil warming and organic carbon content. *Nature* 408: 789–790.

Davies, M. A., et al. 2000. Fluctuating resources in plant communities: a general theory of invasibility. *Journal of Ecology* 88: 528–534.

Davies, R. G., et al. 1999. Successional response of a tropical forest termite assemblage to experimental habitat perturbation. *Journal of Applied Ecology* 36: 946–962.

De Angelis, D. L. 1992. *Dynamics of Nutrient Cycling and Food Webs.* London: Chapman and Hall.

Decaëns, T., et al. 1999. Soil surface macrofaunal communities associated with earthworm casts in grasslands of the eastern plains of Colombia. *Applied Soil Ecology* 13: 87–100.

Degens, B. 1998. Decreases in microbial functional diversity do not result in corresponding changes in decomposition under different moisture regimes. *Soil Biology and Biochemistry* 30: 1989–2000.

Degens, B., and Harris, J. A. 1997. Development of a physiological approach to measuring the catabolic diversity of soil microbial communities. *Soil Biology and Biochemistry* 29: 1309–1320.

Degens, B., et al. 2000. Decreases in organic C reserves in soils can reduce the catabolic diversity of soil microbial communities. *Soil Biology and Biochemistry* 32: 189–196.

De Goede, R. G. M., and Dekker, H. H. 1993. Effects of liming and fertilization on nematode communities in coniferous forests. *Pedobiologia* 37: 193–209.

De Mazancourt, C., and Loreau, M. 2000. Effects of herbivory on primary production, and plant species replacement. *American Naturalist* 155: 734–754.

De Mazancourt, C., et al. 1998. Grazing optimization and nutrient cycling: when do herbivores enhance plant production? *Ecology* 79: 2242–2252.

Demure, Y., et al. 1979. Anhydrobiotic coiling of nematodes. *Journal of Nematology* 11: 189–195.

Denton, C. S., et al. 1999. Low amounts of root herbivory positively influence the rhizosphere microbial community in a temperate grassland soil. *Soil Biology and Biochemistry* 31: 155–165.

Derner, J. D., et al. 1997. Does grazing mediate soil carbon and nitrogen accumulation beneath C4 perennial grasses along an environmental gradient? *Plant and Soil* 191: 157–156.

Derouard, L., et al. 1997. Effects of earthworm introduction on soil processes and plant growth. *Soil Biology and Biochemistry* 29: 541–545.

Derry, A. M., et al. 1999. Functional diversity and community structure of micro-organisms in three Arctic soils as determined by sole-carbon-source utilization. *Biodiversity and Conservation* 8: 205–221.

De Ruiter, P. C., et al. 1993a. Calculation of nitrogen mineralization in soil food webs. *Plant and Soil* 157: 263–273.

———· 1993b. Simulation of nitrogen mineralization in the belowground food webs of two winter wheat fields. *Journal of Applied Ecology* 30: 95–106.

———· 1994. Modeling food webs and nutrient cycling in agroecosystems. *Trends in Ecology and Evolution* 9: 378–383.

———· 1995. Energetics, patterns of interaction strength and stability in real ecosystems. *Science* 269: 1257–1260.

Detling, J. K., et al. 1980. Effects of simulated grazing by belowground herbivores on growth, CO_2 exchange, and carbon allocation patterns of *Bouteloua gracilis*. *Journal of Applied Ecology* 17: 771–778.

Díaz, S. 1995. Elevated CO_2 responsiveness, interactions at the community level and plant functional types. *Journal of Biogeography* 22: 289–295.

Díaz, S., and Cadibo, M. 1997. Plant functional types and ecosystem function in relation to global change. *Journal of Vegetation Science* 8: 463–474.

Díaz, S., et al. 1993. Evidence of a feedback mechanism limiting plant response to elevated carbon dioxide. *Nature* 364: 616–617.

———· 1998. Plant functional types and environmental filters at a regional scale. *Journal of Vegetation Science* 9: 113–122.

Didden, W. A. M. 1993. Ecology of terrestrial Enchytraeidae. *Pedobiologia* 37: 2–29.

Dighton, J. 1978. Effects of synthetic lime aphid honeydew on populations of soil organisms. *Soil Biology and Biochemistry* 10: 369–376.

Dighton, J., and Mason, P. A. 1985. Mycorrhizal dynamics during forest tree development. In D. Moore, L. A. Casselton, D. A. Wood, and J. C. Frankland, eds., *Developmental Biology of Higher Fungi*, 117–139. Cambridge: Cambridge University Press.

Diquélou, S., et al. 1999. Changes in microbial biomass and activity during old field successions in Brittany, France. *Pedobiologia* 43: 470–479.

Doak, D. F., et al. 1998. The statistical inevitability of stability-diversity relationships in community ecology. *American Naturalist* 151: 264–276.

Doube, B. M., et al. 1994. Enhanced root nodulation of subterranean clover (*Trifolium subterraneum*) by *Rhizoctonia leguminosarium* biovar *trifolii* in the presence of the earthworm *Aporrectodea trapezoides* (Lumbricidae). *Biology and Fertility of Soils* 18: 169–174.

Drury, C. F., et al. 1991. Microbial biomass and soil structure associated with corn, grass and legumes. *Soil Science Society of America Journal* 55: 805–811.

Duncan, R. P., and Young, J. R. 2000. Determinants of plant extinction and rarity 145 years after European settlement of Auckland, New Zealand. *Ecology* 81: 3048–3061.

Dyer, M. I., et al. 1986. A model of herbivore feedback on plant productivity. *Mathematical Biosciences* 79: 171–184.

———. 1991. Source-sink carbon relations in two *Panicum coloratum* ecotypes in response to herbivory. *Ecology* 72: 1472–1483.

———. 1993. Herbivory and its consequences. *Ecological Applications* 3: 10–16.

Edwards, C. A., and Fletcher, K. E. 1988. Interactions between earthworms and microorganisms in organic matter breakdown. *Agriculture, Ecosystems and Environment* 24: 235–247.

Egerton-Warburton, L., and Allen, E. B. 2000. Shifts in arbuscular mycorrhizal communities along an anthropogenic nitrogen deposition gradient. *Ecological Applications* 10: 484–496.

Eggleton, P., and Bignall, D. E. 1995. Monitoring the response of tropical insects to changes in the environment: troubles with termites. In R. Harrington and N. E. Stork, eds., *Insects in a Changing Environment*, 434–497. London: Academic.

Ehrlich, P. R., and Ehrlich, A. H. 1981. *Extinction. The Causes and Consequences of the Disappearance of Species.* New York: Random House.

Eissenstat, D. M., and Newman, E. I. 1990. Seedling establishment near large plants: effects of vesicular-arbuscular mycorrhizas on the intensity of plant competition. *Functional Ecology* 4: 95–99.

Eissenstat, D. M., and Yanai, R. D. 1997. The ecology of root lifespan. *Advances in Ecological Research* 27: 1–60.

Elhottová, D., et al. 1997. Rhizosphere microflora of winter wheat plants cultivated under elevated CO_2. *Plant and Soil* 197: 251–259.

Elliott, E. T., et al. 1983. Short term bacterial growth, nutrient uptake and ATP turnover in sterilized, inoculated and C-amended soil: the influence of N availability. *Soil Biology and Biochemistry* 15: 85–91.

Emmerson, M. C., and Raffaelli, D. 2000. Detecting the effects of diversity on measures of ecosystem function: experimental design, null models and empirical observations. *Oikos* 91: 195–203.

REFERENCES

Enquist, B. J., and Niklaus, K. J. 2001. Invariant sealing relations across tree-dominated communities. *Nature* 410: 685–690.

Enríquez, S., et al. 1993. Patterns in decomposition rates among photosynthetic organisms: the importance of detritus C:N:P content. *Oecologia* 94: 457–471.

Entry, J. A., and Backman, C. B. 1995. Influence of carbon and nitrogen on cellulose and lignin degradation in forest soils. *Canadian Journal of Forest Research* 25: 1231–1236.

Eom, A.-H., et al. 2000. Host plant species effects on arbuscular mycorrhizal fungal communities in tallgrass prairie. *Oecologia* 122: 435–444.

Esler, J. J., et al. 2000a. Biological stoichiometery from genes to ecosystems. *Ecology Letters* 3: 540–550.

——— 2000b. Nutritional constraints in terrestrial and freshwater food webs. *Nature* 408: 578–580.

Ettema, C. H. 1998. Soil nematode diversity: species coexistence and ecosystem function. *Journal of Nematology* 30: 159–169.

Ettema, C. H., et al. 1998. Spatiotemporal distributions of bacterivorous nematodes and soil resources in a restored riparian wetland. *Ecology* 79: 2721–2734.

——— 1999a. Riparian soil response to surface nitrogen input: temporal changes in denitrification, labile and microbial C and N pools, and bacterial and fungal respiration. *Soil Biology and Biochemistry* 31: 1609–1624.

——— 1999b. Riparian soil response to surface nitrogen input: the indicator potential of free living soil nematode populations. *Soil Biology and Biochemistry* 31: 1625–1638.

Faber, J. H., and Joose, E. N. G. 1993. Vertical distribution of Collembola in a *Pinus nigra* organic soil. *Pedobiologia* 37: 336–350.

Facelli, J., and Ladd, B. 1996. Germination requirements and responses to leaf litter of four species of eucalypt. *Oecologia* 107: 441–445.

Facelli, J., and Pickett, S. T. A. 1991. Plant litter: its dynamic and effects on plant community structure. *Botanical Reviews* 57: 1–32.

Fenn, M. E., et al. 1993. Microbial N and biomass, respiration and N mineralization in soils beneath two chaparral species along a fire induced age gradient. *Soil Biology and Biochemistry* 25: 457–466.

Findlay, S., et al. 1996. Effects of damage to living plants on leaf litter quality. *Ecological Applications* 6: 269–275.

Finlay. B. J., et al. 1999. Global distribution of free living microbial species. *Ecography* 22: 138–144.

Finzi, A. C., and Canham, A. C. 1998. Non-additive effects of litter mixtures on net mineralization in a southern New England forest. *Forest Ecology and Management* 105: 129–136.

Finzi, A. C., et al. 1998. Canopy tree-soil interactions within temperate forests: species effects on soil carbon and nitrogen. *Ecological Applications* 8: 440–446.

Fisher, M. J., et al. 1994. Carbon storage by introduced deep rooted grasses in the South American savannas. *Nature* 371: 236–238.

Fitter, A. H. 1977. Influence of mycorrhizal infection on competition for phosphorus and potassium by two grasses. *New Phytologist* 79: 119–125.

Fitter, A. H., et al. 1998. Carbon transfer between plants and its control in networks of arbuscular mycorrhizas. *Functional Ecology* 12: 406–412.

Flanagan, P. W., and van Cleve, K. 1983. Nutrient cycling in relation to decomposition and litter quality in Taiga ecosystems. *Canadian Journal of Forest Research* 13: 795–817.

Flannery, T. F. 1994. *The Future Eaters*. Melbourne: Reed Books.

Florence, R. G. 1965. Decline of old growth forests in relation to some soil microbiological processes. *Ecology* 46: 52–64.

Fog, K. 1988. The effect of added nitrogen on the rate of decomposition of added matter. *Biological Reviews* 63: 433–462.

Foissner, W. 1997. Global soil ciliate (Protozoa, Ciliophora) diversity: a probability based approach using large sample collections from Africa, Australia and Antarctica. *Biodiversity and Conservation* 6: 1627–1638.

Foster, B. L., and Gross, K. L. 1997. Partitioning the effects of plant biomass and litter on *Andropogon gerardi* in old field succession. *Ecology* 78: 2091–2104.

———. 1998. Species richness in a successional grassland: effects of nitrogen enrichment and plant litter. *Ecology* 79: 2593–2602.

Francis, R., and Read, D. J. 1984. Direct transfer of carbon between plants connected by vesicular-arbuscular mycelium. *Nature* 307: 53–56.

———. 1994. The contributions of mycorrhizal fungi to the determinants of plant community structure. *Plant and Soil* 159: 11–25.

Franck, V. M., et al. 1997. Decomposition of litter produced under elevated CO_2: dependence on plant species and nutrient supply. *Biogeochemistry* 36: 223–237.

Frank, D. A., and Evans, R. D. 1997. Effects of native grazers on grassland N cycling in Yellowstone National Park. *Ecology* 78: 2238–2248.

Frank, D. A., and Groffman, P. M. 1998. Ungulate vs. landscape control of soil C and N processes in grasslands of Yellowstone National Park. *Ecology* 79: 2229–2241.

Frank, D. A., and McNaughton, S. J. 1991. Stability increase with diversity in plant communities: empirical evidence with the 1988 Yellowstone drought. *Oikos* 62: 360–362.

Frankland, J. C. 1998. Fungal succession—unravelling the unpredictable. *Mycological Research* 102: 1–15.

Fransen, B., et al. 1999. Root morphological and physiological plasticity of perennial grass species and the exploitation of spatial and temporal heterogeneous nutrient patches. *Plant and Soil* 211: 179–189.

Fraser, L. H., and Grime, J. P. 1999. Interacting effects of herbivory and fertility on a synthesized plant community. *Journal of Ecology* 87: 514–515.

Freckman, D. W. 1978. Ecology of anhydrobiotic nematodes. In J. H. Crowe and J. S. Clegy, eds., *Dried Biological Systems*, 345–357. New York: Academic.

———. 1988. Bacterivorous nematodes and organic matter decomposition. *Agriculture, Ecosystems and Environment* 24: 195–217.

Freckman, D. W., and Ettema, C. H. 1993. Assessing nematode communities in agroecosystems of varying human intervention. *Agriculture, Ecosystems and Environment* 45: 239–261.

Freckman, D. W., and Womersley, C. 1983. Physiological adaptations of nematodes in Chihuahuan soils. In P. LeBrun, H. M. Andre, A. DeMedts, C. Gregorie-Wimo, and G. Wauthy, eds., *New Trends in Soil Biology*, 395–403. Louvain: Dieu-Brichart.

Fyles, I. H., et al. 1988. Dynamics of microbial biomass and faunal populations in long-term plots on a Gray Luvisol. *Canadian Journal of Soil Science* 68: 91–100.

Fyles, J. W., and Fyles, J. H. 1993. Interactions of Douglas fir with red alder and salal foliage litter during decomposition. *Canadian Journal of Forest Research* 23: 358–361.

Gallardo, A., and Merino, J. 1999. Control of leaf litter decomposition rate in a Mediterranean shrubland as indicated by N, P and lignin concentrations. *Pedobiologia* 43: 64–72.

Galloway, J. N., et al. 1995. Nitrogen fixation: atmospheric enhancement—environmental response. *Global Biogeochemical Cycles* 9: 235–252.

Gange, A. C. 1999. On the relationship between arbuscular mycorrhizal colonization and plant "benefit." *Oikos* 87: 615–621.

———. 2000. Arbuscular mycorrhizal fungi, collembola and plant growth. *Trends in Ecology and Evolution* 15: 369–372.

Gange, A. C., and Bower, E. 1997. Interactions between insects and mycorrhizal fungi. In A. C. Gange and V. K. Brown, eds., *Multitrophic Interactions in Terrestrial Systems*, 115–131. Oxford: Blackwells.

Gange, A. C., and West, H. M. 1994. Interactions between arbuscular mycorrhizal fungi and foliar-feeding insects in *Plantago lanceolata* L. *New Phytologist* 128: 79–87.

Gange, A. C., et al. 1999. Positive effects of an arbuscular mycorrhizal fungus on aphid life history traits. *Oecologia* 120: 123–131.

Garnier, E., et al. 1997. A problem for biodiversity-productivity studies: how to compare the productivity of multispecific plant mixtures to that of monocultures. *Acta Œcologica* 18: 657–670.

Garrett, S. D. 1963. *Soil Fungi and Soil Fertility*. Oxford: Permagon.

Gaudet, C. L., and Keddy, P. A. 1988. A comparative approach to predicting competitive ability from plant traits. *Nature* 334: 242–243.

Gehring, C. A., and Whitham, T. G. 1991. Herbivore driven mycorrhizal mutualism in insect susceptible pinyon pine. *Nature* 353: 556–557.

———. 1994. Interactions between aboveground herbivores and the mycorrhizal mutualists of plants. *Trends in Ecology and Evolution* 9: 251–255.

Geiger, R. 1950. *The Climate Near the Ground*. Cambridge: Harvard University Press.

Gentry, A. M. 1988. Changes in plant community diversity and floristic composition on environmental and geographic gradients. *Annals of the Missouri Botanical Garden* 75: 1–34.

Gerdol, R., et al. 2000. Response of dwarf shrubs to neighbor removal and nutrient addition and their influence on community structure in a subalpine heath. *Journal of Ecology* 88: 256–266.

Ghilarov, M. S. 1977. Why so many species and so many individuals can exist in the soil. *Ecological Bulletins* 25: 593–597.

Gholz, H. L., et al. 2000. Long term dynamics of pine and hardwood litter in contrasting environments: toward a global model of decomposition. *Global Change Biology* 6: 751–755.

Giardina, C. P., and Ryan, M. G. 2000a. Evidence that decomposition rates of organic matter in mineral soil do not vary with temperature. *Nature* 404: 858–861.

———. 2000b. Soil warming and organic carbon content. *Nature* 408: 790.

Gifford, R. M., et al. 2000. The effects of elevated [CO_2] on the C:N and C:P mass ratios of plant tissues. *Plant and Soil* 224: 1–14.

Giller, K. E., and Cadisch, G. 1995. Future benefits from biological nitrogen fixation: an ecological approach to agriculture. *Plant and Soil* 174: 255–277.

Giller, K. E., et al. 1997. Agricultural intensification, soil biodiversity and ecosystem function. *Applied Soil Ecology* 6: 3–16.

————· 1998. Toxicity of heavy metals to microorganisms and microbial processes in agricultural soils: a review. *Soil Biology and Biochemistry* 30: 1389–1414.

Giller, P. S. 1996. The diversity of soil communities, the "poor man's tropical rainforest." *Biodiversity and Conservation* 5: 135–168.

Goldberg, D. E., and Barton, A. M. 1992. Patterns and consequences of interspecific competition in natural communities: a review of field experiments with plants. *American Naturalist* 139: 771–801.

Goldberg, D. E., and Novoplansky, A. 1997. On the relative importance of competition in unproductive environments. *Journal of Ecology* 85: 409–418.

Goldberg, D. E., et al. 1999. Empirical approaches to quantifying interaction intensity: competition and facilitation along productivity gradients. *Ecology* 80: 1118–1131.

Gorissen, A., and Cotrufo, M. F. 1999. Elevated carbon dioxide effects on nitrogen dynamics in grasses, with emphasis on rhizosphere processes. *Soil Science Society of America Journal* 63: 1695–1702.

Gorissen, A., and Kupyer, T. W. 2000. Fungal species-specific response of ectomycorrhizal Scots pine (*Pinus sylvestris*) to elevated [CO_2]. *New Phytologist* 146: 163–168.

Gotelli, N. J., and Arnett, A. E. 2000. Biogeographic effects of red fire ant invasion. *Ecology Letters* 3: 257–261.

Goverde, M., et al. 2000. Arbuscular mycorrhizal fungi influence life history traits of a lepidopteran larvae. *Oecologia* 125: 362–369.

Grace, J. B. 1998. The factors controlling species density in herbaceous plant communities: an assessment. *Perspectives in Plant Ecology, Evolution and Systematics* 2: 1–28.

Granatstein, D. M., et al. 1987. Long-term tillage and rotation effects on soil microbial biomass, carbon and nitrogen. *Biology and Fertility of Soils* 5: 265–270.

Grant, D. J. 1983. The activities of earthworms and the fate of seeds. In J. E. Satchell, ed., *Earthworm Ecology*, 107–122. London: Chapman and Hall.

Grayston, S. J., et al. 1998. Selective influence of plant species on microbial diversity in the rhizosphere. *Soil Biology and Biochemistry* 30: 369–378.

Greenfield, L. G. 1999. Weight loss and release of mineral nitrogen from decomposing pollen. *Soil Biology and Biochemistry* 31: 353–361.

Gremmen, N. J. M., et al. 1998. Impact of the introduced grass *Agrostis stolonifera* on vegetation and soil faunal communities at Marion Island, sub-Antarctic. *Biological Conservation* 85: 223–231.

Griffiths, B. S. 1994. Soil nutrient flow. In J. F. Darbyshire, ed., *Soil Protozoa*, 65–91. Wallingford: CAB International.

Griffiths, B. S., and Caul, S. 1993. Migration of bacterial-feeding nematodes, but not protozoa, to decomposing grass residues. *Biology and Fertility of Soils* 15: 201–207.

Griffiths, B. S., et al. 1992. The effect of nitrate-nitrogen on bacteria and bacterial-feeding fauna in the rhizosphere of different grass species. *Oecologia* 91: 253–259.

———. 2000. Ecosystem response of pasture soil communities to fumigation-induced microbial diversity reductions: an examination of the biodiversity—ecosystem function relationship. *Oikos* 90: 279–294.

Grime, J. P. 1973a. Control of species density in herbaceous vegetation. *Journal of Environmental Management* 1: 151–167.

———. 1973b. Competitive exclusion in herbaceous vegetation. *Nature* 242: 344–347.

———. 1977. Evidence for the existence of three primary strategies of plants and its relevance to ecological and evolutionary theory. *American Naturalist* 111: 1169–1194.

———. 1979. *Plant Strategies and Vegetation Processes.* Chichester: John Wiley & Sons.

———. 1987. Dominant and subordinate components of plant communities: implications for succession, stability and diversity. In A. J. Gray, M. J. Crawley, and P. J. Edwards, eds., *Colonization, Succession and Stability,* 413–428. Oxford: Blackwells.

———. 1994. The role of plasticity in exploiting environmental heterogeniety. In M. M. Caldwell and R. W. Pearcy, eds., *Exploitation of Environmental Heterogeniety by Plants: Ecophysiological Processes Above- and Belowground,* 1–19. San Diego: Academic.

———. 1997. The hump-backed model: a response to Oksanen. *Journal of Ecology* 85: 97–98.

———. 1998. Benefits of plant diversity to ecosystems: immediate, filter and founder effects. *Journal of Ecology* 86: 902–910.

Grime, J. P., et al. 1987. Floristic diversity in a model system using experimental microcosms. *Nature* 328: 420–422.

———. 1996. Evidence of a causal connection between antiherbivore defence and the decomposition rate of leaves. *Oikos* 77: 489–494.

———· 1997. Integrated screening validates primary axes of specialization in plants. *Oikos* 79: 259–281.

———· 2000. The response of two contrasting limestone grasslands to simulated climate change. *Science* 289: 762–765.

Groffman, P. M., et al. 1996. Grass species and soil type effects on microbial biomass and activity. *Plant and Soil* 183: 61–67.

Grubb, P. J. 1985. Plant populations and vegetation in response to habitat disturbance and competition: problems of generalization. In J. White, ed., *The Population Structure of Vegetation*, 595–621. Dordrecht: Junk.

———· 1998. A reassessment of the strategies of plants which cope with the shortages of resources. *Perspectives in Plant Ecology, Evolution and Systematics* 1: 3–31.

Grubb, P. J., and Tanner, E. V. J. 1976. The montane forests and soils of Jamaica: a reassessment. *Journal of the Arnold Arboretum* 57: 313–368.

Gulledge, J., and Schimel, J. P. 1998. Moisture control over atmospheric CH_4 consumption and CO_2 production in diverse Alaskan soils. *Soil Biology and Biochemistry* 30: 1127–1132.

Guitian, R., and Bardgett, R. D. 2000. Plant and soil microbial responses to defoliation in temperate semi-natural grasslands. *Plant and Soil* 220: 271–277.

Gurevitch, J., et al. 1992. A meta-analysis of competition experiments. *American Naturalist* 140: 539–572.

———· 2000. The interaction between competition and predation: a meta-analysis of field experiments. *American Naturalist* 155: 435–453.

Haimi, J., et al. 1992. Growth increase of birch seedlings under the influence of earthworms—a laboratory study. *Soil Biology and Biochemistry* 24: 1525–1528.

———· 2000. Responses of soil decomposer animals to wood ash fertilization and burning in a coniferous forest stand. *Forest Ecology and Management* 129: 53–61.

Hairston, N. G., et al. 1960. Community structure, population control and competition. *American Naturalist* 94: 421–425.

Halaj, J., and Wise, D. H. 2001. Terrestrial trophic cascades: how much do they trickle? *American Naturalist* 157: 262–281.

Hall, M., and Hedlund, K. 1999. The predatory mite *Hypoaspis aculifer* is attracted to food of its fungivorous prey. *Pedobiologia* 43: 11–17.

Hall, S. J., and Raffaelli, D. G. 1993. Food webs: theory and reality. *Advances in Ecological Research* 24: 187–239.

Halvorson, J. J., et al. 1991. Lupine influence on soil carbon, nitrogen and microbial activity in developing ecosystems at Mt. St. Helens. *Oecologia* 87: 162–170.

Hamilton, W. E., and Sillman, D. Y. 1989. Effects of earthworm middens on the distribution of soil microarthropods. *Biology and Fertility of Soils* 8: 279–284.

Hance, T., et al. 1990. Agriculture and ground beetles: the consequences of crop types and surrounding habitats on activity and species composition. *Pedobiologia* 34: 337–346.

Handley, W. R. C. 1954. *Mull and Mor in Relation to Forest Soils.* London: Her Majesty's Stationery Office.

———. 1961. Further evidence for the importance of residual leaf protein complexes in litter decomposition and the supply of nitrogen for plant growth. *Plant and Soil* 15: 37–73.

Hanlon, R. D. G. 1981. Influence of grazing by Collembola on the activity of senescent fungal colonies grown on media of different nutrient concentration. *Oikos* 36: 362–367.

Hansen, R. A. 1999. Red oak litter promotes a microarthropod functional group that accelerates its decomposition. *Plant and Soil* 209: 37–45.

———. 2000. Effect of habitat complexity and composition on a diverse litter microarthropod assemblage. *Ecology* 81: 1120–1132.

Hansen, R. A., and Coleman, D. C. 1998. Litter complexity and composition are determinants of the diversity and composition of oribatid mites (Acari: Oribatida) in litterbags. *Applied Soil Ecology* 9: 17–23.

Haria, A. H., et al. 1998. Impact of the New Zealand flatworm (Artoposthia triangulata) on soil structure and hydrology in the U.K. *Science of the Total Environment* 215: 259–265.

Harinikumar, K. M., and Bagyaraj, D. J. 1994. Potential of earthworms, ants, millipedes and termites for dissemination of vesicular-arbuscular mycorrhizal fungi in soil. *Biology and Fertility of Soils* 18: 115–118.

Harley, J. L., and Waid, J. S. 1956. The effect of light upon the roots of birch and its surface population. *Plant and Soil* 7: 96–112.

Hart, S. C., and Stark, J. M. 1997. Nitrogen limitation of the microbial biomass in an old growth forest. *Écoscience* 4: 91–98.

Harte, J., and Shaw, R. 1995. Shifting dominance within a montane vegetation community: results of a climate-warming experiment. *Science* 267: 876–880.

Harte, J., et al. 1996. Effects of manipulated soil microclimate on mesofaunal biomass and diversity. *Soil Biology and Biochemistry* 28: 313–322.

Hartnett, D. C., and Wilson, G. W. T. 1999. Mycorrhizae influence plant community structure and diversity in tallgrass prairie. *Ecology* 80: 1187–1195.

Hartwig, U. A., et al. 2000. Due to symbiotic N_2-fixation, five years of elevated atmospheric pCO_2 had no effect on the N-concentration of plant litter in fertile, mixed grassland. *Plant and Soil* 224: 43–50.

Hassall, M., and Rushton, S. P. 1982. The role of coprophagy in the feeding strategies of terrestrial isopods. *Oecologia* 53: 374–381.

Hassall, M., et al. 1987. Effects of terrestrial isopods on the decomposition of woodland leaf litter. *Oecologia* 72: 597–602.

Hassink, J., et al. 1993. Relationships between soil texture, physical protection of organic matter, and C and N mineralization in grassland soils. *Geoderma* 57: 105–128.

Hatch, D. J., and Murray, P. J. 1994. Transfer of nitrogen from damaged roots of white clover (*Trifolium repens* L.) to closely associated roots of intact perennial ryegrass (*Lolium perenne* L.) *Plant and Soil* 166: 181–185.

Hättenschwiler, S., and Vitousek, P. M. 2000. The role of polyphenols in terrestrial ecosystem nutrient cycling. *Trends in Ecology and Evolution* 15: 238–243.

Hättenschwiler, S., et al. 1999. Quality, decomposition and isopod consumption of tree litter produced under elevated CO_2. *Oikos* 85: 271–281.

Hattersley, P. W. 1983. The distribution of C3 and C4 grasses in Australia in relation to climate. *Oecologia* 57: 113–128.

Hawkesworth, D. L. 1991. The fungal dimension of biodiversity: magnitude, significance and conservation. *Mycological Research* 95: 641–655.

Haynes, R. J., and Williams, P. H. 1993. Nutrient cycling and soil fertility in the grazed pasture ecosystem. *Advances in Agronomy* 49: 119–199.

——— 1999. Influence of stock camping behaviour on the soil microbiological and biochemical properties of grazed pastoral soil. *Biology and Fertility of Soils* 28: 253–258.

Heal, O. W., et al. 1997. Plant litter quality and decomposition: an historical overview. In K. E. Giller and G. Cadisch, eds., *Driven by Nature—Plant Litter Quality and Decomposition*, 1–30. Wallingford: CAB International.

Hector, A. 2000. Biodiversity and ecosystem functioning. *Progress in Environmental Science* 2: 155–162.

Hector, A., et al. 1999. Plant diversity and productivity in European grasslands. *Science* 286: 1123–1127.

———· 2000a. No consistent effect of plant diversity on productivity (response). *Science* 289: 1255a.

———· 2000b. Consequences of the reduction of plant diversity for litter decomposition: effects through litter quality and micro-environment. *Oikos* 90: 357–371.

———· 2001. Biodiversity and the functioning of grassland ecosystems: multiple comparisons. In A. Kinzig, D. Tilman, and S. Pacala, eds., *Functional Consequences of Biodiversity: Experimental Progress and Theoretical Extensions*. Princeton: Princeton University Press (in press).

Hedlund, K., and Augustsson, A. 1995. Effects of enchytraeid grazing on fungal growth and respiration. *Soil Biology and Biochemistry* 27: 905–909.

Hedlund, K, and Öhrn, M. S. 2000. Tritrophic interactions in a soil community enhance decomposition rates. *Oikos* 88: 585–591.

Hendriksen, N. B. 1990. Leaf litter selection by detritivore and geophagous earthworms. *Biology and Fertility of Soils* 10: 17–21.

Hendrix, P. F., et al. 1986. Detritus food webs in conventional and no-tillage agroecosystems. *BioScience* 36: 374–380.

Heneghan, L., et al. 1999. Soil microarthropod contributions to decomposition dynamics: tropical-temperate comparisons of a single substrate. *Ecology* 80: 1873–1882.

Herbert, D. A., et al. 1999a. Effects of plant growth characteristics on biogeochemistry and community composition in a changing climate. *Ecosystems* 2: 367–382.

———· 1999b. Hurricane damage to a Hawaiian forest: nutrient supply rate affects resistance and resilience. *Ecology* 80: 908–920.

Hetrick, B. A., et al. 1989. Relationship between mycorrhizal dependence and competitive ability of two tallgrass prairie grasses. *Canadian Journal of Botany* 67: 2608–2615.

Hiernaux, P., and Turner, M. D. 1996. The effect of clipping on growth and nutrient uptake of Sahelian annual rangelands. *Journal of Applied Ecology* 33: 387–399.

Hirschel, G., et al. 1997. Will rising atmospheric CO_2 affect litter quality and *in situ* decomposition rates in native plant communities? *Oecologia* 110: 387–392.

Hobbie, S. E. 1992. Effects of plant species on nutrient cycling. *Trends in Ecology and Evolution* 7: 336–339.

———. 1996. Temperature and plant species control over litter decomposition in Alaskan tundra. *Ecological Monographs* 66: 503–522.

Hobbie, S. E., and Vitousek, P. M. 2000. Nutrient limitation of decomposition in Hawaiian forests. *Ecology* 81: 1867–1877.

Hobbie, S. E., et al. 1999. Plant responses to species removal and experimental warming in Alaskan tussock tundra. *Oikos* 84: 417–434.

Hobbs, N. T., et al. 1991. Fire and grazing in tallgrass prairie: contigent effects on nitrogen budgets. *Ecology* 72: 1374–1382.

Hobbs, R. J., and Mooney, H. A. 1985. Community and population dynamics of serpentine grassland annuals in relation to gopher disturbance. *Oecologia* 67: 342–351.

Hodge, A., et al. 2000. Spatial and physical heterogeneity of N supply from soil does not influence N capture by two grass species. *Functional Ecology* 14: 645–653.

Hodgson, J. G., et al. 1998. Does biodiversity determine ecosystem function? The Ecotron revisited. *Functional Ecology* 12: 843–848.

Hodkinson, I. D., et al. 1996. Can high Arctic soil microarthropods survive elevated summer temperatures? *Functional Ecology* 10: 314–321.

Högberg, P., et al. 1999. Natural ^{13}C abundance reveals trophic status of fungi and host-origin of carbon in mycorrhizal fungi in mixed forests. *Proceedings of the National Academy of Sciences of the U.S.A.* 96: 8534–8539.

Hoitink, H. A. J., and Boehm, M. J. 1999. Biocontrol within the context of microbial communities: a substrate-dependent phenomenon. *Annual Review of Phytopathology* 37: 427–446.

Holdaway, R. N., and Jacomb, C. 2000. Rapid extinction of the moas (Aves: Dinornithiformes): model, test and implications. *Science* 287: 2250–2254.

Holdgate, M. 1996. The ecological significance of biological diversity. *Ambio* 25: 409–416.

Holland, E. A., and Detling, J. K. 1990. Plant response to herbivory and below ground nitrogen cycling. *Ecology* 71: 1040–1049.

Holland, E. A., et al. 1992. Physiological responses of plant populations to herbivory and their consequences for ecosystem nutrient flow. *American Naturalist* 140: 685–706.

Holland, J. N. 1995. Effects of above-ground herbivory on soil microbial biomass in conventional and no-tillage agroecosystems. *Applied Soil Ecology* 2: 275–279.

Holland, J. N., et al. 1996. Herbage-induced changes in plant carbon allocation: assessment of below ground carbon fluxes using carbon-14. *Oecologia* 107: 87–94.

Holmer, L., and Stenlid, J. 1996. Diffuse competition for heterogeneous substrates in soil among six species of wood decomposing basidiomycetes. *Oecologia* 106: 531–538.

Holmgren, M., et al. 2001. El Niño effects on the dynamics of terrestrial ecosystems. *Trends in Ecology and Evolution* 16: 89–94.

Holt, R. D., and Lawton, J. H. 1994. The ecological consequences of shared natural enemies. *Annual Review of Ecology and Systematics* 25: 495–520.

Hooper, D. U. 1998. The role of complementarity and competition in ecosystem responses to variation in plant diversity. *Ecology* 79: 704–719.

Hooper, D. U., and Vitousek, P. M. 1997. The effects of plant composition and diversity on ecosystem processes. *Science* 277: 1302–1305.

———. 1998. Effects of plant composition and diversity on nutrient cycling. *Ecological Monographs* 68: 121–149.

Hooper, D. U., et al. 2000. Interactions between above- and belowground biodiversity in terrestrial ecosystems: patterns, mechanisms and feedbacks. *BioScience* 50: 1049–1061.

Horner, J. D., et al. 1988. The role of carbon based secondary metabolites in decomposition in terrestrial ecosystems. *American Naturalist* 132: 869–883.

House, G. J. 1989. Soil arthropods from weed and crop roots of an agroecosystem in a wheat-soybean-corn rotation: impact of tillage and herbicides. *Agriculture, Ecosystems and Environment* 25: 233–244.

House, G. J., et al. 1984. Nitrogen cycling in conventional and no-tillage agroecosystems in the southern piedmont. *Journal of Soil and Water Conservation* 39: 194–200.

Howarth, F. G. 1983. Ecology of cave arthropods. *Annual Review of Entomology* 28: 365–389.

Hu, S., et al. 2001. Nitrogen limitation of microbial decomposition in a grassland under elevated CO_2. *Nature* 409: 188–191.

Huenneke, L. S., et al. 1990. Effects of soil resources on plant invasion and community structure in Californian serpentine grassland. *Ecology* 71: 478–491.

338

Hughes, J. B., and Roughgarden, J. 2000. Species diversity and biomass stability. *American Naturalist* 155: 618–627.

Huhta, V., and Viberg, K. 1999. Competitive interactions between the earthworm *Dendrobaena octaedra* and the enchytraeid *Cognettia sphagnetorum*. *Pedobiologia* 43: 886–890.

Huhta, V., et al. 1998. Interactions between enchytraeid (*Cognettia sphagnetorum*), microarthropod and nematode populations in forest soil at different moistures. *Applied Soil Ecology* 9: 53–58.

Hungate, B. A., et al. 1996a. Plant species mediate changes in soil microbial N in response to elevated CO_2. *Ecology* 77: 2505–2515.

———· 1996b. Detecting changes in soil carbon in CO_2 enrichment experiments. *Plant and Soil* 187: 135–145.

———· 1997. The fate of carbon in grasslands under carbon dioxide enrichment. *Nature* 388: 576–579.

———· 2000. Soil microbiota in two annual grasslands: responses to elevated atmospheric CO_2. *Oecologia* 124: 589–598.

Hunt, H. W., et al. 1987. The detrital food web in a shortgrass prairie. *Biology and Fertility of Soils* 3: 57–68.

———· 1988. Nitrogen limitation of production and decomposition in prairie, mountain meadow and pine forest. *Ecology* 69: 1009–1016.

Hunter, M. D., and Forkner, R. E. 1999. Hurricane damage influences foliar nutrient phenolics and subsequent herbivory on surviving trees. *Ecology* 80: 2676–2682.

Huston, M. A. 1979. A general model of species diversity. *American Naturalist* 113: 81–101.

———· 1994. *Biological Diversity. The Coexistence of Species on Changing Landscapes*. Cambridge: Cambridge University Press.

———· 1997. Hidden treatments in ecological experiments: reevaluating the ecosystem function of biodiversity. *Oecologia* 110: 449–460.

———· 1999. Local processes and regional patterns: appropriate scales for understanding variation in the diversity of plants and animals. *Oikos* 86: 393–401.

Huston, M. A., and De Angelis, D. L. 1994. Competition and coexistence: the effects of resource transport and supply rates. *American Naturalist* 144: 854–877.

Huston, M. A., and Smith, T. 1987. Plant succession: life history and competition. *American Naturalist* 130: 168–198.

Huston, M. A., et al. 2000. No consistent effect of plant diversity on productivity. *Science* 289: 1255a.

Hutchings, M. J., and De Kroon, H. 1994. Foraging in plants: the role of morphological plasticity in resource acquisition. *Advances in Ecological Research* 25: 159–237.

Hutchinson, G. E. 1961. The paradox of the plankton. *American Naturalist* 45: 137–145.

Hyodo, F., et al. 2000. Role of the mutualistic fungus in lignin degradation of the soil fungus *Macromycetes gilvus* (Isoptera: Macrotermitinae). *Soil Biology and Biochemistry* 32: 653–658.

Hyvönen, R., et al. 1994. Effects of lumbricids and enchytraeids on nematodes in limed and unlimed coniferous mor humus. *Biology and Fertility of Soils* 17: 201–205.

Ingham, E. R., et al. 1986. Trophic interactions and nitrogen cycling in a semi-arid grassland soil. 1. Seasonal dynamics of the natural populations, their interactions and effects on nutrient cycling. *Journal of Applied Ecology* 23: 597–614.

——— 1989. An analysis of food web structure and function in a shortgrass prairie, a mountain meadow and a lodgepole pine forest. *Biology and Fertility of Soils* 8: 29–37.

Ingham, R. E., and Detling, J. K. 1984. Plant-herbivore interactions in a North American mixed-grass prairie. III. Soil nematode populations and root biomass on *Cynmoys ludovicianus* colonies and adjacent uncolonized areas. *Oecologia* 63: 307–313.

Ingham, R. E., et al. 1985. Interactions of bacteria, fungi and their nematode grazers: effects on nutrient cycling and plant growth. *Ecological Monographs* 55: 119–140.

Insam, H. 1990. Are the soil microbial biomass and basal respiration governed by the climatic regime? *Soil Biology and Biochemistry* 22: 525–532.

Insam, H., and Domsch, K. H. 1988. Relationship between soil organic carbon and microbial biomass on chronosequences of reclamation sites. *Microbial Ecology* 15: 177–188.

Insam, H., and Haselwandter, K. 1989. Metabolic quotient of the soil microflora in relation to plant succession. *Oecologia* 79: 174–178.

Insam, H., et al. 1991. Relationship of soil microbial biomass and activity with fertilization practice and crop yield of three Ultisols. *Soil Biology and Biochemistry* 23: 459–464.

Iseman, T. M., et al. 1999. Revegetation and nitrate leaching from Lake States northern hardwood forests following harvest. *Soil Science Society of America Journal* 63: 1424–1429.

Ives, A. R., et al. 1999. Stability and variability in competitive communities. *Science* 286: 542–544.

Jakobsen, I. 1992. Phosphorus transport by external hyphae of vesicular-arbuscular mycorrhizas. In D. J. Read, D. H. Lewis, A. H. Fitter and I. Alexander, eds., *Mycorrhizas in Ecosystems*, 48–54. Wallingford: CAB International.

James, S. W. 1988. The postfire environment and earthworm populations in tallgrass prairie. *Ecology* 69: 476–483.

———. 1992. Seasonal and experimental variation in population structure of earthworms in tall grass prairie. *Soil Biology and Biochemistry* 24: 1445–1449.

James, S. W., and Seastedt, T. R. 1986. Nitrogen mineralization by native and introduced earthworms: effects on big bluestem growth. *Ecology* 67: 1094–1097.

Jane, G. T., and Pracy, L. T. 1974. Observations on two animal exclosures in Hikurangi forest over a period of twenty years (1951–1971). *New Zealand Journal of Forestry* 19: 102–113.

Janzen, D. H. 1970. Herbivores and the number of tree species in tropical forests. *American Naturalist* 104: 501–528.

Janzen, R. A., et al. 1995. A community-level concept of controls on decomposition processes: decomposition of barley straw by *Phanerochaete chrysosporium* or *Phlebia radiata* in pure or mixed culture. *Soil Biology and Biochemistry* 27: 173–179.

Järemo, J., et al. 1996. Plant compensatory growth: herbivory or competition? *Oikos* 77: 238–247.

Jarosz, A. M., and Davelos, A. L. 1995. Effects of disease in wild plant populations and the evolution of plant aggressiveness. *New Phytologist* 129: 371–387.

Jenkinson, D. S., and Ladd, J. N. 1981. Microbial biomass in soil: measurement and turnover. In J. N. Ladd and E. A. Paul, *Soil Biochemistry: Volume 5*, 415–471. New York: Marcel Dekker.

Jenkinson, D. S., et al. 1991. Model estimates of CO_2 emissions from soil in response to global warming. *Nature* 351: 304–306.

———. 1999. Nitrogen deposition and carbon sequestration. *Nature* 400: 629.

Jensen, E. S. 1994. Availability of nitrogen in ^{15}N-labeled mature pea residues to subsequent crops in the field. *Soil Biology and Biochemistry* 26: 465–472.

Jentschke, G., et al. 1995. Soil protozoa and forest tree growth: non-nutritional effects and interaction with mycorrhizae. *Biology and Fertility of Soils* 20: 263–269.

Jobbagy, E. G., and Jackson, R. B. 2000. The vertical distribution of soil organic carbon and its relation to climate and vegetation. *Ecological Applications* 10: 423–436.

Johansson, G. 1992. Release of organic C from growing roots of meadow fescue (*Festuca pratensis* L.). *Soil Biology and Biochemistry* 24: 427–433.

Johnson, C. N. 1996. Interactions between mammals and ectomycorrhizal fungi. *Trends in Ecology and Evolution* 11: 503–507.

Johnson, K. H. 2000. Trophic-dynamic considerations in relating species diversity to ecosystem resilience. *Biological Reviews* 75: 347–376.

Johnson, L. C., et al. 2000. Plant carbon-nutrient interactions control CO_2 exchange in Alaskan wet sedge tundra ecosystems. *Ecology* 81: 453–469.

Joliffe, P. A. 1997. Are mixed populations of plants more productive than pure stands? *Oikos* 80: 595–602.

Jonasson, S., et al. 1996a. Microbial biomass C, N and P in two arctic soils and responses to addition of NPK fertilizer and sugar: implications for plant nutrient uptake. *Oecologia* 106: 507–515.

———. 1996b. Effects of carbohydrate amendments on nutrient partitioning, plant and microbial performance of a grassland-shrub ecosystem. *Oikos* 75: 220–226.

———. 1999. Responses in plants and microbes to changed temperature, nutrient and light regimes in the Arctic. *Ecology* 80: 1828–1843.

Jones, C. G., and Lawton, J. H. Editors. 1995. *Linking Species and Ecosystems.* New York: Academic.

Jones, C. G., et al. 1994. Organisms as ecosystem engineers. *Oikos* 69: 373–386.

Jones, M. D., et al. 1998. A comparison of arbuscular and ectomycorrhizal *Eucalyptus coccifera*: growth response, phosphorus uptake efficiency and external hyphal production. *New Phytologist* 140: 125–134.

Jones, T. H., et al. 1998. Impacts of rising atmospheric carbon dioxide on model terrestrial ecosystems. *Science* 280: 221–223.

Jongmans, A. G., et al. 1997. Rock eating fungi. *Nature* 389: 682–683.

Jonsson, L., et al. 2000. Spatiotemporal distribution of an ectomycorrhizal community in an oligotrophic Swedish *Picea abies* forest subjected to experimental nitrogen addition: above- and below-ground views. *Forest Ecology and Management* 132: 143–156.

———. 2001. Context dependent effects of ectomycorrhizal species richness on tree seedling productivity. *Oikos* (in press).

Joose, E. N. G., and Verhoef, H. A. 1987. Developments in ecophysiological research on soil invertebrates. *Advances in Ecological Research* 16: 175–248.

342

Kaiser, J. 2000. Rift over biodiversity divides ecologists. *Science* 289: 1282–1283.

Kajak, A., and Lukasiewicz, J. 1994. Do semi-natural patches enrich crop fields with preditory epigean arthropods? *Agriculture, Ecosystems and Environment* 49: 149–161.

Kajak, A., et al. 1993. Experimental studies on the effect of epigeic predators on matter decomposition process in managed peat grassland. *Polish Ecological Studies* 17: 289–310.

Kandeler, E., et al. 1998. The response of soil microorganisms and roots to elevated CO_2 and temperature in a terrestrial model ecosystem. *Plant and Soil* 202: 251–262.

Kaneko, N., and Salamanca, N. 1999. Mixed leaf litter effects on decomposition rates and soil arthropod communities in an oak-pine forest stand in Japan. *Ecological Research* 14: 131–138.

Kareiva, P. 1996. Diversity and sustainability on the prairie. *Nature* 279: 673–674.

Kates, R. W., et al. 1990. The great transformation. In B. L. Turner, W. C. Clark, R. W Kates, J. F. Richards, J. T. Matthews, and W. B. Meyer, eds., *The Earth as Transformed by Human Action*, 1–17. Cambridge: Cambridge University Press.

Kaunzinger, C. M. K., and Morin, P. J. 1998. Productivity controls food chain properties in microbial communities. *Nature* 395: 495–497.

Kaye, J. P., and Hart, S. C. 1997. Competition for nitrogen between plants and soil microorganisms. *Trends in Ecology and Evolution* 12: 139–143.

Keddy, P. A. 1989. *Competition*. London: Chapman and Hall.

Keddy, P. A., et al. 1994. Competitive effect and response rankings in 20 wetland plants: are they consistent across three environments? *Journal of Ecology* 82: 635–641.

——— 2000. Effects of low and high nutrients on the competitive hierarchy of 26 shoreline plants. *Journal of Ecology* 88: 413–423.

Kelly, D. 1994. The evolutionary ecology of mast seeding. *Trends in Ecology and Evolution* 9: 465–471.

Kieft, T. L., et al. 1987. Microbial response to a rapid increase in water potential when dry soil is rewetted. *Soil Biology and Biochemistry* 19: 119–126.

Kielland, K., and Bryant, J. P. 1998. Moose herbivory in taiga: effects on biogeochemistry and vegetation dynamics in primary succession. *Oikos* 82: 377–383.

Kielland, K., et al. 1997. Moose herbivory and carbon turnover of early successional stands in interior Alaska. *Oikos* 80: 25–30.

Kiens, E. T., et al. 2000. Differential effects of tropical arbuscular mycorrhizal fungal inocula on root colonization and tree seedling growth: implications for tropical forest diversity. *Ecology Letters* 3: 106–113.

Kinzig, A. P., et al. Editors. 2001. *The Functional Consequences of Biodiversity: Experimental Progress and Theoretical Extensions.* Princeton: Princeton University Press (in press).

Kirchbaum, M. U. F. 1995. The temperature dependence of soil organic matter decomposition and the effect of global warming on soil organic C storage. *Soil Biology and Biochemistry* 27: 753–760.

Kirchmann, H., and Eklund, M. 1994. Microbial biomass in a savanna-woodland and an adjacent arable soil profile in Zimbabwe. *Soil Biology and Biochemistry* 26: 1281–1283.

Kjøller, A., and Struwe, S. 1982. Microfungi in ecosystems: fungal occurrence and activity in litter and soil. *Oikos* 39: 393–422.

Kleb, H., and Wilson, S. D. 1997. Vegetation effects on soil resource heterogeneity in prairie and forest. *American Naturalist* 150: 283–298.

Klemendtsson, L., et al. 1987. Microbial nitrogen transformations in the root environment of barley. *Soil Biology and Biochemistry* 19: 551–558.

Klironomos, J. N., and Kendrick, W. B. 1995. Stimulative effects of arthropods on endomycorrhizas of sugar maple in the presence of decaying litter. *Functional Ecology* 9: 528–536.

Klironomos, J. N., et al. 1992. Feeding preferences of the collembolan *Folsomia candida* in relation to microfungal successions on decaying litter. *Soil Biology and Biochemistry* 24: 685–692.

———. 1996. Below-ground microbial and microfaunal responses to *Artemisia tridentata* grown in elevated atmospheric CO_2. *Functional Ecology* 10: 527–534.

———. 1997. Soil fungal-arthropod responses to *Populus tremuloides* grown under enriched atmospheric CO_2 under field conditions. *Global Change Biology* 3: 473–478.

———. 1999. Reproductive significance of feeding on saprobic and arbuscular mycorrhizal fungi by the collembolan *Folsomia candida*. *Functional Ecology* 13: 756–761.

Klopatek, C. C., et al. 1992. The sustainable biosphere initiative: a commentary from the U.S. Soil Ecology Society. *Bulletin of the Ecological Society of America* 73: 223–228.

Knapp, A. K., and Smith, M. D. 2001. Variation among biomes in temporal dynamics of aboveground primary production. *Science* 291: 481–484.

344

Knapp, A. K., et al. 1999. The keystone role of bison in North American tallgrass prairie. *BioScience* 49: 39–50.

Knops, J. M. H., et al. 1999. Effects of plant species richness on invasion dynamics, disease outbreaks, insect abundances and diversity. *Ecology Letters* 2: 286–293.

Knusden, I. M. B., et al. 1999. Suppressiveness of organically and conventionally managed soils towards brown foot rot of barley. *Applied Soil Ecology* 12: 61–72.

Koopmans, C. J., et al. 1998. Effects of reduced N deposition on litter decomposition and N cycling in two N saturated forests in the Netherlands. *Soil Biology and Biochemstry* 30: 141–151.

Körner, C. 2000. Biosphere responses to CO_2 enrichment. *Ecological Applications* 10: 1590–1619.

Körner, C., and Arnone, J. A. III. 1992. Response to elevated carbon dioxide in artificial tropical ecosystems. *Science* 257: 1672–1675.

Korsaeth, A., et al. 2001. Modeling the competition for nitrogen between plants and microflora as a function of soil heterogeneity. *Soil Biology and Biochemistry* 33: 215–226.

Kristensen, H. L., and McCarty, G. W. 1999. Mobilization and immobilization of nitrogen in heath soil under *Calluna*, after heather beetle infestation and nitrogen fertilization. *Applied Soil Ecology* 13: 187–198.

Kronzucker, H. J., et al. 1997. Conifer root discrimination against soil nitrate and the ecology of forest succession. *Nature* 385: 59–61.

Kruckelmann, H. W. 1975. Effects of fertilizers, soils, soil tillage and plant species on the frequency of endogone chlamydospores and mycorrhizal infection in arable soils. In F. E. Sanders, P. Mosse, and P. B. Tinker, eds., *Endomycorrhizas*, 511–525. London: Academic.

Kuikman, P. J., and van Veen, J. A. 1989. The impact of protozoa on the availability of bacterial nitrogen to plants. *Biology and Fertility of Soils* 8: 13–18.

Laakso, J., and Setälä, H. 1997. Nest mounds of red wood ants (*Formica aquilonia*): hot spots for litter dwelling earthworms. *Oecologia* 111: 565–569.

———. 1998. Composition and trophic structure of detrital food web in ant nest mounds of *Formica aquilonia* and in the surrounding forest soil. *Oikos* 81: 266–278.

———. 1999a. Population- and ecosystem-effects of predation on microbial-feeding nematodes. *Oecologia* 120: 279–286.

———. 1999b. Sensitivity of primary production to changes in the architecture of belowground food webs. *Oikos* 87: 57–64.

345

Lamont, B. B. 1995. Testing the effects of composition/structure on its functioning. *Oikos* 74: 283–295.

Lavelle, P. 1994. Faunal activities and soil processes: adaptive strategies that determine ecosystem function. In: *XV ISSS Congress Proceedings, Acapulco, Mexico, Volume 1: Introductory Conferences*, pp. 189–220.

———. 1997. Faunal activities and soil processes: adaptive strategies that determine ecosystem function. *Advances in Ecological Research* 27: 93–132.

Lavelle, P., and Martin, A. 1992. Small-scale and large-scale effects of endogeic earthworms on soil organic matter dynamics in soils of the humid tropics. *Soil Biology and Biochemistry* 24: 1491–1498.

Lavelle, P., et al. 1994. Soil fauna and land use in the humid tropics. In D. J. Greenland and I. Szabolcs, eds., *Soil Resilience and Sustainable Land Use*, 291–308. Wallingford: CAB International.

———. 1995. Mutualism and biodiversity in soils. *Plant and Soil* 170: 23–33.

———. 1997. Soil function in a changing world: the role of invertebrate ecosystem engineers. *European Journal of Soil Biology* 33: 159–193.

Lawrence, K. L., and Wise, D. H. 2000. Spider predation on forest floor Collembola and evidence for indirect effects on decomposition. *Pedobiologia* 44: 33–39.

Lawton, J. H. 1994. What do species do in ecosystems? *Oikos* 71: 367–374.

Lawton, J. H., and McNeill, S. 1979. Between the devil and the deep blue sea: on the problems of being a herbivore. In R. M. Anderson, B. D. Turner, and L. R. Taylor, eds., *Population Dynamics*, 223–244. London: Blackwells.

Lawton, J. H., et al. 1996. Carbon flux and diversity of nematodes and termites in Cameroon forest soils. *Biodiversity and Conservation* 5: 261–273.

———. 1998a. Biodiversity inventories, indicator taxa and effects of habitat modification in tropical forests. *Nature* 391: 72–76.

———. 1998b. Biodiversity and ecosystem function: getting the Ecotron experiment in its correct context. *Functional Ecology* 12: 848–852.

Ledgard, S. F., et al. 1985. Assessing nitrogen transfer from legumes to associated grasses. *Soil Biology and Biochemistry* 17: 575–577.

Lehman, C. L., and Tilman, D. 2000. Biodiversity, stability, and productivity in competitive communities. *American Naturalist* 156: 534–552.

Le Tacon, F., et al. 1992. Variations in field response of forest trees to nursery ectomycorrhizal inoculation in Europe. In D. J. Read, D. H. Lewis, A. H. Fitter, and I Alexander, eds., *Mycorrhizas in Ecosystems*, 119–134. Wallingford: CAB International.

Lethbridge, G., and Davidson, M. S. 1983. Microbial biomass as a source of nitrogen for cereals. *Soil Biology and Biochemistry* 15: 375–376.

Letourneau, D. K., and Dyer, L. A. 1998. Experimental tests in lowland tropical forest shows top-down effects through four trophic levels. *Ecology* 79: 1678–1687.

Levine, J. M. 2000. Species diversity and biological invasions: relating local process to community pattern. *Science* 288: 852–854.

Ley, R. E., and D'Antonio, C. M. 1998. Exotic grass invasion alters potential rates of N fixation in Hawaiian woodlands. *Oecologia* 113: 179–187.

Likens, G. E., et al. 1968. Nitrification: important to nutrient losses from a cutover forested ecosystem. *Science* 163: 1205–1206.

Lindahl, B., et al. 1999. Translocation of ^{32}P between interacting mycelia of a wood-decomposing fungus and ectomycorrhizal fungi in microcosm systems. *New Phytologist* 144: 183–193.

Linkel, L. L., et al. 1992. Microbial community analysis in incompletely or destructively sampled systems. *Microbial Ecology* 24: 227–242.

Lloyd, J. 1999. The CO_2 dependence of photosynthesis, plant growth responses to elevated CO_2 concentrations and their interactions with soil nutrient status, II. Temperature and boreal forest productivity and the combined effects of increasing CO_2 concentrations and increased nitrogen deposition at a global scale. *Functional Ecology* 13: 439–459.

Lockwood, J. L. 1981. Exploitation competition. In D. T. Wicklow and G. C. Carroll, eds., *The Fungal Community: Its Organization and Role in the Ecosystem*, 319–349. New York: Marcel Dekker.

Lodge, D. 1993. Biological invasions: lessons for ecology. *Trends in Ecology and Evolution* 8: 133–137.

Loranger, G., et al. 1998. Impact of earthworms on the diversity of microarthropods in a vertisol (Martinique). *Biology and Fertility of Soils* 27: 21–26.

Loreau, M. 1995. Consumers as maximizers of matter and energy flow in ecosystems. *American Naturalist* 145: 22–42.

——— 1998. Biodiversity and ecosystem functioning: a mechanistic model. *Proceedings of the National Academy of Sciences, U.S.A.* 95: 5632–5636.

———— 2000. Biodiversity and ecosystem function: recent theoretical advances. *Oikos* 91: 3–17.

———— 2001. Microbial diversity, producer-decomposer interactions and ecosystem processes: a theoretical model. *Proceedings of the Royal Society of London Series B* (in press).

Loreau, M., and Behera, N. 1999. Phenotypic diversity and stability of ecosystem processes. *Theoretical Population Biology* 56: 29–47.

Lovett, G. M., and Ruesink, A. E. 1995. Carbon and nitrogen mineralization from decomposing gypsy moth frass. *Oecologia* 104: 133–138.

Lovett, G. M., and Rueth, H. 1999. Soil nitrogen transformations in beech and maple stands along a nitrogen deposition gradient. *Ecological Applications* 9: 1330–1334.

Luizão, F. J., et al. 1998. Rain forest on Maracá Island, Roraima, Brazil: soil and litter process response to artificial gaps. *Forest Ecology and Management* 102: 291–303.

Luizao, R. C. C., et al. 1992. Seasonal variation of microbial biomass—the effects of clearfelling a tropical rainforest and establishment of pasture in the central Amazon. *Soil Biology and Biochemistry* 24: 805–813.

Luken, J. O., and Fonda, R. W. 1983. Nitrogen accumulation in a chronosequence of red alder communities along the Hoh River, Olympic National Park, Washington. *Canadian Journal of Forest Research* 13: 1228–1237.

Luo, Y., and Reynolds, J. F. 1999. Validity of extrapolating field CO_2 experiments to predict carbon sequestration in natural ecosystems. *Ecology* 80: 1568–1583.

Lüscher, A. 1996. Differences between legumes and non-legumes of permanent grasslands in their response to free air carbon dioxide enrichment. In C. Körner and F. A. Bazzaz, eds., *Biological Diversity in a CO_2 Rich World*, 150–160. San Diego: Academic.

Lussenhop, J., et al. 1998. Response of soil biota to elevated atmospheric CO_2 in poplar model systems. *Oecologia* 113: 247–251.

Lutze, J. L., et al. 2000. Litter quality and decomposition in *Danthonia richardsonii* swards in response to CO_2 and nitrogen supply over four years of growth. *Global Change Biology* 6: 13–24.

MacArthur, R. H. 1964. Environmental factors affecting bird species diversity. *American Naturalist* 98: 387–397.

MacGillivray, C. W., et al. 1995. Testing predictions of the resistance and resilience of vegetation subjected to extreme events. *Functional Ecology* 9: 640–649.

Mack, M. C., and D'Antonio, C. M. 1998. Impacts of biological invasions on disturbance regimes. *Trends in Ecology and Evolution* 13: 195–198.

Mao, D. M., et al. 1992. Effect of afforestation on microbial biomass and activity in soils of tropical China. *Soil Biology and Biochemistry* 24: 865–872.

Maraun, M., et al. 1999. Middens of the earthworm *Lumbricus terrestris* (Lumbricidae): microhabitats for micro- and mesofauna in forest soil. *Pedobiologia* 43: 276–287.

Marilley, L., and Aragno, M. 1999. Phylogenetic diversity of bacterial communities differing in degree of proximity of *Lolium perenne* and *Trifolium repens* roots. *Applied Soil Ecology* 13: 127–136.

Martens, R. 1990. Contribution of rhizodeposits to the maintenance and growth of soil microbial biomass. *Soil Biology and Biochemistry* 22: 141–147.

Martikainen, E., and Huhta, V. 1990. Interactions between nematodes and predatory mites in raw humus soil: a microcosm experiment. *Revue d'Ecologie et de Biologie du Sol* 27: 13–20.

Martikainen, P. J. 1996. Microbial processes in boreal forests as affected by forest management practices and atmospheric stress. In G. Stotzky and J.-M. Bollag, eds., *Soil Biochemistry, Volume 9*, 195–232. New York: Marcel Dekker.

Martin, A. 1991. Short- and long-term effects of the endogenic earthworm *Millsonia anomala* (Omodeo) (Megascolecidae: Oligochaeta) of tropical savannas, on soil organic matter. *Biology and Fertility of Soils* 11: 234–238.

Martin, A., et al. 1987. Les mucus intestinaex de ver de terre moteur de leurs interactions avec la microflorie. *Revue d'Écologie et de Biologie du Sol* 24: 549–558.

Martinez, N. D. 1992. Constant connectance in food webs. *American Naturalist* 139: 1208–1218.

Masters, G. J. 1995. The impact of root herbivory on aphid performance: field and laboratory evidence. *Acta Œcologica* 16: 135–142.

Masters, G. J., and Brown, V. K. 1992. Plant-mediated interactions between two spatially separated insects. *Functional Ecology* 6: 175–179.

———. 1997. Host-plant mediated interactions between spatially separated herbivores: effects on community structure. In A. C. Gange and V. K. Brown, eds., *Multitrophic Interactions in Terrestrial Systems*, 217–236. Oxford: Blackwells.

Masters, G. J., et al. 1993. Plant mediated interactions between above- and below-ground herbivores. *Oikos* 66: 148–151.

Mawdsley, J. L., and Bardgett, R. D. 1997. Continuous defoliation of perennial ryegrass (*Lolium perenne*) and white clover (*Trifolium repens*) and associated changes in the microbial population of an upland soil. *Biology and Fertility of Soils* 24: 52–58.

May, R. M. 1973. *Stability and Complexity in Model Ecosystems.* Princeton: Princeton University Press.

——. 1988. How many species are there on Earth? *Science* 241: 1441–1449.

McCann, K., et al. 1998. Weak trophic interactions and the balance of nature. *Nature* 395: 794–798.

McCulloch, D. G., et al. 1998. Fire and insects in northern and boreal forests of North America. *Annual Review of Entomology* 43: 107–127.

McGill, W. B., et al. 1986. Dynamics of soil microbial biomass and water soluble organic carbon in Breton L. after 50 years of cropping to two rotations. *Canadian Journal of Soil Science* 66: 1–19.

McGrady-Steed, J., et al. 1997. Biodiversity regulates ecosystem predictability. *Nature* 390: 162–165.

McIlwee, A. P., and Johnson, C. N. 1998. The contribution of fungus to the diets of three mycophagous marsupials in *Eucalyptus* forest, revealed by stable isotope analysis. *Functional Ecology* 12: 223–231.

McIntosh, P. D., et al. 1998. Effect of exclosures on soils, biomass, plant nutrients and vegetation, on unfertilized steeplands, upper Waitaki district, South Island, New Zealand. *New Zealand Journal of Ecology* 22: 209–217.

McIntyre, S., et al. 1999. Disturbance response in vegetation—towards a global perspective on functional traits. *Journal of Vegetation Science* 10: 621–630.

McKane, R. B., et al. 1997. Climatic effects on tundra carbon storage inferred from experimental data and a model. *Ecology* 78: 1170–1187.

McLean, M. A., and Parkinson, D. 1998a. Impacts of the epigeic earthworm *Dendrobaena octaedra* on microfungal community structure in pine forest floor: a mesocosm study. *Applied Soil Ecology* 8: 61–75.

——. 1998b. Impacts of the epigeic earthworm *Dendrobaena octaedra* on oribatid mite community diversity and microarthropod abundances in pine forest floor: a mesocosm study. *Applied Soil Ecology* 7: 125–136.

——. 2000a. Field evidence of the effects of the epigeic earthworm *Dendrobaena octaedra* on the microfungal community in pine forest floor. *Soil Biology and Biochemistry* 32: 351–360.

———. 2000b. Introduction of the epigeic earthworm *Dendrobaena octaedra* changes the oribatid community and microarthropod abundances in a pine forest. *Soil Biology and Biochemistry* 32: 1671–1681.

McLean, M. A., et al. 1996. Does selective grazing by mites and collembola affect fungal community structure? *Pedobiologia* 40: 97–105.

McNaughton, S. J. 1977. Diversity and stability of ecological communities: a comment on the role of empiricism in ecology. *American Naturalist* 111: 515–525

———. 1979. Grazing as an optimization process: grass-ungulate relationships in the Serengeti. *American Naturalist* 113: 691–703.

———. 1985. Ecology of a grazing system: the Serengeti. *Ecological Monographs* 55: 259–294.

———. 1986. On plants and herbivores. *American Naturalist* 128: 765–770.

McNaughton, S. J., et al. 1988. Large mammals and process dynamics in African ecosystems. *BioScience* 38: 794–800.

———. 1989. Ecosystem-level patterns of primary productivity and herbivory in terrestrial habitats. *Nature* 341: 142–144.

———. 1998. Root biomass and productivity in a grazing system: the Serengeti. *Ecology* 79: 587–592.

McTiernan, K. B., et al. 1997. Respiration and nutrient release from tree leaf litter mixtures. *Oikos* 78: 527–538.

Mearns, L. O., et al. 1984. Extreme high-temperature events: changes in their probabilities with changes in mean temperature. *Journal of Climate and Applied Meteorology* 23: 1601–1613.

Menge, B. A. 1995. Indirect effects in marine rocky intertidal interaction webs: patterns and importance. *Ecological Monographs* 65: 21–74.

Menge, B. A., and Sutherland, J. P. 1976. Species diversity gradients: synthesis of the roles of predation, competition and spatial heterogeniety. *American Naturalist* 110: 351–369.

Merrill, E. H., et al. 1994. Responses of bluebunch wheatgrass, Idaho fescue and nematodes to ungulate grazing in Yellowstone National Park. *Oikos* 69: 231–240.

Michalzik, B., et al. 1999. Aphids on Norway spruce and their effects on forest floor solution chemistry. *Forest Ecology and Management* 118: 1–10.

Michelsen, A., et al. 1996. Leaf ^{15}N abundance of subarctic plants provides field evidence that ericoid, ectomycorrhizal and non- and arbuscular mycorrhizal species access different sources of soil nitrogen. *Oecologia* 105: 53–63.

Mikola, J. 1999. Effects of microbivore species composition and basal resource enrichment on trophic-level biomasses in an experimental microbial based food web. *Oecologia* 117: 396–403.

Mikola, J., and Setälä, H. 1998a. Productivity and trophic level biomasses in a microbial-based soil food web. *Oikos* 82: 158–168.

———. 1998b. No evidence of trophic cascades in an experimental microbial-based soil food web. *Ecology* 79: 153–164.

———. 1998c. Relating species diversity to ecosystem functioning: mechanistic backgrounds and experimental approach with a decomposer food web. *Oikos* 83: 180–194.

Mikola, J., et al. 2000. Linking above-ground and below-ground effects in autotrophic microcosms: effects of shading and defoliation on plant and soil properties. *Oikos* 89: 577–587.

———. 2001a. Response of soil food web structure to defoliation of different plant species combinations in an experimental grassland community. *Soil Biology and Biochemistry* 33: 205–214.

———. 2001b. Effects of defoliation intensity on soil food web properties in an experimental grassland community. *Oikos* 92: 333–343.

Milchunas, D. G., and Lauenroth, W. K. 1993. Quantitative effects of grazing on vegetation and soils over a global range of environments. *Ecological Monographs* 63: 327–366.

Milchunas, D. G., et al. 1998. Livestock grazing: animal and plant biodiversity of shortgrass steppe and the relationship to ecosystem functioning. *Oikos* 83: 65–74.

Miller, R. F., and Rose, J. A. 1992. Growth and carbon allocation of *Agropyron desertorum* following summer defoliation. *Oecologia* 89: 482–486.

Moen, J., and Oksanen, L. 1991. Ecosystem trends. *Nature* 353: 510.

Molvar, E. M., et al. 1997. Moose herbivory, browse quality and nutrient cycling in an Alaskan treeline community. *Oecologia* 94: 472–479.

Monz, C. A., et al. 1994. The response of mycorrhizal colonization to elevated CO_2 and climate change in *Pascopyrum smithii* and *Boutelea gracilis*. *Plant and Soil* 165: 75–80.

Moore, J. C., and de Ruiter, P. C. 2000. Invertebrates in detrital food webs along gradients of productivity. In P. F. Hendrix and

D. C. Coleman, eds., *Invertebrates as Webmasters in Ecosystems*, 161–184. Wallingford: CAB International.

Moore, J. C., and Hunt, H. W. 1988. Resource compartmentation and the stability of real ecosystems. *Nature* 333: 261–263.

Moore, J. C., et al. 1988. Arthropod regulation of micro- and mesobiota in below-ground detrital food webs. *Annual Review of Entomology* 33: 419–439.

———. 1993. Influence of productivity on the stability of real and model ecosystems. *Science* 261: 906–908.

Moore, T. R., et al. 1999. Litter decomposition rates in Canadian forests. *Global Change Biology* 5: 75–82.

Moran, N. A., and Whitham, T. G. 1990. Interspecific competition between root-feeding and leaf-galling aphids mediated by host-plant resistance. *Ecology* 71: 1050–1058.

Morris, W. F., and Wood, D. M. 1989. The role of lupine in Mt. St. Helens: facilitation or inhibition? *Ecology* 70: 697–703.

Mortimer, S. R., et al. 1999. Insect and nematode herbivory below ground: interactions and role in succession. In H. Olff, V. K. Brown and R. H. Drent, eds., *Herbivores: Between Plants and Predators*, 205–238. Oxford: Blackwell Science.

Mosimann, J. E., and Martin, P. S. 1975. Simulating overkill by Paleoindians. *American Scientist* 63: 304–313.

Mulder, C. P. H., and Keall, S. N. 2001. Burrowing seabirds and reptiles: impacts on seeds, seedlings and soils in an island forest in New Zealand. *Oecologia* (in press).

Mulder, C. P. H., et al. 1999. Insects affect relationships between plant species richness and ecosystem processes. *Ecology Letters* 2: 237–246.

Muller, P. E. 1884. Studier over skovjord, som bidrag til skovdyrkningens theori. II. Om muld og mor i egeskove og paa heder. *Tidsskrift for Skovbrug* 7: 1–232.

Müller-Schärer, H., and Brown, V. K. 1995. Direct and indirect effects of above- and below-ground insect herbivory on plant density and performance of *Tripleurospermum perforatum* during early plant succession. *Oikos* 72: 36–41.

Murdoch, W. W., et al. 1972. Diversity and pattern in plants and insects. *Ecology* 53: 819–829.

Myrold, D. D., et al. 1989. Relationships between soil microbial properties and aboveground stand characteristics of coniferous forests in Oregon. *Biogeochemistry* 8: 265–281.

Nadelhoffer, K., et al. 1992. Microbial processes and plant nutrient availability in arctic soil. In F. S. Chapin III, R. L. Jefferies, J. F. Reynolds, G. R. Shaver, and J. Svoboda, eds., *Arctic Ecosystems in a Changing Climate. An Ecophysiological Perspective*, 281–300. San Diego: Academic.

———. 1999. Nitrogen deposition makes a minor contribution to carbon sequestration in temperate forests. *Nature* 398: 145–148.

Naeem, S., and Li, S. 1997. Biodiversity enhances ecosystem reliability. *Nature* 390: 507–509.

Naeem, S., et al. 1994. Declining biodiversity can alter the performance of ecosystems. *Nature* 368: 734–737.

———. 1996. Biodiversity and productivity in a model assemblage of plant species. *Oikos* 76: 259–264.

———. 2000a. Producer-decomposer co-dependency influences biodiversity effects. *Nature* 403: 762–764.

———. 2000b. Plant diversity increases resistance to invasion in the absence of covarying extrinsic factors. *Oikos* 91: 97–108.

Nakatsubo, T., et al. 1997. Comparative study of the mass loss rate of moss litter in boreal and subalpine forests in relation to temperature. *Ecological Research* 12: 47–54.

Näsholm, T., et al. 1998. Boreal forest plants take up organic nitrogen. *Nature* 392: 914–916.

———. 2000. Uptake of organic nitrogen in the field by four agriculturally important plant species. *Ecology* 81: 1155–1161.

Neill, C., et al. 1996. Forest- and pasture-derived carbon contributions to carbon stocks and microbial respiration of tropical pasture soils. *Oecologia* 107: 113–119.

———. 1997. Soil carbon and nitrogen stocks following forest clearing for pasture in the southwestern Brazilian Amazon. *Ecological Applications* 7: 1216–1225.

Newbery, D. McC., et al. 1997. Phosphorus dynamics in a lowland African rain forest: the influence of ectomycorrhizal trees. *Ecological Monographs* 67: 367–409.

———. 2000. Does proximity to conspecific adults influence the establishment of ectomycorrhizal trees in rain forest? *New Phytologist* 147: 401–409.

Newell, K. 1984. Interactions between two decomposer basidiomycetes and a collembolan under sitka spruce: grazing and its potential effects on fungal distribution and litter composition. *Soil Biology and Biochemistry* 16: 235–239.

Newman, E. I. 1988. Mycorrhizal links between plants: their functioning and ecological significance. *Advances in Ecological Research* 18: 243–270.

Newsham, K. K., et al. 1994. Root pathogenic and arbuscular mycorrhizal fungi determine fecundity of asymptomatic plants in the field. *Journal of Ecology* 82: 805–814.

———. 1995. Multi-functionality and biodiversity in arbuscular mycorrhizas. *Trends in Ecology and Evolution* 10: 407–411.

Newton, P. C. D., et al. 1995. Plant growth and soil processes in temperate grassland communities at elevated CO_2. *Journal of Biogeography* 22: 235–240.

Nicolai, V. 1988. Phenolic and mineral contents of leaves influences decomposition in European forest systems. *Oecologia* 75: 575–579.

Niemelä, J. 1993. Interspecific competition in ground beetle assemblages (Carabidae): what have we learned? *Oikos* 66: 325–335.

Niemelä, J., et al. 1996. The importance of small scale heterogeniety in boreal forests: variation in diversity of forest floor invertebrates across the successional gradient. *Ecography* 19: 352–368.

———. 1997. Establishment and interactions of carabid populations: an experiment with native and exotic species. *Ecography* 20: 643–652.

Nieminen, J. K., and Setälä, H. 1998. Enclosing decomposer food web: implications for community structure and function. *Biology and Fertility of Soils* 26: 50–57.

Nijs, I., and Roy, J. 2000. How important are species richness, species evenness, and interspecific differences to productivity? A mathematical model. *Oikos* 88: 57–66.

Niklaus, P. A., and Körner, C. 1996. Responses of soil microbiota of a late successional alpine grassland to long term CO_2 enrichment. *Plant and Soil* 184: 219–229.

Nilsson, M.-C. 1994. Separation of allelopathy and resource competition by the boreal dwarf shrub *Empetrum hermaphroditum* Hagerup. *Oecologia* 98: 1–7.

Nilsson, M.-C., et al. 1993. Allelopathic effects by *Empetrum hermaphroditum* Hagerup on development and nitrogen uptake by roots and mycorrhizas of *Pinus sylvestris*. *Canadian Journal of Botany* 71: 620–628.

———. 2000. Characterization of the differential interference effects of two boreal dwarf shrub species. *Oecologia* 123: 122–128.

Nohrstedt, H. O., et al. 1989. Changes in carbon content, respiration rate, ATP content and microbial biomass in nitrogen-fertilized pine forest soils in Sweden. *Canadian Journal of Forest Research* 19: 323–328.

Norby, R. J. 1994. Issues and perspectives for investigating root responses to elevated atmospheric carbon dioxide. *Plant and Soil* 165: 9–20.

Nordgren, A. 1992. A method for determining microbially available N and P in an organic soil. *Biology and Fertility of Soils* 13: 195–199.

Northup, B. K., et al. 1999. Grazing impacts on the spatial distribution of soil microbial biomass around tussock grasses in a tropical grassland. *Applied Soil Ecology* 13: 259–270.

Northup, R. A., et al. 1995. Polyphenol control of nitrogen release from pine litter. *Nature* 377: 227–229.

———. 1998. Polyphenols as regulators of plant-litter-soil interactions in northern California's pygmy forest: a positive feedback? *Biogeochemistry* 42: 189–220.

Odell, R. T., et al. 1984. Changes in organic carbon and nitrogen of Morrow Plot soils under different management, 1904–1973. *Soil Science* 137: 160–171.

Odum, E. P. 1969. The strategy of ecosystem development. *Science* 164: 262–270.

Oechel, W. C., et al. 1993. Recent changes of Arctic tundra ecosystems from a net carbon dioxide sink to a source. *Nature* 361: 520–523.

Ohtonen, R., et al. 1997. Ecological theories in soil biology. *Soil Biology and Biochemistry* 29: 1613–1620.

———. 1999. Ecosystem properties and microbial community changes in primary succession on a glacier forefront. *Oecologia* 119: 239–246.

Okano, S., et al. 1991. Negative relationship between microbial biomass and root amount in topsoil of a renovated grassland. *Soil Science and Plant Nutrition* 37: 47–53.

Oksanen, J. 1996. Is the humped back relationship between species richness and biomass an artefact due to plot size? *Journal of Ecology* 84: 293–295.

Oksanen, L. 1988. Ecosystem organisation: mutualism and cybernetics or plain Darwinian struggle for existence? *American Naturalist* 131: 424–444.

Oksanen, L., and Oksanen, T. 2000. The logic and realism of the hypothesis of exploitation ecosystems. *American Naturalist* 155: 703–723.

Oksanen, L., et al. 1981. Exploitation ecosystems in gradients of primary productivity. *American Naturalist* 118: 240–261

O'Lear, H. A., and Seastedt, T. R. 1994. Landscape patterns of litter decomposition in alpine tundra. *Oecologia* 99: 95–101.

Olff, H., and Ritchie, M. E. 1998. Effects of herbivores on grassland plant diversity. *Trends in Ecology and Evolution* 13: 261–265.

Olsson, M. O., and Falkengren-Grerup, U. 2000. Potential nitrification as an indicator of preferential uptake of ammonium or nitrate by plants in an oak wood understory. *Annals of Botany* 85: 299–305.

Orchard, V. A., and Cook, F. J. 1983. Relationship between soil respiration and soil moisture. *Soil Biology and Biochemistry* 15: 447–453.

Ostertag, R., and Hobbie, S. E. 1999. Early stages of root and leaf decomposition in Hawaiian forests: effects of nutrient availability. *Oecologia* 121: 564–573.

Ostfeld, R. S., and Keesing, F. 2000. Pulsed resources and community dynamics of consumers in terrestrial ecosystems. *Trends in Ecology and Evolution* 15: 232–237.

Otrosina, W. J., et al. 1999. Root infecting fungi associated with a decline of longleaf pine in the southeastern United States. *Plant and Soil* 217: 145–150.

Øvreås, L. 2000. Population and community approaches for analyzing microbial diversity in natural environments. *Ecology Letters* 3: 236–251.

Owensby, C. E., et al. 1993. Biomass production in tall grass prairie ecosystems exposed to ambient and elevated CO_2. *Ecological Applications* 3: 644–653.

Pace, M. L., et al. 1999. Trophic cascades revealed in diverse ecosystems. *Trends in Ecology and Evolution* 14: 483–488.

Packer, A., and Clay, K. 2000. Soil pathogens and spatial patterns of seedling mortality in a temperate tree. *Nature* 404: 278–281.

Paine, R. T. 1966. Food web complexity and species diversity. *American Naturalist* 100: 65–75.

——— 1992. Food web analysis through field measurement of per capita interaction strength. *Nature* 355: 73–75.

Palm, C. A., and Sanchez, P. A. 1991. Nitrogen release from the leaves of some tropical legumes as affected by their lignin and polyphenolic contents. *Soil Biology and Biochemistry* 23: 83–88.

Paoletti, M. 1988. Soil invertebrates in cultivated and uncultivated soil in northeastern Italy. *Redia* 71: 501–563.

Papavizas, G. C. 1985. *Trichoderma* and *Gliocladium*: biology, ecology and potential for biocontrol. *Annual Review of Phytopathology* 23: 23–54.

Papavizas, G. C., and Lumsden, R. D. 1980. Biological control of soilborne fungal propagules. *Annual Review of Phytopathology* 18: 389–413.

Paquin, P., and Coderre, D. 1997. Changes in soil macroarthropod communities in relation to forest maturation through three successional stages in the Canadian boreal forest. *Oecologia* 112: 104–111.

Parmelee, R. W., et al. 1989. Decomposition and nitrogen dynamics of surface weed residues in no-tillage agroecosystems under drought conditions: influence of resource quality on the decomposer community. *Soil Biology and Biochemistry* 21: 97–103.

————. 1993. Effects of pine roots on microorganisms, fauna, and nitrogen availability in two horizons on a coniferous forest spodosol. *Biology and Fertility of Soils* 15: 113–119.

Parsons, W. F. J., et al. 1994. Root gap dynamics in lodgepole pine forest: nitrogen transformation in gaps of different size. *Ecological Applications* 4: 354–362.

Pastor, J., and Post, W. M. 1988. Response of northern forests to CO_2-induced climate change. *Nature* 334: 55–58.

Pastor, J., et al. 1988. Moose, microbes and the boreal forest. *BioScience* 38: 770–777.

————. 1993. Moose browsing and soil fertility in the boreal forests of Isle Royale National Park. *Ecology* 74: 467–480.

Paterson, E., et al. 1996. Effect of elevated atmospheric CO_2 concentration on C-partitioning and rhizosphere C-flow for three plant species. *Soil Biology and Biochemistry* 28: 195–201.

Paustian, K., et al. 1990. Carbon and nitrogen budgets of four agroecosystems with annual and perennial crops, with and without N fertilization. *Journal of Applied Ecology* 27: 60–84.

Peltzer, D. A., et al. 1998. Competition intensity along a productivity gradient in a low diversity grassland. *American Naturalist* 151: 465–476.

Peñuelas, J., and Estiarte, M. 1998. Can elevated CO_2 affect secondary metabolism and ecosystem function? *Trends in Ecology and Evolution* 13: 21–24.

Peoples, M. B., et al. 1995. Biological nitrogen fixation: an efficient source of nitrogen for agricultural production? *Plant and Soil* 174: 3–28.

Pereira, A. P., et al. 1998. Leaf litter decomposition in relation to litter physico-chemical properties, fungal biomass, arthropod colonization, and geographic origin of plant species. *Pedobiologia* 42: 316–327.

Perez-Moreno, J., and Read, D. J. 2000. Mobilization and transfer of nutrients from litter to tree seedlings via the vegetative mycelia of mycorrhizal plants. *New Phytologist* 145: 301–309.

358

Perry, D. A., et al. 1989. Ectomycorrhizal mediation of competition between coniferous tree species. *New Phytologist* 112: 501–511.

Persson, T., et al. 1980. Trophic structure, biomass dynamics and carbon metabolism of soil organisms in a Scots pine forest. *Ecological Bulletins* 32: 419–459.

Petchey, O. L. 2000. Species diversity, species extinction, and ecosystem function. *American Naturalist* 155: 696–702.

Petchey, O. L., et al. 1999. Environmental warming alters food web structure and ecosystem function. *Nature* 402: 69–72.

Petersen, H., and Luxton, M. 1982. A comparative analysis of soil fauna populations and their role in decomposition processes. *Oikos* 39: 287–388.

Phillips, R. E., and Phillips, S. H. Editors. 1984. *No Tillage Agriculture. Principles and Practices.* New York: Van Nostrand Reinhold.

Pickett, S. T. A., et al. 1987. Models, mechanisms and pathways of succession. *Botanical Reviews* 53: 336–371.

Pietikäinen, J., and Fritze, J. 1995. Clear-cutting and prescribed burning in coniferous forest: comparison of effects on soil fungal and total microbial biomass, respiration activity and nitrification. *Soil Biology and Biochemistry* 27: 101–109.

Pietikäinen, J., et al. 2000a. Does short term heating of forest humus change its properties as a substrate for microbes? *Soil Biology and Biochemistry* 32: 277–288.

———. 2000b. Charcoal as a habitat for microbes and its effect on the microbial community of the underlying humus. *Oikos* 89: 231–242.

Pimm, S. L. 1982. *Food Webs.* London: Chapman and Hall.

Pimm, S. L., et al. 1991. Food web patterns and their consequences. *Nature* 350: 669–674.

———. 1995. The future of biodiversity. *Science* 269: 347–350.

Podger, F. D., and Newhook, F. J. 1971. *Phytophthora cinnamoni* in indigenous plant communities in New Zealand. *New Zealand Journal of Botany* 9: 625–638.

Polis, G. A. 1991. Complex trophic interactions in deserts: an empirical critique of food web theory. *American Naturalist* 138: 123–155.

———. 1994. Food webs, trophic cascades and community structure. *Australian Journal of Ecology* 19: 121–136.

———. 1999. Why are parts of the world green? Multiple factors control productivity and the distribution of biomass. *Oikos* 86: 3–15.

Polis, G. A., and Hurd, S. D. 1996. Linking marine and terrestrial food webs: allochthonous input from the ocean supports high secondary productivity on small islands and coastal land communities. *American Naturalist* 147: 396–423.

Polis, G. A., et al. 1997a. Towards an integration of landscape and food web ecology: the dynamics of spatially subsidized food webs. *Annual Review of Ecology and Systematics* 28: 289–316.

———· 1997b. El Niño effects on the dynamics and control of an island ecosystem in the Gulf of California. *Ecology* 78: 1884–1897.

Ponsard, S., and Arditi, R. 2000. What can stable isotopes ($\delta^{15}N$ and $\delta^{13}C$) tell us about the food web of soil macro-invertebrates? *Ecology* 81: 852–864.

Poorter, H., and Bergkotte, M. 1992. Chemical composition of 24 wild species differing in relative growth rate. *Plant, Cell and Environment* 15: 221–229.

Poorter, H., and Remkes, C. 1990. Leaf area ratio and net assimilation rate of 24 wild species differing in relative growth rate. *Oecologia* 83: 553–559.

Post, D. M., et al. 2000. Ecosystem size determines food chain length in lakes. *Nature* 405: 1047–1049.

Post, E., et al. 1999. Ecosystem consequences of wolf behavioral response to climate. *Nature* 401: 905–907.

Post, W. M., et al. 1992. Aspects of the interaction between vegetation and soil under global change. *Water, Air and Soil Pollution* 64: 345–363.

Potvin, C., and Vasseur, L. 1997. Long term CO_2 enrichment of a pasture community: species richness, dominance and succession. *Ecology* 78: 666–677.

Power, M. E. 1992. Top-down and bottom-up forces in food webs: do plants have primacy? *Ecology* 73: 733–746.

Power, M. E., et al. 1996. Challenges in the quest for keystones. *BioScience* 46: 609–620.

Pregitzer, K. S., et al. 2000. Interactive effects of atmospheric CO_2 and soil N availability on fine roots of *Populus tremuloides*. *Ecological Applications* 10: 18–23.

Prentice, K. C., and Fung, I. Y. 1990. The sensitivity of terrestrial carbon storage to climate change. *Nature* 346: 48–51.

Prieur-Richard, A.-H., et al. 2000. Plant community diversity and invasion by exotics: invasion of Mediterranean old fields by *Conyza bonariensis* and *Conyza canadensis*. *Ecology Letters* 2: 412–422.

Priha, O., et al. 1999. Comparing microbial biomass, denitrification enzyme activity, and numbers of nitrifiers in the rhizospheres of *Pinus sylvestris*, *Picea abies* and *Betula pendula* seedlings by microscale methods. *Biology and Fertility of Soils* 30: 14–19.

Proulx, M., and Mazumder, A. 1998. Reversal of grazing impact on plant species richness in nutrient poor vs. nutrient rich ecosystems. *Ecology* 79: 2581–2592.

Raghubanshi, A. S. 1991. Dynamics of soil C, N and P in a dry tropical forest in India. *Biology and Fertility of Soils* 12: 55–59.

Rai, B., and Upadhyay, R. S. 1983. Competitive saprophytic colonization of pigeon-pea substrate by *Fusarium udum* in relation to environmental factors, chemical treatments and microbial antagonism. *Soil Biology and Biochemistry* 15: 187–191.

Rainey, P. B., and Travisano, M. 1998. Adaptive radiation in a heterogeneous environment. *Nature* 394: 69–72.

Rainey, P. B., et al. 2000. The emergence and maintenance of diversity: insights from experimental bacterial populations. *Trends in Ecology and Evolution* 15: 243–247.

Ramsay, A. 1983. Bacterial biomass in ornithogenic soils in Antarctica. *Polar Biology* 1: 221–225.

Ramsay, A., and Stannard, R. E. 1986. Numbers and viability of bacteria in ornithogenic soils in Antarctica. *Polar Biology* 5: 195–198.

Randlett, D. L., et al. 1996. Elevated carbon dioxide and leaf litter chemistry: influences on microbial respiration and net nitrogen mineralization. *Soil Science Society of America Journal* 60: 1571–1577.

Rapoport, E. H. 1982. *Areography: Geographical Strategies of Species.* New York: Permagon.

Rapson, G. L., et al. 1997. The humped relationship between species richness and biomass: testing its sensitivity to sample quadrat size. *Journal of Ecology* 85: 99–100.

Rayner, A. D. M. 1994. Pattern generating processes in fungal communities. In K. Ritz, J. Dighton, and K. E. Giller, eds., *Beyond the Biomass. Compositional and Functional Analysis of Soil Microbial Communities,* 247–258. Chichester: Wiley.

Rayner, A. D. M., and Todd, N. K. 1979. Population and community structure and dynamics in decaying wood. *Advances in Botanical Research* 7: 334–420.

Read, D. J., et al. 1985. Mycorrhizal mycelia and nutrient cycling in plant communities. In A. H. Fitter, D. J. Read and M. B. Usher, eds., *Ecological Interactions in Soil,* 193–217. Oxford: Blackwells.

Recher, H. F. 1969. Bird species diversity and habitat diversity in Australia and North America. *American Naturalist* 103: 75–80.

Reddell, P., and Spain, A. V. 1991. Transmission of infective *Frankia* (Actinomycetales) propagules in casts of the endogeic earthworm *Pontoscolex corethurus* (Oligochaeta: Glossoscolecidae). *Soil Biology and Biochemistry* 23: 775–778.

Reich, P. B., et al. 1998. Leaf structure (specific leaf area) modulates photosynthesis-nitrogen relations: evidence from within and across species and functional groups. *Functional Ecology* 12: 948–958.

Reiners, W. A., et al. 1994. Tropical rain forest conversion to pasture: changes in vegetation and soil properties. *Ecological Applications* 4: 363–377.

Rejmánek, M. 1996. A theory of seed plant invasiveness: the first sketch. *Biological Conservation* 78: 171–181.

Rhoades, C. C., et al. 2000. Soil carbon differences among forest, agriculture and secondary vegetation in lower montane Ecuador. *Ecological Applications* 10: 497–505.

Rhoades, D. F. 1985. Offensive-defensive interactions between herbivores and plants: their relevance in herbivore population dynamics and ecological theory. *American Naturalist* 125: 205–238.

Rice, C. W., et al. 1994. Soil microbial response in tallgrass prairie to elevated CO_2. *Plant and Soil* 165: 65–75.

Rice, E. L., and Pancholy, S. K. 1973. Inhibition of nitrification by climax ecosystems. II. Additional evidence and possible role of tannins. *American Journal of Botany* 60: 691–702.

Richardson, D. M., et al. 2000. Plant invasions—the role of mutualisms. *Biological Reviews* 75: 65–93.

Riedell, W. E. 1989. Western corn rootworm damage in maize: greenhouse technique and plant response. *Crop Science* 29: 412–415.

Rillig, M. C., and Allen, M. F. 1998. Arbuscular mycorrhizae of *Gutierrezia sarothrae* and elevated carbon dioxide: evidence for shifts in C allocation to and within the mycobiont. *Soil Biology and Biochemistry* 30: 2001–2008.

Roberts, M. S., and Cohen, F. M. 1995. Recombination and migration rates in natural populations of *Baccillus subtilus* and *Baccillus mojavensis*. *Evolution* 49: 1081–1094.

Robinson, C. H., et al. 1993. Nutrient and carbon dioxide release by interacting species of straw-decomposing fungi. *Plant and Soil* 151: 139–142.

———. 1999. Decomposition of roots from high Arctic plants: a microcosm study. *Soil Biology and Biochemistry* 31: 1101–1108.

Rogers, H. H., et al. 1996. Root to shoot ratio as influenced by CO_2. *Plant and Soil* 187: 229–248.

Rohde, K. 1992. Latitudinal gradients in species diversity: the search for primary cause. *Oikos* 65: 514–527.

Rohde, K., et al. 1993. Rapoport's rule does not apply to marine telosts and cannot explain latitudinal gradients in species richness. *American Naturalist* 143: 1–16.

Rønn, R., et al. 1996. Spatial distribution and successional pattern of microbial activity and microfaunal populations on decomposing barley roots. *Journal of Applied Ecology* 33: 662–672.

Rosenzweig, M. L. 1995. *Species Diversity in Time and Space.* Cambridge: Cambridge University Press.

Roser, D. J., et al. 1993. Microbiology of ornithogenic soils from the Windmill Islands, Budd Coast, continental Antarctica: microbial biomass distribution. *Soil Biology and Biochemistry* 25: 165–175.

Ross, D. J. 1987. Soil microbial biomass estimated by the fumigation-incubation procedure: seasonal fluctuations and influence of soil moisture content. *Soil Biology and Biochemistry* 19: 397–404.

Ross, D. J., et al. 1995. Elevated CO_2 and temperature effects on soil carbon and nitrogen cycling in ryegrass / white clover turves of an Endoaquept soil. *Plant and Soil* 176: 37–49.

———. 1999. Land use change: effects on soil carbon, nitrogen and phosphorus pools and fluxes in three adjacent ecosystems. *Soil Biology and Biochemistry* 31: 803–813.

Rossow, J. J., et al. 1997. Effects of above-ground browsing by mammals on mycorrhizal infection in an early successional taiga ecosystem. *Oecologia* 110: 94–98.

Rotty, R. M., and Marland, G. 1986. Fossil fuel consumption: recent amounts, patterns and trends of CO_2. In J. R. Trabalka and D. E. Reichle, eds., *The Changing Global Cycle: A Global Analysis*, 474–490. New York: Springer-Verlag.

Rouhier, H., et al. 1996. Carbon fluxes in the rhizosphere of sweet chestnut (*Castanea sativa*) grown under two atmospheric CO_2 concentrations: ^{14}C partitioning after pulse labelling. *Plant and Soil* 180: 101–111.

Ruess, L., et al. 1999. Simulated climate change affecting microorganisms, nematode density and biodiversity in subarctic soils. *Plant and Soil* 212: 63–73.

Ruess, R. W. 1988. The interaction of defoliation and nutrient uptake in *Sporobolus kentrophyllus*, a short-grass species from the Serengeti plains. *Oecologia* 77: 550–556.

Ruess, R. W., and McNaughton, S. J. 1987. Grazing and the dynamics and energy related microbial processes in the Serengeti grassland. *Oikos* 49: 101–110.

Ruess, R. W., et al. 1998. Regulation of fine root dynamics by mammalian browsers in early successional Alaskan taiga forests. *Ecology* 79: 2706–2720.

Runion, G. B., et al. 1994. Effects of CO_2 enrichment on microbial populations in the rhizosphere and phyllosphere of cotton. *Agricultural and Forest Meteorology* 70: 117–130.

Rusek, J. 1998. Biodiversity of Collembola and their functional role in the ecosystem. *Biodiversity and Conservation* 7: 1207–1219.

Rutherford, P. M., and Juma, N. G. 1992. Effect of glucose amendment on microbial biomass, fertilizer ^{15}N-recovery and distribution in a barley-soil system. *Biology and Fertility of Soils* 12: 228–232.

Rygiewicz, P. T., et al. 2000. Morphotype community structure of ectomycorrhizas on Douglas fir (*Pseudotsuga menziesii* Mirb. Franco) seedlings grown under elevated CO_2 and temperature. *Oecologia* 124: 299–308.

Saetre, P. 1998. Decomposition, microbial community structure and earthworm effects along a birch-spruce soil gradient. *Ecology* 79: 834–846.

Saetre, P., and Bååth, E. 2000. Spatial variation and patterns of the soil microbial community structure in a mixed spruce-birch stand. *Soil Biology and Biochemistry* 32: 909–917.

Saggar, S., et al. 1999. Changes in soil microbial biomass, metabolic quotient and organic matter turnover under *Hieracium* (*H. pilosella* L.). *Biology and Fertility of Soils* 30: 232–238.

Sala, O. E., et al. 2000. Global biodiversity scenarios for the year 2100. *Science* 287: 1770–1774.

Sanchez-Piñero, F., and Polis, G. A. 2000. Bottom-up dynamics of allochthonous input: direct and indirect effects of seabirds on islands. *Ecology* 81: 3117–3232.

Sankaran, M., and McNaughton, S. J. 1999. Determinants of biodiversity regulate compositional stability of communities. *Nature* 401: 691–693.

Santos, P. F., et al. 1981. The role of mites and nematodes in early stages of buried litter decomposition in a desert. *Ecology* 63: 664–669.

Sanyal, D., and Kulshrestha, G. 1999. Effects of repeated metalochlor applications on its persistence in field soil and degradation kinetics in mixed microbial cultures. *Biology and Fertility of Soils* 30: 124–131.

Sarathchandra, S. U., et al. 1989. Soil microbial biomass: influence of simulated temperature changes on size, activity and nutrient content. *Soil Biology and Biochemistry* 21: 987–993.

Sarig, S., et al. 1994. Annual plant growth and soil characteristics under desert halophyte canopy. *Acta Œcologica* 15: 521–527.

Satchell, J. E. 1967. Lumbricidae. In A. Burgess and F. Raw, eds., *Soil Biology*, 259–322. London: Academic.

———. 1983. Earthworm microbiology. In J. E. Satchell, ed., *Earthworm Ecology*, 351–364. New York: Chapman and Hall.

Schaefer, M., and Schauermann, J. 1990. The fauna of beech forests: comparison between a mull and moder soil. *Pedobiologia* 34: 299–304.

Schenck, U., et al. 1995. Effects of CO_2 enrichment and intraspecific competition on biomass partitioning, nitrogen content and microbial biomass carbon in soil of perennial ryegrass and white clover. *Journal of Experimental Botany* 46: 987–993.

Scheu, S. 1990. Changes in microbial nutrient status during secondary succession and its modification by earthworms. *Oecologia* 84: 351–358.

———. 1992. Changes in the lumbricid coenosis during secondary succession from a wheat field to a beechwood on limestone. *Soil Biology and Biochemistry* 24: 1641–1646.

———. 1994. There is an earthworm mobilizable nitrogen pool in soil. *Pedobiologia* 38: 243–249.

Scheu, S., and Falca, M. 2000. The soil food web of two beech forests (*Fagus sylvatica*) of contrasting humus type: stable isotope analysis of a macro- and a mesofauna dominated community. *Oecologia* 123: 285–296.

Scheu, S., and Parkinson, D. 1994. Effects of invasion of an aspen forest (Canada) by *Dendrobaena octaedra* (Lumbricidae) on plant growth. *Ecology* 75: 2348–2361.

Scheu, S., and Schaefer, M. 1998. Bottom-up control of the soil macrofauna community in a beechwood on limestone: manipulation of food resources. *Ecology* 79: 1573–1585.

Scheu, S., et al. 1999. Links between the detritivore and herbivore system: effects of earthworms and Collembola on plant growth and aphid development. *Oecologia* 119: 541–551.

Schimel, D. 1995. Terrestrial ecosystems and the carbon cycle. *Global Change Biology* 1: 77–91.

Schimel, D., et al. 1994. Climatic, edaphic and biotic controls over storage and turnover of carbon in soils. *Global Biogeochemical Cycles* 8: 279–193.

———. 2000. Contribution of increasing CO_2 and climate to carbon storage by ecosystems in the United States. *Science* 287: 2004–2006.

Schimel, J. P., and Clein, J. S. 1996. Microbial responses to freeze-thaw cycles in tundra and taiga soils. *Soil Biology and Biochemistry* 28: 1061–1066.

Schimel, J. P., et al. 1998. The role of balsam poplar secondary chemicals in controlling soil nutrient dynamics through succession in the Alaskan taiga. *Biogeochemistry* 42: 221–234.

Schippers, B., et al. 1987. Interactions of deleterious and beneficial rhizosphere microorganisms and the effect of cropping practices. *Annual Review of Phytopathology* 25: 339–358.

Schippers, P., et al. 1999. Competition under high and low nutrient levels among three grassland species occupying different positions in a successional sequence. *New Phytologist* 143: 547–559.

Schläpfer, F., and Schmid, B. 1999. Ecosystem effects of biodiversity: a clarification of hypotheses and exploration of empirical studies. *Ecological Applications* 9: 893–912.

Schlesinger, W. H. 1990. Evidence from chronosequence studies for a low carbon-storage potential of soils. *Nature* 348: 232–234.

Schlesinger, W. H., and Pilmanis, A. M. 1998. Plant-soil interactions in deserts. *Biogeochemistry* 42: 169–187.

Schlesinger, W. H., et al. 1990. Biological feedbacks in global desertification. *Science* 247: 1043–1048.

——— 1996. On the spatial distribution of soil nutrients in desert ecosystems. *Ecology* 77: 364–374.

——— 1998. The biogeochemistry of phosphorus after the first century of soil development on Rakata Island, Krakatau, Indonesia. *Biogeochemistry* 40: 37–55.

Schmit, J. P. 1999. Resource consumption and competition by unit restricted fungal decomposers of patchy substrates. *Oikos* 87: 509–519.

Schmitz, O. J., et al. 2000. Trophic cascades in terrestrial ecosystems: a review of the effects of carnivore removal on plants. *American Naturalist* 155: 141–153.

Schneider, S. H., 1992. The climatic response to greenhouse gases. *Advances in Ecological Research* 22: 1–32.

Schnürer, J., and Rosswall, T. 1987. Mineralization of nitrogen from ^{15}N labeled fungi, soil microbial biomass and roots and its uptake by barley plants. *Plant and Soil* 102: 71–78.

Schnürer, J., et al. 1986. Effects of soil moisture on soil microorganisms and nematodes: a field experiment. *Microbial Ecology* 12: 217–230.

Schoener, T. W. 1974. Resource partitioning in ecological communities. *Science* 185: 27–39.

————· 1983. Field experiments on interspecific competition. *American Naturalist* 122: 240–285.

————· 1989. Food webs from the small to the large. *Ecology* 70: 1559–1589.

Schoener, T. W., and Spiller, D. A. 1996. Devastation of prey diversity by experimentally introduced predators in the field. *Nature* 381: 691–694.

Scholle, G., et al. 1992. Effects of mesofauna exclusion on the microbial biomass in two moder profiles. *Biology and Fertility of Soils* 12: 253–260.

Schortemeyer, M., et al. 2000. Effects of elevated atmospheric CO_2 concentration on C and N pools and rhizosphere processes in a Florida scrub oak community. *Global Change Biology* 6: 383–391.

Schulze, E.-D., and Mooney, H. A. Editors. 1993. *Biodiversity and Ecosystem Function.* Berlin: Springer.

Schwartz, M. W., et al. 2000. Linking biodiversity to ecosystem function: implications for conservation biology. *Oecologia* 122: 297–305.

Scowcroft, P. G., et al. 2000. Decomposition of *Metrosideros polymorpha* leaf litter along elevational gradients in Hawaii. *Global Change Biology* 6: 73–85.

Seastedt, T. R. 1984. The role of microarthropods in decomposition and mineralization procresses. *Annual Review of Entomology* 29: 25–46.

————· 1985. Maximization of primary and secondary productivity by grazers. *American Naturalist* 126: 559–564.

————· 2000. Soil fauna and controls of carbon dynamics: comparisons of rangelands and forests across latitudinal gradients. In P. F. Hendrix and D. C. Coleman, eds., *Invertebrates as Webmasters in Ecosystems*, 293–312. Wallingford: CAB International.

Seastedt, T. R., and Adams, G. A. 2001. Effects of mobile tree islands on alpine tundra soil. *Ecology* 82: 8–17.

Seastedt, T. R., and Crossley, D. A. 1980. Effects of microarthropods on the seasonal dynamics of nutrients in forest litter. *Soil Biology and Biochemistry* 12: 337–342.

Seastedt, T. R., et al. 1988a. Interactions among soil invertebrates, microbes and plant growth in the tallgrass prairie. *Agriculture, Ecosystems and Environment* 24: 219–228.

————· 1988b. Maximization of densities of soil animals by foliage herbivory: empirical evidence, graphical and conceptual models. *Oikos* 51: 243–248.

———. 1994. Controls of plant and soil carbon in a semihumid temperate grassland. *Ecological Applications* 4: 344–353.

Setälä, H. 1995. Growth of birch and pine seedlings in relation to grazing by soil fauna on ectomycorrhizal fungi. *Ecology* 76: 1844–1851.

Setälä, H., and Huhta, V. 1991. Soil fauna increase *Betula pendula* growth: laboratory experiments with coniferous forest floor. *Ecology* 72: 665–671.

Setälä, H., and Marshall, V. 1994. Stumps as a habitat for Collembola during succession from clear-cuts to old-growth Douglas-fir forests. *Pedobiologia* 38: 307–326.

Setälä, H., et al. 1996. Influence of body size of soil fauna on litter decomposition and ^{15}N uptake by poplar in a pot trial. *Soil Biology and Biochemistry* 28: 1661–1675.

———. 1997. Conditional outcomes in the relationship between pine and ectomycorrhizal fungi in relation to biotic and abiotic environment. *Oikos* 80: 112–122.

———. 1999. Influence of ectomycorrhiza on the structure of detrital food webs in pine rhizosphere. *Oikos* 87: 113–122.

———. 2000. Sensitivity of soil processes in northern forest soils: are management practices a threat? *Forest Ecology and Management* 133: 5–11.

Sgardelis, S. P., and Usher, M. B. 1994. Responses of soil Cryptostigmata across the boundary between a farm woodland and an arable field. *Pedobiologia* 38: 36–49.

Shaver, G., et al. 1992. Global change and the carbon balance of Arctic ecosystems. *BioScience* 42: 433–441.

Shvarts, E. A., et al. 1997. Do shrews have an impact on soil invertebrates in Eurasian forests? *Écoscience* 4: 158–162.

Siemann, E. 1998. Experimental tests of effects of plant productivity and diversity on grassland arthropod diversity. *Ecology* 79: 2057–2070.

Siemann, E., et al. 1998. Experimental tests of the dependence of arthropod diversity on plant diversity. *American Naturalist* 152: 738–750.

Sievering, H. 1999. Nitrogen deposition and carbon sequestration. *Nature* 400: 629–630.

Sih, A., et al. 1985. Predation, competition and prey communities: a review of field experiments. *Annual Review of Ecology and Systematics* 16: 269–311.

Silver, W. L., and Vogt, K. A. 1993. Fine root dynamics following single and multiple disturbances in a subtropical wet forest ecosystem. *Journal of Ecology* 81: 729–738.

REFERENCES

Simard, S. W., et al. 1997. Net transfer of carbon between ecto-mycorrhizal tree species in the field. *Nature* 388; 579–582.

Singh, J. P., et al. 1989. Microbial biomass acts as a source of plant nutrients in a dry tropical forest and savanna. *Nature* 338: 499–500.

Sirotnak, J. M., and Huntly, N. J. 2000. Direct and indirect effects of herbivores on nitrogen dynamics: voles in riparian areas. *Ecology* 81: 78–87.

Skujins, J., and Klubek, B. 1982. Soil biological properties of a montane forest sere: corroboration of Odum's postulates. *Soil Biology and Biochemistry* 14: 505–513.

Smith, H. G. 1996. Diversity of Antarctic terrestrial protozoa. *Biodiversity and Conservation* 5: 1379–1394.

Smith, M. D., et al. 1999. Interacting effects of mycorrhizal symbiosis and competition on plant diversity in tallgrass prairie. *Oecologia* 121: 574–582.

Smith, S. D., et al. 2000. Elevated CO_2 increases productivity and invasive species success in an arid ecosystem. *Nature* 408: 79–82.

Smith, S. E., and Read, D. J. 1997. *Mycorrhizal Symbiosis.* London: Academic.

Smolander, A., et al. 1994. Microbial biomass C and N, and respiratory activity in soil of repeatedly limed and N- and P-fertilized Norway spruce stands. *Soil Biology and Biochemistry* 26: 957–962.

———. 1998. Nitrogen and carbon transformations before and after clear-cutting in repeatedly N-fertilized and limed forest soil. *Soil Biology and Biochemistry* 30: 477–490.

Söderström, B., et al. 1983. Decrease in soil microbial activity and biomass owing to nitrogen amendments. *Canadian Journal of Microbiology* 23: 1500–1506.

Sohlenius, B. 1996. Structure and composition of the nematode fauna in pine forests under the influence of clear-cutting—effects of slash removal and field layer vegetation. *European Journal of Soil Biology* 32: 1–14.

———. 1997. Fluctuations of nematode populations in pine forest soil: influence by clear-cutting. *Fundamental and Applied Nematology* 20: 103–114.

Sohlenius, B., and Boström, S. 1999. Effect of climate change on soil factors and metazoan microfauna (nematodes, tardigrades and rotifers) in a Swedish tundra soil—a soil transplantation experiment. *Applied Soil Ecology* 12: 113–128.

Southwood, T. R. E., et al. 1979. The relationships between plant and insect diversities in succession. *Biological Journal of the Linnean Society* 12: 327–348.

Spain, A. V., and McIvor, J. G. 1988. The nature of herbaceous vegetation associated with termitaria in north-eastern Australia. *Journal of Ecology* 76: 181–191.

Spain, A. V., et al. 1992. Stimulation of plant growth by tropical earthworms. *Soil Biology and Biochemistry* 24: 1629–1633.

Sparling, G. P., et al. 1994. Microbial C and N in revegetated wheatbelt soils in Western Australia: estimation in soil, humus and leaf litter using the ninhydrin method. *Soil Biology and Biochemistry* 26: 1179–1184.

Spehn, E. M., et al. 2000. Plant diversity and soil heterotrophic activity in experimental grassland systems. *Plant and Soil* 224: 217–230.

Srivastava, S. C., and Singh, J. S. 1991. Microbial C, N and P in dry tropical forest soils: effects of alternative land uses and nutrient flux. *Soil Biology and Biochemistry* 23: 117–124.

Stadler, B., and Michalzik, B. 1998. Linking aphid honeydew, throughfall, and forest floor solution chemistry of Norway spruce. *Ecology Letters* 1: 13–16.

Stahl, P. D., and Christensen, M. 1992. *In vitro* mycelial competition among members of a soil microfungal community. *Soil Biology and Biochemistry* 24: 309–316.

Stamp, N. E., and Bowers, M. D. 1996. Consequences for plantain chemistry and growth when herbivores are attacked by predators. *Ecology* 77: 535–549.

Starfield, A. M., and Chapin, F. S. III. 1996. Model of transient changes in Arctic and boreal vegetation in response to climate and land use change. *Ecological Applications* 6: 842–864.

Stark, J. M., and Hart, S. C. 1997. High rates of nitrification and nitrate turnover in undisturbed coniferous soils. *Nature* 385: 61–64.

Stark, S., et al. 2000. The effect of reindeer grazing on decomposition, mineralization and soil biota in a dry oligotrophic Scots pine forest. *Oikos* 90: 301–310.

Steinberger, Y., and Sarig, S. 1993. Response by soil nematode populations and the soil microbial biomass to a rain episode in the hot, dry Negev desert. *Biology and Fertility of Soils* 16: 188–192.

Steltzer, H., and Bowman, W. D. 1998. Differential influence of plant species on soil nitrogen transformations within moist meadow alpine tundra. *Ecosystems* 1: 464–474.

Stevens, G. C. 1989. The latitudinal gradient in geographical ranges: how can so many species coexist in the tropics? *American Naturalist* 133: 240–256.

Stevens, M. H. H., and Carson, W. P. 1999a. The significance of assemblage-level thinning for species richness. *Journal of Ecology* 87: 490–502.

———· 1999b. Plant density determines species richness along an experimental fertility gradient. *Ecology* 80: 455–465.

Stohlgren, T. J., et al. 1999. Exotic plant species invade hotspots of native plant diversity. *Ecological Monographs* 69: 25–46.

Stout, J. D. 1980. The role of protozoa in nutrient cycling and energy flow. *Advances in Microbial Ecology* 4: 1–50.

Strong, D. R. 1992. Are trophic cascades all wet? Differentiation and donor control in speciose ecosystems. *Ecology* 73: 747–754.

———· 1999. Predator control in terrestrial ecosystems: the underground food chain of bush lupine. In H. Olff, V. K. Brown, and R. H. Drent, eds., *Herbivores: Between Plants and Predators*, 577–602. Oxford: Blackwell Science.

Strong, D. R., et al. 1999. Model selection for a subterranean trophic cascade: root-feeding caterpillars and entomopathogenic nematodes. *Ecology* 80: 2750–2761.

Suding, K. N., and Goldberg, D. E. 1999. Variation in the effects of vegetation and litter on recruitment across productivity gradients. *Journal of Ecology* 87: 436–449.

Suominen, O. 1999. Impact of cervid browsing and grazing on the terrestrial gastropod fauna in the boreal forests of Fennoscandia. *Ecography* 22: 651–658.

Suominen, O., et al. 1999. Moose, trees and ground living invertebrates: indirect interactions in Swedish pine forest. *Oikos* 84: 215–226.

Swift, M. J., et al. 1979. *Decomposition in Terrestrial Ecosystems*. Oxford: Blackwells.

Symstad, A., et al. 1998. Species loss and ecosystem functioning: effects of species identity and community composition. *Oikos* 81: 389–397.

Tanner, E. J. V. 1977. Four montane rain forests in Jamaica: a quantitative assessment of the floristics, the soils and the foliar mineral levels, and a discussion of their interrelationships. *Journal of Ecology* 65: 883–918.

Taylor, B. R., and Parkinson, D. 1988a. Does repeated wetting and drying accelerate decay of leaf litter? *Soil Biology and Biochemistry* 20: 647–656.

———· 1988b. Does repeated freezing and thawing accelerate decay of leaf litter? *Soil Biology and Biochemistry* 20: 657–665.

Taylor, B. R., et al. 1989. Nitrogen and lignin content as predictors of decomposition rate: a microcosm test. *Ecology* 70: 97–104.

Taylor, P. H., and Gaines, S. D. 1999. Can Rapoport's rule be rescued? Modeling causes of the latitudinal gradient in species richness. *Ecology* 80: 2474–2482.

Tearle, P. V. 1987. Cryptogramic carbohydrate release and microbial response during spring freeze-thaw cycles in Antarctic fellfield fines. *Soil Biology and Biochemistry* 19: 381–390.

Terwilliger, J., and Pastor, J. 1999. Small mammals, ectomycorrhizae and conifer succession in beaver meadows. *Oikos* 85: 83–94.

Teuben, A. 1991. Nitrogen availability and interactions between soil arthropods and microorganisms during decomposition of coniferous litter: a field study. *Biology and Fertility of Soils* 10: 256–266.

Theenhaus, A., et al. 1999. Contramensural interactions between two collembolan species: effects on population development and on soil processes. *Functional Ecology* 13: 238–246.

Theodose, T. A., and Bowman, W. D. 1997. Nutrient availability, plant abundance and species diversity in two alpine tundra communities. *Ecology* 78: 1861–1872.

Thompson, K. 1987. The resource ratio hypothesis and the meaning of competition. *Functional Ecology* 1: 297–303.

Thompson, K., et al. 1994. Seeds in soils and worm casts from a neutral grassland. *Functional Ecology* 8: 29–35.

———. 1995. Native and alien species: more of the same? *Ecography* 18: 390–402.

Tian, G., et al. 1993. Biological effects of plant residues with contrasting chemical compositions under humid tropical conditions: effects on soil fauna. *Soil Biology and Biochemistry* 25: 731–737.

Tiedje, J. M., et al. 1999. Opening the black box of soil microbial diversity. *Applied Soil Ecology* 13: 109–122.

Tiessen, H., et al. 1994. The role of soil organic matter in sustaining soil fertility. *Nature* 371: 783–785.

Tilman, D. 1982. *Resource Competition and Community Structure.* Princeton: Princeton University Press.

———. 1988. *Plant Strategies and the Dynamics and Structure of Plant Communities.* Princeton: Princeton University Press.

———. 1996. Biodiversity: population versus ecosystem stability. *Ecology* 77: 350–363.

———. 1999. The ecological consequences of changes in biodiversity: the search for general principles. *Ecology* 80: 1455–1474.

Tilman, D., and Wedin, D. 1991. Plant traits and resource reduction for five grasses growing on a nitrogen gradient. *Ecology* 72: 585–700.

Tilman, D., et al. 1996. Productivity and sustainability influenced by biodiversity in grassland ecosystems. *Nature* 379: 718–720.

———. 1997a. The influence of functional diversity and composition on ecosystem processes. *Science* 277: 1300–1302.

———. 1997b. Biodiversity and ecosystem properties. *Science* 278: 1866–1867.

———. 1997c. Plant diversity and ecosystem productivity: theoretical considerations. *Proceedings of the National Academy of Sciences, U.S.A.* 94: 1857–1861.

———. 2001. Experimental and observational studies of diversity, productivity and stability. In A. Kinzig, D. Tilman, and S. Pacala, eds., *Functional Consequences of Biodiversity: Experimental Progress and Theoretical Extensions*. Princeton: Princeton University Press (in press).

Timmermann, A., et al. 1999. Increased El Niño frequency in a climate model forced by future greenhouse warming. *Nature* 398: 694–696.

Tiunov, A. V., et al. 2001. Microflora, Protozoa and Nematoda in Lumbricus terrestris burrows: a laboratory experiment. *Pedobiologia* 45: 46–60.

Tiwari, S. C., and Mishra, R. R. 1993. Fungal abundance and diversity in earthworm casts and in uningested soil. *Biology and Fertility of Soils* 16: 131–134.

Todd, N. K., and Rayner, A. D. M. 1978. Genetic structure of a natural population of *Coriolus versicolor* (L. ex Fr) Quél. *Genetical Research* 32: 55–65.

Torsvik, V., et al. 1994. Use of DNA analysis to determine the diversity of soil communities. In K. Ritz, J. Dighton, and K. E. Giller, eds., *Beyond the Biomass. Compositional and Functional Analysis of Soil Microbial Communities*, 39–48. Chichester: Wiley.

Toyota, K., and Kimura, M. 1994. Earthworms disseminate a soil-borne pathogen, *Fusarium oxysporum* f. sp. *raphani*. *Biology and Fertility of Soils* 18: 32–36.

Tracy, B. F., and Frank, D. A. 1998. Herbivore influence on soil microbial biomass and nitrogen mineralization in a northern grassland ecosystem: Yellowstone National Park. *Oecologia* 114: 556–562.

Trenbath, B. R. 1974. Biomass productivity of mixtures. *Advances in Agronomy* 26: 177–210.

Treonis, A. M., and Lussenhop, J. F. 1997. Rapid responses of soil protozoa to elevated CO_2. *Biology and Fertility of Soils* 25: 60–62.

Treonis, A. M., et al. 2000. The use of anhydrobiosis by soil nematodes in the Antarctic dry valleys. *Functional Ecology* 14: 460–467.

Tresender, K., and Vitousek, P. M. 2001. Potential ecosystem-level effects of genetic variation among populations of Metrosideros polymorpha from a soil fertility gradient in Hawaii. *Oecologia* 126: 266–275.

Trigo, D., et al. 1999. Mutualism between earthworms and soil microflora. *Pedobiologia* 43: 866–873.

Troumbis, A. Y., and Memtis, D. 2000. Observational evidence that diversity may increase productivity in Mediterranean shrublands. *Oecologia* 125: 101–108.

Troumbis, A. Y., et al. 2000. Hidden diversity and productivity patterns in mixed Mediterranean grasslands. *Oikos* 90: 549–559.

Trumbore, S. E., et al. 1996. Rapid exchange between soil carbon and atmospheric carbon dioxide driven by temperature change. *Science* 272: 393–396.

Tunlid, A., and White, D. C. 1992. Biochemical analysis of biomass, community structure, nutritional status and metabolic activity of microbial communities in soil. In G. Stotzky and J. M. Bollag, eds., *Soil Biochemistry—Volume 7*, 229–262. New York: Marcel Dekker.

Tuomi, J., et al. 1990. The Panglossian paradigm and delayed inducible accumulaton of foliar phenolics in mountain birch. *Oikos* 59: 399–410.

Turkington, R., et al. 1998. The effects of NPK fertilization for nine years on boreal forest vegetation in northwestern Canada. *Journal of Vegetation Science* 9: 333–346.

Unger, P. W., and Cassell, D. K. 1991. Tillage implement disturbance effects on soil properties related to soil and water conservation: a literature review. *Soil and Tillage Research* 19: 363–382.

Vallis, I. 1983. Uptake by grass and transfer to soil of nitrogen from ^{15}N-labeled legume materials applied to a Rhodes Grass pasture. *Australian Journal of Agricultural Research* 34: 367–376.

Valone, T. J., and Brown, J. H. 1995. Effects of competition, colonization and extinction on rodent species diversity. *Science* 267: 880–883.

Van Breemen, N., et al. 2000. Mycorrhizal weathering: a true case of mineral plant nutrition? *Biogeochemistry* 49: 53–67.

Vance, E. D., and Chapin, F. S. III. 2001. Substrate limitations to microbial activity in taiga forest floors. *Soil Biology and Biochemistry* 33: 173–188.

Vance, E. D., and Nadkarni, N. M. 1990. Microbial biomass and activity in canopy organic matter and the forest floor of a tropical cloud forest. *Soil Biology and Biochemistry* 22: 677–684.

Van de Koppel, J., et al. 1996. Patterns of herbivory along a productivity gradient: an empirical and theoretical investigation. *Ecology* 77: 736–745.

Van der Heijden, M., et al. 1998a. Different arbuscular mycorrhizal fungal species are potential determinants of plant community structure. *Ecology* 79: 2082–2091.

———. 1998b. Mycorrhizal fungal diversity determines plant biodiversity, ecosystem variability and productivity. *Nature* 396: 69–72.

———. 1999. "Sampling effect", a problem in biodiversity manipulation? A reply to David A. Wardle. *Oikos* 87: 408–410.

Vandermeer, J. 1990. *The Ecology of Intercropping*. Cambridge: Cambridge University Press.

Van der Putten, W. H. 2001. Interactions of plants, soil pathogens, and their natural antagonists in natural ecosystems. In M. J. Jeger and N. J. Spence, eds., *Biotic Interactions in Plant-Pathogen Associations*. Wallingford: CAB International (in press).

Van der Putten, W. H., and Peters, B. A. M. 1997. How soil-borne pathogens may affect plant competition. *Ecology* 78: 1785–1795.

Van der Putten, W. H., et al. 1993. Plant specific soil borne diseases contribute to succession in foredune vegetation. *Nature* 363: 53–56.

———. 2000. Plant species diversity as a driver of early succession in abandoned fields: a multisite approach. *Oecologia* 124: 91–99.

Van der Wal, R., et al. 2000. Effects of resource competition and herbivory on plant performance along a natural productivity gradient. *Journal of Ecology* 88: 317–330.

Van Straalen, N., and Verhoef, H. A. 1997. The development of a bioindicator system for soil acidity based on arthropod food preferences. *Journal of Applied Ecology* 34: 217–232.

Van Veen, J. A., et al. 1991. Carbon fluxes in plant-soil systems at elevated CO_2 levels. *Ecological Applications* 1: 175–181.

Van Wensem, J., et al. 1993. Litter degradation stage as a prime factor for isopod interaction with mineralization processes. *Soil Biology and Biochemistry* 25: 1175–1183.

Van Wesemael, B., and Veer, M. A. C. 1992. Soil organic matter accumulation, litter decomposition and humus forms under mediterranean-type forests in southern Tuscany, Italy. *Journal of Soil Science* 43: 133–144.

Van Wijnen, H. H., et al. 1999. The impact of herbivores on nitrogen mineralization rate: consequences for salt marsh succession. *Oecologia* 118: 225–231.

Väre, H., et al. 1996. The effect and extent of heavy grazing by reindeer in oligotrophic pine heaths in northeastern Fennoscandia. *Ecography* 19: 245–253.

Vedder, B., et al. 1996. Impact of faunal complexity on microbial biomass and N turnover in field mesocosms from a spruce forest soil. *Biology and Fertility of Soils* 22: 22–30.

Vegter, J. J. 1987. Phenology and seasonal resource partitioning in forest floor Collembola. *Oikos* 48: 175–185.

Veldkamp, E. 1994. Organic carbon turnover in three tropical soils under pasture after deforestation. *Soil Science Society of America Journal* 58: 175–180.

Verhoef, H. A., and Brussaard, L. 1990. Decomposition and nitrogen mineralization in natural and agro-ecosystems: the contribution of soil animals. *Biogeochemistry* 11: 175–211.

Verville, J. H., et al. 1998. Response of tundra CH_4 and CO_2 flux to manipulation of temperature and vegetation. *Biogeochemistry* 41: 215–235.

Vinton, M. A., and Burke, I. C. 1995. Interactions between individual plant species and soil nutrient status in shortgrass steppe. *Ecology* 76: 1116–1133.

Visser, S., et al. 1984. Effects of topsoil storage on microbial activity, primary production and decomposition potential. *Plant and Soil* 82: 41–50.

Vitousek, P. M. 1982. Nutrient cycling and nutrient use efficiency. *American Naturalist* 119: 553–572.

Vitousek, P. M., and Farrington, H. 1997. Nutrient limitation and soil development: experimental test of a biogeochemical theory. *Biogeochemistry* 37: 63–75.

Vitousek, P. M., and Hooper, D. U. 1993. Biological diversity and terrestrial ecosystem biogeochemistry. In E.-D. Schulze and H. A. Mooney, eds., *Biodiversity and Ecosystem Function*, 3–14. Berlin: Springer-Verlag.

Vitousek, P. M., and Matson, P. A. 1984. Mechanisms of nitrogen retention in forest ecosystems: a field experiment. *Science* 225: 51–52.

Vitousek, P. M., and Walker, L. R. 1989. Biological invasion by *Myr-*

ica faya in Hawai'i: plant demography, nitrogen fixation, ecosystem effects. *Ecological Monographs* 59: 247–265.

Vitousek, P. M., et al. 1979. Nitrate losses from perturbed ecosystems. *Science* 204: 469–474.

———· 1987. Biological invasion by *Myrica faya* alters ecosystem development in Hawaii. *Science* 238: 802–804.

———· 1993. Nutrient limitation to plant growth during primary succession in Hawaii Volcanoes National Park. *Biogeochemistry* 23: 197–215.

———· 1994. Litter decomposition on the Mauna Loa environmental matrix, Hawai'i: patterns, mechanisms and models. *Ecology* 75: 418–429.

———· 1997a. Human domination of the Earth's ecosystems. *Science* 277: 494–499.

———· 1997b. Introduced species: a significant component of human-induced global change. *New Zealand Journal of Ecology* 21: 1–16.

———· 1997c. Human alteration of the global nitrogen cycle: sources and consequences. *Ecological Applications* 7: 737–750.

Vohland, K., and Scroth, G. 1999. Distribution patterns of the litter macrofauna in agroforestry and monoculture plantations in central Amazonia as affected by plant species and management. *Applied Soil Ecology* 13: 57–68.

Vose, J. M., et al. 1995. Effects of elevated CO_2 and N fertilization on soil respiration from ponderosa pine (*Pinus ponderosa*) in open top chambers. *Canadian Journal of Forest Research* 25: 1243–1251.

Vossbrink, C. R., et al. 1979. Abiotic and biotic factors in litter decomposition in a semiarid grassland. *Ecology* 60: 265–271.

Vreeken-Buijs, M. J., et al. 1997. The effects of bacterivorous mites and amoebae on mineralization in a detrital based below-ground food web: microcosm experiment and simulation of interactions. *Pedobiologia* 41: 481–493.

Vtorov, I. P. 1993. Feral pig removal: effects on soil microarthropods in a Hawaiian rain forest. *Journal of Wildlife Management* 57: 875–880.

Wacquant, J. P., et al. 1989. Evidence for a periodic excretion of nitrogen by roots of grass-legume associations. *Plant and Soil* 116: 57–68.

Wagner, D., et al. 1997. Harvester ants, soil biota and soil chemistry. *Oecologia* 112: 232–236.

Waide, R. B., et al. 1999. The relationship between productivity and species richness. *Annual Review of Ecology and Systematics* 30: 257–301.

Wali, M. K. 1999. Ecological succession and the rehabilitation of disturbed terrestrial ecosystems. *Plant and Soil* 213: 195–220.

Walker, B. H. 1992. Biodiversity and ecological redundancy. *Conservation Biology* 6; 18–23.

Walker, B. H., et al. 1999. Plant attribute diversity, resilience and ecosystem function: the nature and significance of dominant and minor species. *Ecosystems* 2: 95–113.

Walker, J., et al. 1983. Soil weathering, vegetation succession and canopy dieback. *Pacific Science* 37: 471–481.

Walker, L. R. 1989. Soil nitrogen changes during primary succession on a floodplain in Alaska, U.S.A. *Arctic and Alpine Research* 21: 341–349.

Walker, L. R., and Chapin, F. S. III 1987. Interactions among processes controlling successional change. *Oikos* 50: 131–135.

Walker, T. R., and Syers, J. K. 1976. The fate of phosphorus during pedogenesis. *Geoderma* 15: 1–19.

Wall, D. H., and Moore, J. C. 1999. Interactions underground. *BioScience* 49: 109–117.

Wall, D. H., and Virginia, R. A. 1999. Controls on biodiversity: insights from extreme environments. *Applied Soil Ecology* 13: 137–150.

Wall-Freckman, D., and Huang, S. P. 1998. Response of the soil nematode community in a shortgrass steppe to long-term and short-term grazing. *Applied Soil Ecology* 9: 39–44.

Wallis, F. P., and James, I. L. 1972. Introduced animal effects and erosion phenomena in the northern Urewera forests. *New Zealand Journal of Forestry* 17: 21–36.

Wallwork, J. A. 1983. Oribatids in forest ecosystems. *Annual Review of Entomology* 28: 109–130.

Walter, D. E., and Behan-Pelletier, V. 1999. Mites in forest canopies: filling the size-distribution shortfall? *Annual Review of Entomology* 44: 1–19.

Walter, D. E., et al. 1991. Missing links: a review of methods used to estimate trophic links in food webs. *Agriculture, Ecosystems and Environment* 34: 399–405.

Walter, H. 1985. *Vegetation of the Earth and Ecological Systems of the Geobiosphere.* Berlin: Springer-Verlag.

Ward, D., et al. 1990. 16S rRNA sequences reveal numerous uncultured microorganisms in a natural community. *Nature* 345: 63–65.

Wardle, D. A. 1991. Free lunch? *Nature* 352: 482.

——— · 1992. A comparative assessment of factors which influence microbial biomass carbon and nitrogen levels in soils. *Biological Reviews* 67: 321–358.

——— · 1993. Changes in the microbial biomass and metabolic quotient during leaf litter succession in some New Zealand forest and scrubland ecosystems. *Functional Ecology* 7: 346–355.

——— · 1995. Impact of disturbance on detritus food-webs in agro-ecosystems of contrasting tillage and weed management practices. *Advances in Ecological Research* 26: 105–185.

——— · 1998a. Controls of temporal variability of the soil microbial biomass: a global scale synthesis. *Soil Biology and Biochemistry* 30: 1627–1637.

——— · 1998b. A more reliable design for biodiversity study? *Nature* 394: 30.

——— · 1999a. Biodiversity, ecosystems and interactions that transcend the interface. *Trends in Ecology and Evolution* 14: 125–127.

——— · 1999b. Is "sampling effect" a problem for experiments investigating biodiversity—ecosystem function relationships? *Oikos* 87: 403–407.

——— · 2001. Experimental demonstration that plant diversity reduces invisibility—evidence of an ecological mechanism or a consequence of "sampling effect"? *Oikos* (in press).

Wardle, D. A., and Barker, G. M. 1997. Competition and herbivory in establishing grassland communities: implications for plant biomass, species diversity and soil microbial activity. *Oikos* 80: 470–480.

Wardle, D. A., and Ghani, A. 1995. A critique of the microbial metabolic quotient (qCO_2) as a bioindicator of disturbance and ecosystem development. *Soil Biology and Biochemistry* 27: 1601–1610.

Wardle, D. A., and Giller, K. E. 1996. The quest for a contemporary ecological dimension to soil biology. *Soil Biology and Biochemistry* 28: 1549–1554.

Wardle, D. A., and Greenfield, L. G. 1991. Mineral nitrogen release from plant root nodules. *Soil Biology and Biochemistry* 23: 827–832.

Wardle, D. A., and Lavelle, P. 1997. Linkages between soil biota, plant litter quality and decomposition. In K. E. Giller and G. Cadisch, eds., *Driven by Nature—Plant Litter Quality and Decomposition*, 107–124. Wallingford: CAB International.

Wardle, D. A., and Nicholson, K. S. 1996. Synergistic effects of grassland plant species on soil microbial biomass and activity: implications for ecosystem-level effects of enriched plant diversity. *Functional Ecology* 10: 410–416.

Wardle, D. A., and Parkinson, D. 1992. The influence of the herbicide glyphosate on interspecific interactions between four fungal species. *Mycological Research* 96: 180–186.

Wardle, D. A., and Yeates, G. W. 1993. The dual importance of competition and predation as regulatory forces in terrestrial ecosystems: evidence from decomposer food-webs. *Oecologia* 93: 303–306.

Wardle, D. A., et al. 1993. Interspecific competitive interactions between pairs of fungal species in natural substrates. *Oecologia* 94: 165–172.

——— 1995a. Ecological effects of the invasive weed species *Senecio jacobaea* L. (ragwort) in a New Zealand pasture. *Agriculture, Ecosystems and Environment* 56: 19–28.

——— 1995b. Development of the decomposer food-web, trophic relationships and ecosystem properties during a three-year primary succession of sawdust. *Oikos* 73: 155–166.

——— 1997a. Biodiversity and plant litter: experimental evidence which does not support the view that enhanced species richness improves ecosystem function. *Oikos* 79: 247–258.

——— 1997b. The influence of island area on ecosystem properties. *Science* 277: 1296–1299.

——— 1997c. Biodiversity and ecosystem properties. *Science* 278: 1867–1869.

——— 1998a. Can comparative approaches based on plant ecophysiological traits predict the nature of biotic interactions and individual plant species effects in ecosystems? *Journal of Ecology* 86: 405–420.

——— 1998b. Trophic relationships in the soil microfood-web: predicting the responses to a changing global environment. *Global Change Biology* 4: 713–727.

——— 1998c. The charcoal effect in boreal forests: mechanisms and ecological consequences. *Oecologia* 115: 419–426.

——— 1999a. The regulation and functional significance of soil biodiversity in agroecosystems. In D. Wood and J. M. Lenné, eds., *Agrobiodiversity: Characterization, Utilization and Management,* 87–121. Wallingford: CAB International.

——— 1999b. Plant removals in perennial grassland: vegetation dynamics, decomposers, soil biodiversity and ecosystem properties. *Ecological Monographs* 69: 535–568.

——— 1999c. Effects of agricultural intensification on soil-associated arthropod population dynamics, community structure, diversity and temporal variability over a seven year period. *Soil Biology and Biochemistry* 31: 1691–1706.

————· 1999d. Response of soil microbial biomass dynamics, activity and plant litter decomposition to agricultural intensification over a seven-year period. *Soil Biology and Biochemistry* 31: 1671–1706.

————· 2000. Stability of ecosystem properties in response to above-ground functional group richness and composition. *Oikos* 89: 11–23.

————· 2001. Introduced browsing mammals in New Zealand natural forests: aboveground and belowground consequences. *Ecological Monographs* (in press).

Wasilewska, L. 1979. The structure and function of soil nematode communities in natural ecosystems and agrocoenoses. *Polish Ecological Studies* 5: 97–145.

————· 1994. The effect of age of meadows on succession and diversity in soil nematode communities. *Pedobiologia* 38: 1–11.

————· 1995. Differences in the development of soil nematode communities in single- and multi-species grass experimental treatments. *Applied Soil Ecology* 2: 53–64.

Webb, N. R., et al. 1998. The effects of experimental temperature elevation on populations of cryptostigmatid mites in high Arctic soils. *Pedobiologia* 42: 298–308.

Webber, J. F., and Hedger, J. N. 1986. Comparison of interactions between *Ceratocytis ulmi* and elm bark saprobes *in vitro* and *in vivo*. *Transactions of the British Mycological Society* 86: 93–101.

Wedin, D. A., and Tilman, D. 1990. Species effects on nitrogen cycling: a test with perennial grasses. *Oecologia* 84: 433–441.

————· 1996. Influence of nitrogen loading and species composition on the carbon balance of grasslands. *Science* 274: 1720–1723.

West, A. W., et al. 1988. Dynamics of microbial C, N-flush and ATP, and enzyme activities of gradually dried soils from a climosequence. *Australian Journal of Soil Research* 26: 519–530.

West, H. M. 1995. Soil mycorrhizal status modifies the response of mycorrhizal and non-mycorrhizal *Senecio vulgaris* L. to infection by the rust *Puccinia lagenophorae* Cooke. *New Phytologist* 129: 107–116.

————· 1997. Arbuscular mycorrhizal fungi and foliar pathogens: consequences for host and pathogen. In A. C. Gange and V. K. Brown, eds., *Multitrophic Interactions in Terrestrial Systems*, 79–89. Oxford: Blackwells.

Wheatley, R., et al. 1990. Microbial biomass and mineral N transformations in soil planted with barley, ryegrass, pea or turnip. *Plant and Soil* 127: 157–167.

381

Whipps, J. M., and Lynch, J. M. 1986. The influence of the rhizosphere on crop productivity. *Advances in Microbial Ecology* 9: 187–244.

White, C. S. 1986. Volatile and water-soluble inhibitors of nitrogen mineralization and nitrification in a ponderosa pine ecosystem. *Biology and Fertility of Soils* 2: 97–104.

———· 1990. Comments on "effects of terpenoids on nitrification in soil." *Soil Science Society of America Journal* 54: 296–298.

White, T. A., et al. 2000. Sensitivity of three grassland communities to simulated extreme climate and rainfall events. *Global Change Biology* 6: 671–684.

———· 2001. Impacts of extreme climatic events on competition during grassland invasions. *Global Change Biology* 7: 1–13.

White, T. R. C. 1978. The importance of relative shortage of food in animal ecology. *Oecologia* 33: 71–86.

Whitford, W. G. 1989. Abiotic controls on the functional structure on soil food webs. *Biology and Fertility of Soils* 8: 1–6.

Whitford, W. G., and DiMarco, R. 1995. Variability in soils and vegetation associated with harvester ant (*Pogonomyrmex rugosus*) nests on a Chihuahuan Desert watershed. *Biology and Fertility of Soils* 20: 169–173.

Whittaker, R. H. 1975. *Communities and Ecosystems*. Second Edition. Macmillan: New York.

Wicklow, D. T. 1981. Interference competition and the organization of fungal communities. In D. T. Wicklow and G. C. Carroll, eds., *The Fungal Community: Its Organization and Role in the Ecosystem*, 351–375. New York: Marcel Dekker.

Widden, P. 1984. The effects of temperature on competition for spruce needles among sympatric species of Trichoderma. *Mycologia* 76: 783–883.

———· 1986. Microfungal community structure from forest soils in southern Quebec, using discriminant function and factor analysis. *Canadian Journal of Botany* 64: 1402–1412.

Widden, P., and Hsu, D. 1987. Competition between *Trichoderma* species: effects of temperature and litter type. *Soil Biology and Biochemistry* 19: 89–93.

Widden, P., and Parkinson, D. 1975. The effects of a forest fire on soil microfungi. *Soil Biology and Biochemistry* 7: 125–138.

Wijesinghe, D. K., and Hutchings, M. J. 1997. The effects of spatial scale of environmental heterogeneity on the growth of a clonal plant: an experimental study with Glechoma hederacea. *Journal of Ecology* 85: 17–28.

Williams, B. L., et al. 2000. Influence of synthetic sheep urine on the microbial biomass, activity and community structure in two pastures in the Scottish uplands. *Plant and Soil* 225: 175–185.

Williams, M. 1990. Deforestation. In B. L. Turner, W. Clark, R. Kates, J. Richards, J. Matthews, and W. Meyer, eds., *The Earth as Transformed by Human Action*, 179–201. Cambridge: Cambridge University Press.

Williams, R. J., and Martinez, N. 2000. Simple rules yield complex food webs. *Nature* 404: 160–163.

Williamson, M., and Fitter, A. 1996. The characters of successful invaders. *Biological Conservation* 78: 163–170.

Wilsey, B. J., and Potvin, C. 2000. Biodiversity and ecosystem functioning: importance of species eveness in an old field. *Ecology* 81: 887–892.

Wilson, S. D. 1999. Plant interactions during secondary succession. In L. R. Walker, ed., *Ecosystems of the World. 16. Ecosystems of Disturbed Ground*, 611–632. Amsterdam: Elsevier.

Wilson, S. D., and Tilman, D. 1991. Components of plant competition along an experimental gradient of nitrogen availability. *Ecology* 72: 1050–1065.

Winchester, N. N., et al. 1999. Arboreal specificity, diversity and abundance of canopy-dwelling oribatid mites (Acari: Oribatida). *Pedobiologia* 43: 391–400.

Winkler, H., and Kampichler, C. 2000. Local and regional richness in communities of surface-dwelling grassland Collembola: indication of species saturation. *Ecography* 23: 385–392.

Winter, J. P., et al. 1990. Soil microarthropods in long-term no-tillage and conventional tillage corn production. *Canadian Journal of Soil Science* 70: 641–653.

Wiser, S. K., et al. 1998. Community structure and forest invasion by an exotic herb over 23 years. *Ecology* 79: 2071–2081.

Wolter, C., and Scheu, S. 1999. Changes in bacterial numbers and hyphal lengths during the gut passage through *Lumbricus terrestris* (Lumbricidae, Oligochaeta). *Pedobiologia* 43: 891–900.

Wolters, V., et al. 2000. Effects of global change on above- and belowground biodiversity in terrestrial ecosystems: implications for ecosystem functioning. *BioScience* 50: 1089–1098.

Wood, C. W., and Edwards, J. H. 1992. Agroecosystem management effects on soil carbon and nitrogen. *Agriculture, Ecosystems and Environment* 39: 123–138.

Wood, C. W., et al. 1994. Free air CO_2 enrichment effects on soil carbon and nitrogen. *Agricultural and Forest Meteorology* 70: 103–116.

Woodward, F. I. 1987. *Climate and Plant Distribution*. Cambridge: Cambridge University Press.

REFERENCES

Wootton, J. T. 1994. The nature and consequences of indirect effects in ecological communities. *Annual Review of Ecology and Systematics* 25: 443–466.

———. 1998. Effects of disturbance on species diversity: a multitrophic perspective. *American Naturalist* 152: 803–825.

Wouts, W. M., et al. 1999. Criconematidae (Nematoda: Tylenchida) from the New Zealand region: genera *Ogma* Southern, 1914 and *Blandicephalanema* Mehta & Raski, 1971. *Nematology* 1: 561–570.

Wright, D. H., and Coleman, D. C. 1993. Patterns of survival and extinction of nematodes in isolated soil. *Oikos* 67: 563–572.

Wu, J., and Levin, S. A. 1994. A spatial patch dynamic approach to pattern and process in an annual grassland. *Ecological Monographs* 64: 447–464.

Wyman, R. L. 1998. Experimental assessment of salamanders as predators of detrital food webs: effects on invertebrates, decomposition and the carbon cycle. *Biodiversity and Conservation* 7: 641–650.

Xiong, S., and Nilsson, C. 1999. The effects of plant litter on vegetation: a meta-analysis. *Journal of Ecology* 87: 984–994.

Yachi, S., and Loreau, M. 1999. Biodiversity and ecosystem productivity in a fluctuating environment: the insurance hypothesis. *Proceedings of the National Academy of Sciences, U.S.A.* 96: 1463–1468.

Yarie, J., and Van Cleve, K. 1996. Effects of carbon, fertilizer and drought on foliar chemistry of tree species in interior Alaska. *Ecological Applications* 6: 815–827.

Yeates, G. W. 1979. Soil nematodes in terrestrial ecosystems. *Journal of Nematology* 11: 213–229.

———. 1981. Soil nematode populations depressed in the presence of earthworms. *Pedobiologia* 22: 191–195.

———. 2001. Nematodes—roundworms of vertebrates, eelworms of plants—the most numerous multicellular animals of earth. In D. P. Gordon, ed., *The New Zealand Inventory of Biodiversity: A Species 2000 Review*. Christchurch: Canterbury University Press (in press).

Yeates, G. W., and Bird, A. F. 1994. Some observations on the influence of agricultural practices on the nematode faunae of some South Australian soils. *Fundamental and Applied Nematology* 17: 133–145.

Yeates, G. W., and Bongers, T. 1999. Nematode biodiversity in agroecosystems. *Agriculture, Ecosystems and Environment* 74: 113–135.

Yeates, G. W., and Orchard, V. A. 1993. Response of pasture soil faunal populations and decomposition processes to elevated carbon dioxide and temperature—a climate chamber experiment. *Australasian Grasslands Invertebrate Ecology Proceedings* 6: 148–154.

Yeates, G. W., et al. 1997. Response of the fauna of a grassland soil to doubling of atmospheric carbon dioxide level. *Biology and Fertility of Soils* 25: 305–317.

———. 1999a. Increase in ^{14}C-carbon translocation to the soil microbial biomass when five species of plant-parasitic nematodes infect roots of white clover. *Nematology* 1: 295–300.

———. 1999b. Responses of nematode populations, community structure, diversity and temporal variability to agricultural intensification over a seven year period. *Soil Biology and Biochemistry* 31: 1721–1733.

———. 2000. Changes in soil fauna and soil conditions under *Pinus radiata* agroforestry regimes during a 25-year rotation. *Biology and Fertility of Soils* 31: 391–406.

Young, I. M., and Ritz, K. 1998. Can there be a contemporary ecological dimension to soil biology without a habitat? *Soil Biology and Biochemistry* 30: 1229–1232.

Youssef, R. A., et al. 1989. Distribution of microbial biomass across the rhizosphere of barley (*Hordeum vulgare* L.) in soils. *Biology and Fertility of Soils* 7: 341–345.

Zackrisson, O., et al. 1996. Key ecological function of charcoal from wildfire in the boreal forest. *Oikos* 77: 10–19.

———. 1999. Nutritional effects of seed fall during mast years in boreal forest. *Oikos* 84: 17–26.

Zagal, E. 1994. Influence of light intensity on the distribution of carbon and consequent effects on mineralization of soil nitrogen in a barley (*Hordeum vulgare*)—soil system. *Plant and Soil* 160: 21–31.

Zak, D. R., et al. 1990a. Carbon and nitrogen cycling during old field succession: constraints on plant and microbial biomass. *Biogeochemistry* 11: 111–129.

———. 1990b. The vernal dam: plant-microbe competition for nitrogen in northern hardwood forests. *Ecology* 71: 651–656.

———. 1993. Elevated atmospheric CO_2 and feedback between carbon and nitrogen cycles. *Plant and Soil* 151: 105–117.

———. 1994. Plant production and soil microorganisms in late-successional ecosystems: a continental-scale study. *Ecology* 75: 2333–2347.

———. 1996. Soil microbial communities beneath *Populus grandi-*

dentata grown under elevated atmospheric CO_2. *Ecological Applications* 6: 257–162.

———· 2000a. Atmospheric CO_2 and the composition and function of soil microbial communities. *Ecological Applications* 10: 47–59.

———· 2000b. Elevated atmospheric CO_2, fine roots and the response of soil microorganisms: a review and hypothesis. *New Phytologist* 147: 201–222.

Zak, J. C., et al. 1994. Functional diversity of soil microbial communities: a quantitative approach. *Soil Biology and Biochemistry* 26: 1101–1108.

Zhang, Q., and Zak, J. C. 1995. Effects of gap size on litter decomposition and microbial activity in a subtropical forest. *Ecology* 76: 2196–2204.

———· 1998. Potential physiological activities of fungi and bacteria in relation to plant litter decomposition along a gap size gradient in a natural subtropical forest. *Microbial Ecology* 35: 172–179.

Zheng, D., et al. 1997. Soil food webs and ecosystem processes: decomposition in donor-controlled and Lotka-Volterra systems. *American Naturalist* 149: 125–148.

———· 1999. How do soil organisms affect total organic nitrogen storage and substrate nitrogen to carbon ratios in soils? A theoretical analysis. *Oikos* 86: 430–442.

Zimmerman, J. K., et al. 1995. Nitrogen immobilization by woody debris and the recovery of tropical wet forest by hurricane damage. *Oikos* 72: 314–322.

Zimov, S. A., et al. 1995. Steppe-tundra transition: a herbivore-driven biome shift at the end of the Pleistocene. *American Naturalist* 146: 765–794.

Zullini, A., and Peretti, E. 1986. Lead pollution and moss-inhabiting nematodes of an industrial area. *Water, Air and Soil Pollution* 27: 403–410.

Index

accelerated decomposition, 70–72
alien organisms. *See* invasive
organisms
ants: physical structures created by,
42; response of plant communities to, 173–174
arbuscular mycorrhizae. *See*
mycorrhizae

bacterial-based energy channel,
35–37, 61–63, 86, 92; and land
use, 261–264; effects of nitrogen
deposition on, 279–289; effects
of soil moisture on, 287–288; effects of temperature on, 286
bacterial to fungal ratios, 29–30;
effects of climate change on,
285–286; effects of defoliation
on, 107; effects of land use on,
261–264; effects of moisture on,
287–288; effects of nitrogen deposition on, 278–280; effects of
soil animals on, 38–39; effects of
plant traits on, 92; and mull and
mor theory, 86; response of soil
processes to, 44–45; and vegetation succession, 99
body size of soil biota: and latitudinal diversity gradients, 208–209;
response of plant growth to,
147–148; response of soil processes to, 47–49; and soil food
web stability, 51–52
bottom-up forces: and biological
diversity, 188; and carbon dioxide enrichment, 274–275; in
food web theory, 9–16; microbial
regulation by, 17–23; and nitrogen deposition, 279–280; soil
faunal regulation by, 23–27; in
soil food webs, 35–37

canopy gaps: in forests, 80–81
carbon dioxide enrichment: effects
on climate, 281; effects on carbon sequestration, 275–277; effects on plant litter quality, 267–
271; effects on soil biota, 273–
275; effects on vegetation composition, 271–273; interactions
with nitrogen deposition, 280
carbon sequestration: effects of
biotic composition on, 302–303;
effects of carbon dioxide enrichment on, 275–277; effects of climate change on, 291–292;
effects of foliar herbivores on,
124, 129–130; effects of land use
on, 257, 263, 264; effects of nitrogen deposition on, 280–281;
effects of plant species on, 72–
73; and mull and mor theory,
86; and plant traits, 92; and vegetation succession, 97
charcoal: from fire, 83
chronosequences, 101–102
climate change: effects on carbon
sequestration, 291–292; effects
on decomposition rates, 289–
291; effects on plant litter quality, 282–283, 290–291; effects on
soil organisms, 283–289, 291; effects on vegetation, 281–283,
290–291; and experimental
warming studies, 285
comparative plant ecology: and foliar herbivory, 119–122; and
plant traits, 89–90
compensatory growth responses: of
microbes to faunal grazing, 29–
30; of plants to herbivory, 109
competition: along primary productivity gradients, 9–14, 91–97;
among fungi, 20–23, 191; among
plants as mediated by mycorrhizae, 165–166; among soil fauna, 26–27; between plants and
microbes, 16, 66–67, 141, 147
competitive exclusion: among soil
organisms, 191–193; 203–205
competitive saprophytic ability, 21
complexity of food webs, 50–51

green world hypotheses, 9–14
guano, 135–136

habitat complexity, 3, 67–68, 194–
199; effects on soil biodiversity,
202–205
Hairston, Smith, and Slobodkin hypothesis, 9–14, 35–37
herbivores. *See* foliar herbivores,
root herbivores
hidden treatments: and diversity-
ecosystem function experiments,
212–219, 232; and primary productivity gradients, 15
honeydew production: by aphids,
116; by scale insects, 116–117
hump-backed relationship between
productivity and diversity, 95,
187, 189–191; and the diversity-
ecosystem function relationship,
211–213; and latitudinal gradients, 208
hurricanes: damage to forests by,
81–82
hyphal connections: between
plants with shared mycorrhizae,
165–166

indirect interactions, 296–298
induced plant defenses, 111–112
insurance effect of biodiversity,
234–236
interaction strengths: in food webs,
50–51
intercropping, 214, 259–260
intensification of land use, 257–
264
intermediate disturbance hypothesis, 187, 191–193
invasive organisms, 245–253; effects of carbon dioxide enrichment on, 271; effects of plant
diversity on, 249; in soils, 251–
253

land use: and abandonment, 264–
265; change of, 253–264
latitudinal gradients of diversity,
205–209
litter mixing experiments, 198–
199, 202–203, 224–225

litter quality: effects of carbon dioxide enrichment on, 267–271;
effects of foliar herbivores on,
110–113, 117–121; effects on microflora, 60, 63; effects of plant
phenotypic plasticity on, 87; effects of plant traits on, 85; effects on soil fauna, 26–27, 63;
effects of soil food webs on, 151–
152; effects of temperature on,
282–283, 290–291; effects of vegetation succession on, 97–102;
litter transformers, 8, 37–38; as
drivers of soil processes, 46; effects of plant species on, 63–64;
processing of carbon dioxide-
enriched litter by, 270–271

macroecology: in relation to soil
biota, 205–209
mast seeding, 77–78
megafauna: extinctions of, 242–
243
microarthropods, 8, 37–38, 63–64;
diversity of, 196–199, 208, 224;
effects of carbon dioxide enrichment on, 273–275; effects of climate on, 283, 285; responses of
mycorrhizae to, 167–168
microbial biomass: effects of carbon dioxide enrichment on, 273;
effects of land use on, 259–260,
262; effects of nitrogen deposition on, 278–279; effects of nutrient supply on, 140–141; effects
of plant species on, 58–60; effects of soil moisture on, 287–
288; effects of substrate quality
on, 57–58; effects of temperature on, 283; substrate limitation
of, 17–20; temporal variability of,
78–79
microbial loop, 146
microfood-webs, 8, 16–37; in comparison with foliage-based food
webs, 36–37; as drivers of soil
processes, 44–46; effects of carbon dioxide enrichment on,
273–275; effects of land use on,
261–264, 258–259; effects of nitrogen deposition on, 279–280;